Computational Molecular Evolution

Oxford Series in Ecology and Evolution
Edited by Paul H. Harvey and Robert M. May

The Comparative Method in Evolutionary Biology
Paul H. Harvey and Mark D. Pagel
The Cause of Molecular Evolution
John H. Gillespie
Dunnock Behaviour and Social Evolution
N. B. Davies
Natural Selection: Domains, Levels, and Challenges
George C. Williams
Behaviour and Social Evolution of Wasps: The Communal Aggregation Hypothesis
Yosiaki Itô
Life History Invariants: Some Explorations of Symmetry in Evolutionary Ecology
Eric L. Charnov
Quantitative Ecology and the Brown Trout
J. M. Elliott
Sexual Selection and the Barn Swallow
Anders Pape Møller
Ecology and Evolution in Anoxic Worlds
Tom Fenchel and Bland J. Finlay
Anolis Lizards of the Caribbean: Ecology, Evolution and Plate Tectonics
Jonathan Roughgarden
From Individual Behaviour to Population Ecology
William J. Sutherland
Evolution of Social Insect Colonies: Sex Allocation and Kin Selection
Ross H. Crozier and Pekka Pamilo
Biological Invasions: Theory and Practice
Nanako Shigesada and Kohkichi Kawasaki
Cooperation Among Animals: An Evolutionary Perspective
Lee Alan Dugatkin
Natural Hybridization and Evolution
Michael L. Arnold
Evolution of Sibling Rivalry
Douglas Mock and Geoffrey Parker
Asymmetry, Developmental Stability, and Evolution
Anders Pape Møller and John P. Swaddle
Metapopulation Ecology
Ilkka Hanski
Dynamic State Variable Models in Ecology: Methods and Applications
Colin W. Clark and Marc Mangel
The Origin, Expansion, and Demise of Plant Species
Donald A. Levin
The Spatial and Temporal Dynamics of Host-Parasitoid Interactions
Michael P. Hassell
The Ecology of Adaptive Radiation
Dolph Schluter
Parasites and the Behavior of Animals
Janice Moore
Evolutionary Ecology of Birds
Peter Bennett and Ian Owens
The Role of Chromosomal Change in Plant Evolution
Donald A. Levin
Living in Groups
Jens Krause and Graeme Ruxton
Stochastic Population Dynamics in Ecology and conservation
Russell Lande, Steiner Engen and Bernt-Erik Sæther
The structure and Dynamics of Geographic Ranges
Kevin J. Gaston
Animal Signals
John Maynard Smith and David Harper
Evolutionary Ecology: The Trinidadian Guppy
Anne E. Magurran
Infectious Disease and Primate Socioecology
Charles L. Nunn and Sonia M. Altizer
Computational Molecular Evolution
Ziheng Yang

Computational Molecular Evolution

ZIHENG YANG

University College London, UK

OXFORD
UNIVERSITY PRESS

OXFORD
UNIVERSITY PRESS

Great Clarendon Street, Oxford OX2 6DP

Oxford University Press is a department of the University of Oxford.
It furthers the University's objective of excellence in research, scholarship,
and education by publishing worldwide in

Oxford New York

Auckland Cape Town Dar es Salaam Hong Kong Karachi
Kuala Lumpur Madrid Melbourne Mexico City Nairobi
New Delhi Shanghai Taipei Toronto

With offices in

Argentina Austria Brazil Chile Czech Republic France Greece
Guatemala Hungary Italy Japan Poland Portugal Singapore
South Korea Switzerland Thailand Turkey Ukraine Vietnam

Oxford is a registered trade mark of Oxford University Press
in the UK and in certain other countries

Published in the United States
by Oxford University Press Inc., New York

© Oxford University Press 2006

The moral rights of the author have been asserted
Database right Oxford University Press (maker)

First published 2006
Reprinted 2007 (with corrections)

All rights reserved. No part of this publication may be reproduced,
stored in a retrieval system, or transmitted, in any form or by any means,
without the prior permission in writing of Oxford University Press,
or as expressly permitted by law, or under terms agreed with the appropriate
reprographics rights organization. Enquiries concerning reproduction
outside the scope of the above should be sent to the Rights Department,
Oxford University Press, at the address above

You must not circulate this book in any other binding or cover
and you must impose the same condition on any acquirer

British Library Cataloguing in Publication Data

Data available

Library of Congress Cataloging in Publication Data

Data available

Typeset by Newgen Imaging Systems (P) Ltd., Chennai, India
Printed in Great Britain
on acid-free paper by
Biddles Ltd., King's Lynn

ISBN 978-0-19-856699-1
ISBN 978-0-19-856702-8 (Pbk.)

10 9 8 7 6 5 4 3 2

To my parents

Preface

Studies of evolution at the molecular level aim to address two major questions: reconstruction of the evolutionary relationships among species and investigation of the forces and mechanisms of the evolutionary process. The first is the realm of systematics, and is traditionally studied using morphological characters and fossils. The great utility and easy availability of molecular data have made molecules the most common type of data used for phylogeny reconstruction in most species groups. The second question concerning the mechanisms of molecular evolution is studied by estimating the rates of nucleotide and amino acid substitutions, and by testing models of mutation and selection using sequence data.

Both areas of research have experienced phenomenal growth in the past few decades, due to the explosive accumulation of genetic sequence data, improved computer hardware and software, and development of sophisticated statistical methods suitable for addressing interesting biological questions. By all indications, this growth is bound to continue, especially on the front of data generation. Phylogenetic analysis has entered the genomic age, with large data sets consisting of hundreds of species or sequences analysed routinely. The debate of morphology versus molecules is largely over; the values of both kinds of data are well appreciated by most researchers. The philosophical debate concerning parsimony versus likelihood is ongoing but appeared to become less acrimonious. Much exciting progress has been made to develop and implement powerful statistical methods and models, which are now used routinely in analysis of real data sets.

The time appears ripe to summarize the methodological advancements in the field, and this book is such an attempt. I make no effort to be comprehensive in the coverage. There is hardly such a need now, thanks to recent publication of Joseph Felsenstein's (2004) treatise, which has discussed almost everything relevant to phylogenies. Instead I take the view that molecular evolutionary analysis, including reconstruction of phylogenies and inference of the evolutionary process, is a problem of statistical inference (Cavalli-Sforza and Edwards 1967). Thus well-established statistical methods such as likelihood and Bayesian are described as standard. Heuristic and approximate methods are discussed from such a viewpoint and are often used to introduce the central concepts, because of their simplicity and intuitive appeal, before more rigorous methods are described. I include some discussions of implementation issues so that the book can serve as a reference for researchers developing methods of data analysis.

The book is written for upper-level undergraduate students, research students, and researchers in evolutionary biology, molecular systematics, and population genetics. It is hoped that biologists who have used software programs to analyse their own data will find the book particularly useful in helping them understand the workings of the methods. The book emphasizes essential concepts but includes detailed mathematical derivations, so it can be read by statisticians, mathematicians, and computer scientists, who would like to work in this exciting area of computational biology.

The book assumes an elementary knowledge of genetics, as provided, for example, by Chapter 1 of Graur and Li (2000). Knowledge of basic statistics or biostatistics is assumed, and calculus and linear algebra is needed in some parts of the book. Likelihood and Bayesian statistics are introduced using simple examples and then used in more sophisticated analyses. Readers who would like a systematic and comprehensive treatment of these methods should consult many of the excellent textbooks in probability theory and mathematical statistics, for example DeGroot and Schervish (2002) at the elementary level, and Davison (2003), Stuart et al. (1999), and Leonard and Hsu (1999) at more advanced levels.

The book is organized as follows. Part I consists of two chapters and introduces Markov-process models of sequence evolution. Chapter 1 discusses models of nucleotide substitution and calculation of the distance between a pair of sequences. This is perhaps the simplest phylogenetic analysis, and I take the opportunity to introduce the theory of Markov chains and the maximum likelihood method, which are used extensively later in the book. As a result, this chapter is probably most challenging for the biologist reader. Chapter 2 describes Markov-process models of amino acid and codon substitution and their use in calculation of the distance between two protein sequences and in estimation of synonymous and nonsynonymous substitution rates between two protein-coding DNA sequences. Part II deals with methods of phylogeny reconstruction. Parsimony and distance methods are discussed briefly (Chapter 3), while likelihood and Bayesian methods are covered in depth (Chapters 4 and 5). Chapter 5 is an expanded version of the chapter in *Mathematics in Phylogeny and Evolution* edited by Olivier Gascuel (Oxford University Press, 2005). Chapter 6 provides a review of studies that compare different phylogeny reconstruction methods and covers testing of trees. Part III discusses a few applications of phylogenetic methods to study the evolutionary process, such as testing the molecular clock and using the clock to estimate species divergence times (Chapter 7), and applications of models of codon substitution to detect natural selection affecting protein evolution (Chapter 8). Chapter 9 discusses basic techniques of computer simulation. Chapter 10 includes a discussion of current challenges and future perspectives of the field. A brief review of major phylogenetics software packages is included in Appendix C. Sections marked with an asterisk * are technical and may be skipped.

Example data sets used in the book and small C programs that implement algorithms discussed in the book are posted at the web site for the book: http://abacus.gene.ucl.ac.uk/CME/. It will also include a list of errors discovered since publication of the book. Please report errors you discover to the author at z.yang@ucl.ac.uk.

I am grateful to a number of colleagues who read earlier versions of chapters of this book and provided constructive comments and criticisms: Hiroshi Akashi (chapters 2 and 8), Adam Eyre-Walker (chapter 2), Jim Mallet (chapters 4 and 5), Konrad Scheffler (chapters 1, 5, and 8), Elliott Sober (chapter 6), Mike Steel (chapter 6), Jeff Thorne (chapter 6), Simon Whelan (chapter 1), and Anne Yoder (chapter 6). Special thanks go to Karen Cranston, Ligia Mateiu and Fengrong Ren, who read the whole book and provided many detailed suggestions. Jessica Vamathevan and Richard Emes were victims of my experiments when I tested some difficult passages. Needless to say, all errors that remain are mine. Thanks are also due to Ian Sherman and Stefanie Gehrig at Oxford University Press for initiating this project and for valuable support and patience throughout it.

Ziheng Yang
London
March 2006

Contents

	PART I: MODELLING MOLECULAR EVOLUTION	1
1	**Models of nucleotide substitution**	**3**
1.1	Introduction	3
1.2	Markov models of nucleotide substitution and distance estimation	4
	1.2.1 The JC69 model	4
	1.2.2 The K80 model	10
	1.2.3 HKY85, F84, TN93, etc.	11
	1.2.4 The transition/transversion rate ratio	17
1.3	Variable substitution rates across sites	18
1.4	Maximum likelihood estimation	22
	1.4.1 The JC69 model	22
	1.4.2 The K80 model	25
	*1.4.3 Profile and integrated likelihood methods	27
1.5	Markov chains and distance estimation under general models	30
	1.5.1 General theory	30
	1.5.2 The general time-reversible (GTR) model	33
1.6	Discussions	37
	1.6.1 Distance estimation under different substitution models	37
	1.6.2 Limitations of pairwise comparison	37
1.7	Exercises	38
2	**Models of amino acid and codon substitution**	**40**
2.1	Introduction	40
2.2	Models of amino acid replacement	40
	2.2.1 Empirical models	40
	2.2.2 Mechanistic models	43
	2.2.3 Among-site heterogeneity	44
2.3	Estimation of distance between two protein sequences	45
	2.3.1 The Poisson model	45

xii • Contents

	2.3.2	Empirical models	46
	2.3.3	Gamma distances	46
	2.3.4	Example: distance between cat and rabbit p53 genes	47
2.4	Models of codon substitution		48
2.5	Estimation of synonymous and nonsynonymous substitution rates		49
	2.5.1	Counting methods	50
	2.5.2	Maximum likelihood method	58
	2.5.3	Comparison of methods	61
	*2.5.4	Interpretation and a plethora of distances	62
*2.6	Numerical calculation of the transition-probability matrix		68
2.7	Exercises		70

PART II: PHYLOGENY RECONSTRUCTION 71

3 Phylogeny reconstruction: overview 73
3.1	Tree concepts		73
	3.1.1	Terminology	73
	3.1.2	Topological distance between trees	77
	3.1.3	Consensus trees	79
	3.1.4	Gene trees and species trees	80
	3.1.5	Classification of tree-reconstruction methods	81
3.2	Exhaustive and heuristic tree search		82
	3.2.1	Exhaustive tree search	82
	3.2.2	Heuristic tree search	83
	3.2.3	Branch swapping	84
	3.2.4	Local peaks in the tree space	87
	3.2.5	Stochastic tree search	89
3.3	Distance methods		89
	3.3.1	Least-squares method	90
	3.3.2	Neighbour-joining method	92
3.4	Maximum parsimony		93
	3.4.1	Brief history	93
	3.4.2	Counting the minimum number of changes given the tree	94
	3.4.3	Weighted parsimony and transversion parsimony	95
	3.4.4	Long-branch attraction	98
	3.4.5	Assumptions of parsimony	99

4 Maximum likelihood methods 100
4.1	Introduction		100
4.2	Likelihood calculation on tree		100
	4.2.1	Data, model, tree, and likelihood	100
	4.2.2	The pruning algorithm	102
	4.2.3	Time reversibility, the root of the tree and the molecular clock	106

	4.2.4	Missing data and alignment gaps	107
	4.2.5	An numerical example: phylogeny of apes	108
4.3	Likelihood calculation under more complex models		109
	4.3.1	Models of variable rates among sites	110
	4.3.2	Models for combined analysis of multiple data sets	116
	4.3.3	Nonhomogeneous and nonstationary models	118
	4.3.4	Amino acid and codon models	119
4.4	Reconstruction of ancestral states		119
	4.4.1	Overview	119
	4.4.2	Empirical and hierarchical Bayes reconstruction	121
	*4.4.3	Discrete morphological characters	124
	4.4.4	Systematic biases in ancestral reconstruction	126
*4.5	Numerical algorithms for maximum likelihood estimation		128
	4.5.1	Univariate optimization	129
	4.5.2	Multivariate optimization	131
	4.5.3	Optimization on a fixed tree	134
	4.5.4	Multiple local peaks on the likelihood surface for a fixed tree	135
	4.5.5	Search for the maximum likelihood tree	136
4.6	Approximations to likelihood		137
4.7	Model selection and robustness		137
	4.7.1	LRT, AIC, and BIC	137
	4.7.2	Model adequacy and robustness	142
4.8	Exercises		144
5	**Bayesian methods**		145
5.1	The Bayesian paradigm		145
	5.1.1	Overview	145
	5.1.2	Bayes's theorem	146
	5.1.3	Classical versus Bayesian statistics	151
5.2	Prior		158
5.3	Markov chain Monte Carlo		159
	5.3.1	Monte Carlo integration	160
	5.3.2	Metropolis–Hastings algorithm	161
	5.3.3	Single-component Metropolis–Hastings algorithm	164
	5.3.4	Gibbs sampler	166
	5.3.5	Metropolis-coupled MCMC (MCMCMC or MC^3)	166
5.4	Simple moves and their proposal ratios		167
	5.4.1	Sliding window using the uniform proposal	168
	5.4.2	Sliding window using normal proposal	168
	5.4.3	Sliding window using the multivariate normal proposal	169
	5.4.4	Proportional shrinking and expanding	170
5.5	Monitoring Markov chains and processing output		171
	5.5.1	Validating and diagnosing MCMC algorithms	171

		5.5.2	Potential scale reduction statistic	173
	5.6		Bayesian phylogenetics	174
		5.5.3	Processing output	174
		5.6.1	Brief history	174
		5.6.2	General framework	175
		5.6.3	Summarizing MCMC output	175
		5.6.4	Bayesian versus likelihood	177
		5.6.5	A numerical example: phylogeny of apes	180
	5.7		MCMC algorithms under the coalescent model	181
		5.7.1	Overview	181
		5.7.2	Estimation of θ	181
	5.8		Exercises	184

6 Comparison of methods and tests on trees 185

6.1	Statistical performance of tree-reconstruction methods	186
	6.1.1 Criteria	186
	6.1.2 Performance	188
6.2	Likelihood	190
	6.2.1 Contrast with conventional parameter estimation	190
	6.2.2 Consistency	191
	6.2.3 Efficiency	192
	6.2.4 Robustness	196
6.3	Parsimony	198
	6.3.1 Equivalence with misbehaved likelihood models	198
	6.3.2 Equivalence with well-behaved likelihood models	201
	6.3.3 Assumptions and Justifications	204
6.4	Testing hypotheses concerning trees	206
	6.4.1 Bootstrap	207
	6.4.2 Interior branch test	210
	6.4.3 Kishino-Hasegawa test and modifications	211
	6.4.4 Indexes used in parsimony analysis	213
	6.4.5 Example: phylogeny of apes	214
*6.5	Appendix: Tuffley and Steel's likelihood analysis of one character	215

PART III: ADVANCED TOPICS 221

7 Molecular clock and estimation of species divergence times 223

7.1	Overview	223
7.2	Tests of the molecular clock	225
	7.2.1 Relative-rate tests	225
	7.2.2 Likelihood ratio test	226

	7.2.3	Limitations of the clock tests	227
	7.2.4	Index of dispersion	228
7.3	Likelihood estimation of divergence times		228
	7.3.1	Global-clock model	228
	7.3.2	Local-clock models	230
	7.3.3	Heuristic rate-smoothing methods	231
	7.3.4	Dating primate divergences	233
	*7.3.5	Uncertainties in fossils	235
7.4	Bayesian estimation of divergence times		245
	7.4.1	General framework	245
	7.4.2	Calculation of the likelihood	246
	7.4.3	Prior on rates	247
	7.4.4	Uncertainties in fossils and prior on divergence times	248
	7.4.5	Application to primate and mammalian divergences	252
7.5	Perspectives		257

8 Neutral and adaptive protein evolution 259
8.1	Introduction		259
8.2	The neutral theory and tests of neutrality		260
	8.2.1	The neutral and nearly neutral theory	260
	8.2.2	Tajima's D statistic	262
	8.2.3	Fu and Li's D and Fay and Wu's H statistics	264
	8.2.4	McDonald-Kreitman test and estimation of selective strength	265
	8.2.5	Hudson–Kreitman–Aquade test	267
8.3	Lineages undergoing adaptive evolution		268
	8.3.1	Heuristic methods	268
	8.3.2	Likelihood method	269
8.4	Amino acid sites undergoing adaptive evolution		271
	8.4.1	Three strategies	271
	8.4.2	Likelihood ratio test of positive selection under random-sites models	273
	8.4.3	Identification of sites under positive selection	276
	8.4.4	Positive selection in the human major histocompatability (MHC) locus	276
8.5	Adaptive evolution affecting particular sites and lineages		279
	8.5.1	Branch-site test of positive selection	279
	8.5.2	Other similar models	281
	8.5.3	Adaptive evolution in angiosperm phytochromes	282
8.6	Assumptions, limitations, and comparisons		284
	8.6.1	Limitations of current methods	284
	8.6.2	Comparison between tests of neutrality and tests based on d_N and d_S	286
8.7	Adaptively evolving genes		286

xvi • Contents

9	Simulating molecular evolution	293
9.1	Introduction	293
9.2	Random number generator	294
9.3	Generation of continuous random variables	295
9.4	Generation of discrete random variables	296
	9.4.1 Discrete uniform distribution	296
	9.4.2 Binomial distribution	297
	9.4.3 General discrete distribution	297
	9.4.4 Multinomial distribution	298
	9.4.5 The composition method for mixture distributions	298
	*9.4.6 The alias method for sampling from a discrete distribution	299
9.5	Simulating molecular evolution	302
	9.5.1 Simulating sequences on a fixed tree	302
	9.5.2 Generating random trees	305
9.6	Exercises	306
10	Perspectives	308
10.1	Theoretical issues in phylogeny reconstruction	308
10.2	Computational issues in analysis of large and heterogeneous data sets	309
10.3	Genome rearrangement data	309
10.4	Comparative genomics	310

Appendices		311
A:	Functions of random variables	311
B:	The delta technique	313
C:	Phylogenetics software	316
References		319
Index		353

Part I

Modelling molecular evolution

1

Models of nucleotide substitution

1.1 Introduction

Calculation of the distance between two sequences is perhaps the simplest phylogenetic analysis, yet it is important for two reasons. First, calculation of pairwise distances is the first step in distance-matrix methods of phylogeny reconstruction, which use cluster algorithms to convert a distance matrix into a phylogenetic tree. Second, Markov-process models of nucleotide substitution used in distance calculation form the basis of likelihood and Bayesian analysis of multiple sequences on a phylogeny. Indeed, joint multiple sequence analysis under the same model can be viewed as a natural extension of pairwise distance calculation. Thus besides discussing distance estimation, this chapter introduces the theory of Markov chains used in modelling nucleotide substitution in a DNA sequence. It also introduces the method of maximum likelihood (ML). Bayesian estimation of pairwise distances and Bayesian phylogenetics are introduced in Chapter 5.

The distance between two sequences is defined as the expected number of nucleotide substitutions per site. If the evolutionary rate is constant over time, the distance will increase linearly with the time of divergence. A simplistic distance measure is the proportion of different sites, sometimes called the p distance. If 10 sites are different between two sequences, each 100 nucleotides long, then $p = 10\% = 0.1$. This raw proportion works fine for very closely related sequences but is otherwise a clear underestimate of the number of substitutions that have occurred. A variable site may result from more than one substitution, and even a constant site, with the same nucleotide in the two sequences, may harbour back or parallel substitutions (Fig. 1.1). Multiple substitutions at the same site or *multiple hits* cause some changes to be hidden. As a result, p is not a linear function of evolutionary time. Thus the raw proportion p is usable only for highly similar sequences, with $p < 5\%$, say.

To estimate the number of substitutions, we need a probabilistic model to describe changes between nucleotides. Continuous-time Markov chains are commonly used for this purpose. The nucleotide sites in the sequence are normally assumed to be evolving independently of each other. Substitutions at any particular site are described by a Markov chain, with the four nucleotides to be the *states* of the chain. The main feature of a Markov chain is that it has no memory: 'given the present, the future does not depend on the past'. In other words, the probability with which

4 · 1 Models of nucleotide substitution

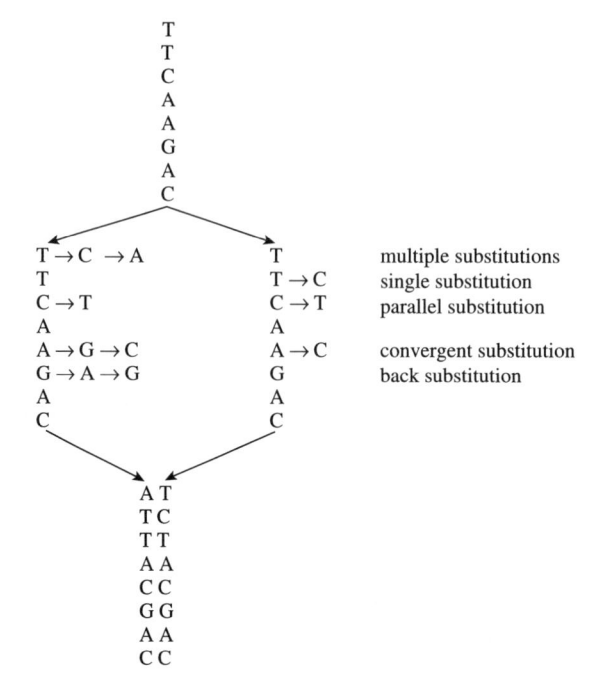

```
              T
              T
              C
              A
              A
              G
              A
              C
         ←  ─────  →
    T→C →A          T                  multiple substitutions
    T                T→C               single substitution
    C→T              C→T               parallel substitution
    A                A
    A→G→C            A→C               convergent substitution
    G→A→G            G                 back substitution
    A                A
    C                C
         ↘        ↙
              A T
              T C
              T T
              A A
              C C
              G G
              A A
              C C
```

Fig. 1.1 Illustration of multiple substitutions at the same site or multiple hits. An ancestral sequence diverged into two sequences and has since accumulated nucleotide substitutions independently along the two lineages. Only two *differences* are observed between the two present-day sequences, so that the proportion of different sites is $\hat{p} = 2/8 = 0.25$, while in fact as many as 10 *substitutions* (seven on the left lineage and three on the right lineage) occurred so that the true distance is $10/8 = 1.25$ substitutions per site. Constructed following Graur and Li (2000).

the chain jumps into other nucleotide states depends on the current state, but not on how the current state is reached. This is known as the *Markovian property*. Besides this basic assumption, we often place further constraints on substitution rates between nucleotides, leading to different models of nucleotide substitution. A few commonly used models are summarized in Table 1.1 and illustrated in Fig. 1.2. These are discussed below.

1.2 Markov models of nucleotide substitution and distance estimation

1.2.1 The JC69 model

The JC69 model (Jukes and Cantor 1969) assumes that every nucleotide has the same rate λ of changing into any other nucleotide. We use q_{ij} to denote the instantaneous

Table 1.1 Substitution-rate matrices for commonly used Markov models of nucleotide substitution

	From	To: T	C	A	G
JC69 (Jukes and Cantor 1969)	T	·	λ	λ	λ
	C	λ	·	λ	λ
	A	λ	λ	·	λ
	G	λ	λ	λ	·
K80 (Kimura 1980)	T	·	α	β	β
	C	α	·	β	β
	A	β	β	·	α
	G	β	β	α	·
F81 (Felsenstein 1981)	T	·	π_C	π_A	π_G
	C	π_T	·	π_A	π_G
	A	π_T	π_C	·	π_G
	G	π_T	π_C	π_A	·
HKY85 (Hasegawa et al. 1984, 1985)	T	·	$\alpha\pi_C$	$\beta\pi_A$	$\beta\pi_G$
	C	$\alpha\pi_T$	·	$\beta\pi_A$	$\beta\pi_G$
	A	$\beta\pi_T$	$\beta\pi_C$	·	$\alpha\pi_G$
	G	$\beta\pi_T$	$\beta\pi_C$	$\alpha\pi_A$	·
F84 (Felsenstein, DNAML program since 1984)	T	·	$(1+\kappa/\pi_Y)\beta\pi_C$	$\beta\pi_A$	$\beta\pi_G$
	C	$(1+\kappa/\pi_Y)\beta\pi_T$	·	$\beta\pi_A$	$\beta\pi_G$
	A	$\beta\pi_T$	$\beta\pi_C$	·	$(1+\kappa/\pi_R)\beta\pi_G$
	G	$\beta\pi_T$	$\beta\pi_C$	$(1+\kappa/\pi_R)\beta\pi_A$	·
TN93 (Tamura and Nei 1993)	T	·	$\alpha_1\pi_C$	$\beta\pi_A$	$\beta\pi_G$
	C	$\alpha_1\pi_T$	·	$\beta\pi_A$	$\beta\pi_G$
	A	$\beta\pi_T$	$\beta\pi_C$	·	$\alpha_2\pi_G$
	G	$\beta\pi_T$	$\beta\pi_C$	$\alpha_2\pi_A$	·
GTR (REV) (Tavaré 1986; Yang 1994b; Zharkikh 1994)	T	·	$a\pi_C$	$b\pi_A$	$c\pi_G$
	C	$a\pi_T$	·	$d\pi_A$	$e\pi_G$
	A	$b\pi_T$	$d\pi_C$	·	$f\pi_G$
	G	$c\pi_T$	$e\pi_C$	$f\pi_A$	·
UNREST (Yang 1994b)	T	·	q_{TC}	q_{TA}	q_{TG}
	C	q_{CT}	·	q_{CA}	q_{CG}
	A	q_{AT}	q_{AC}	·	q_{AG}
	G	q_{GT}	q_{GC}	q_{GA}	·

The diagonals of the matrix are determined by the requirement that each row sums to 0. The equilibrium distribution is $\pi = (1/4, 1/4, 1/4, 1/4)$ under JC69 and K80, and $\pi = (\pi_T, \pi_C, \pi_A, \pi_G)$ under F81, F84, HKY85, TN93, and GTR. Under the general unrestricted (UNREST) model, it is given by the equations $\pi Q = 0$ under the constraint $\sum_i \pi_i = 1$.

6 • *I Models of nucleotide substitution*

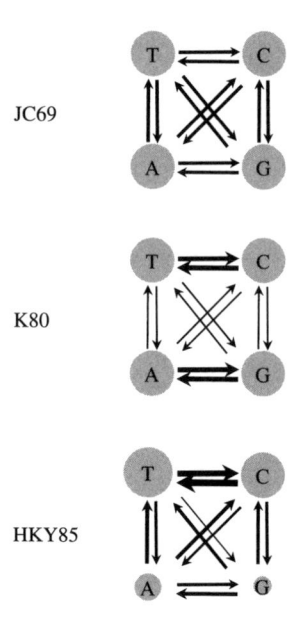

Fig. 1.2 Relative substitution rates between nucleotides under three Markov-chain models of nucleotide substitution: JC69 (Jukes and Cantor 1969), K80 (Kimura 1980), and HKY85 (Hasegawa et al. 1985). The thickness of the lines represents the substitution rates while the sizes of the circles represent the steady-state distribution.

rate of substitution from nucleotide i to nucleotide j, with $i, j =$ T, C, A, or G. Thus the *substitution-rate matrix* is

$$Q = \{q_{ij}\} = \begin{bmatrix} -3\lambda & \lambda & \lambda & \lambda \\ \lambda & -3\lambda & \lambda & \lambda \\ \lambda & \lambda & -3\lambda & \lambda \\ \lambda & \lambda & \lambda & -3\lambda \end{bmatrix}, \qquad (1.1)$$

where the nucleotides are ordered T, C, A, and G. Each row of the matrix sums to 0. The total rate of substitution of any nucleotide i is 3λ, which is $-q_{ii}$.

Note that $q_{ij}\Delta t$ gives the probability that any given nucleotide i will change to a different nucleotide j in an infinitely small time interval Δt. To characterize the Markov chain, we need a similar probability over any time $t > 0$. This is the *transition probability*; $p_{ij}(t)$ is the probability that a given nucleotide i will become j time t later. The matrix $P(t) = \{p_{ij}(t)\}$ is known as the *transition-probability matrix*. As will be discussed later in Section 1.5,

$$P(t) = e^{Qt}. \qquad (1.2)$$

1.2 Markov models of nucleotide substitution

Calculation of this matrix exponential is discussed later. For the moment, we simply give the solution

$$P(t) = e^{Qt} = \begin{bmatrix} p_0(t) & p_1(t) & p_1(t) & p_1(t) \\ p_1(t) & p_0(t) & p_1(t) & p_1(t) \\ p_1(t) & p_1(t) & p_0(t) & p_1(t) \\ p_1(t) & p_1(t) & p_1(t) & p_0(t) \end{bmatrix}, \quad \text{with} \quad \begin{cases} p_0(t) = \frac{1}{4} + \frac{3}{4}e^{-4\lambda t}, \\ p_1(t) = \frac{1}{4} - \frac{1}{4}e^{-4\lambda t}. \end{cases}$$

(1.3)

Imagine a long sequence with nucleotide i at every site; we let every site evolve for a time period t. Then the proportion of nucleotide j in the sequence will be $p_{ij}(t)$, for $j = $ T, C, A, G.

The two different elements of the transition-probability matrix, $p_0(t)$ and $p_1(t)$, are plotted in Fig. 1.3. A few features of the matrix are worth noting. First, every row of $P(t)$ sums to 1, because the chain has to be in one of the four nucleotide states at time t. Second, $P(0) = I$, the identity matrix, reflecting the case of no evolution ($t = 0$). Third, rate λ and time t occur in the transition probabilities only in the form of a product λt. Thus if we are given a source sequence and a target sequence, it will be impossible to tell whether one sequence has evolved into the other at rate λ over time t or at rate 2λ over time $t/2$. In fact, the sequences will look the same for any combination of λ and t as long as λt is fixed. With no external information about either the time or the rate, we can estimate only the distance, but not time and rate individually.

Lastly, when $t \to \infty$, $p_{ij}(t) = 1/4$, for all i and j. This represents the case where so many substitutions have occurred at every site that the target nucleotide is random, with probability 1/4 for every nucleotide, irrespective of the starting nucleotide. The probability that the chain is in state j when $t \to \infty$ is represented by π_j and the distribution $(\pi_T, \pi_C, \pi_A, \pi_G)$ is known as the *limiting distribution* of the chain. For the JC69 model, $\pi_j = 1/4$ for every nucleotide j. If the states of the chain are already in the limiting distribution, the chain will stay in that distribution, so the limiting

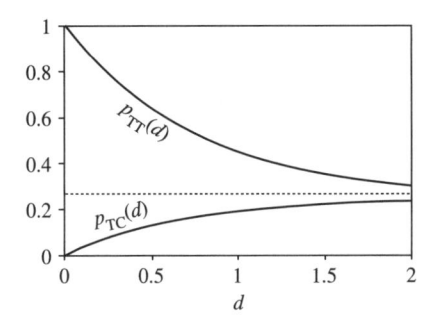

Fig. 1.3 Transition probabilities under the JC69 model (equation 1.3) plotted against distance $d = 3\lambda t$, measured in the expected number of substitutions per site.

distribution is also the *steady-state distribution* or *stationary distribution*. In other words, if a long sequence starts with T at every site, the proportions of the four nucleotides T, C, A, and G will drift away from $(1, 0, 0, 0)$ and approach the limiting distribution $(1/4, 1/4, 1/4, 1/4)$, as the sequence evolves. If the sequence starts with equal proportions of the four nucleotides, the sequence will continue to have equal proportions of the four nucleotides as the sequence evolves. The Markov chain is said to be stationary, or nucleotide substitutions are said to be in equilibrium. This is an assumption made in almost all models in phylogenetic analysis and is clearly violated if the sequences in the data have different base compositions.

How does the Markov-chain model correct for multiple hits and recover the hidden changes illustrated in Fig. 1.1? This is achieved through the calculation of the transition probabilities using equation (1.2), which accommodates all the possible paths the evolutionary process might have taken. In particular, the transition probabilities for a Markov chain satisfy the following equation, known as the Chapman–Kolmogorov theorem (e.g. Grimmett and Stirzaker 1992, p. 239)

$$p_{ij}(t_1 + t_2) = \sum_k p_{ik}(t_1) p_{kj}(t_2). \qquad (1.4)$$

The probability that nucleotide i will become nucleotide j time $t_1 + t_2$ later is a sum over all possible states k at any intermediate time point t_1 (Fig. 1.4).

We now consider estimation of the distance between two sequences. From equation (1.1), the total substitution rate for any nucleotide is 3λ. If the two sequences are separated by time t, for example, if they diverged from a common ancestor time $t/2$ ago, the distance between the two sequences will be $d = 3\lambda t$. Suppose x out of n sites are different between the two sequences, so that the proportion of different sites is $\hat{p} = x/n$. (The hat is used to indicate that the proportion is an estimate from the data.) To derive the expected probability p of different sites, consider one sequence as the ancestor of the other. By the symmetry of the model (equation 1.3), this is

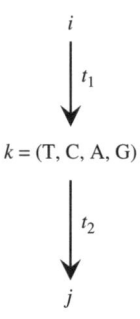

$$i \atop \downarrow t_1$$

$$k = (T, C, A, G)$$

$$\downarrow t_2 \atop j$$

Fig. 1.4 Illustration of the Chapman–Kolmogorov theorem. The transition probability from any nucleotide i to any nucleotide j over time $t_1 + t_2$ is a sum over all possible states k at any intermediate time point t_1.

equivalent to considering the two sequences as descendants of an extinct common ancestor. From equation (1.3), the probability that the nucleotide in the descendant sequence is different from the nucleotide in the ancestral sequence is

$$p = 3p_1(t) = \tfrac{3}{4} - \tfrac{3}{4}e^{-4\lambda t} = \tfrac{3}{4} - \tfrac{3}{4}e^{-4d/3}. \tag{1.5}$$

By equating this to the observed proportion $\hat{p}$, we obtain an estimate of distance as

$$\hat{d} = -\tfrac{3}{4}\log\left(1 - \tfrac{4}{3}\hat{p}\right), \tag{1.6}$$

where the base of the logarithm is the constant e. If $\hat{p} > 3/4$, the distance formula will be inapplicable; two random sequences should have about 75% different sites, and if $\hat{p} > 3/4$, the estimated distance is infinite. To derive the variance of $\hat{d}$, note that $\hat{p}$ is a binomial proportion with variance $\hat{p}(1-\hat{p})/n$. Considering $\hat{d}$ as a function of $\hat{p}$ and using the so-called delta technique (see Appendix B), we obtain

$$\text{var}(\hat{d}) = \text{var}(\hat{p}) \times \left|\frac{d\hat{d}}{d\hat{p}}\right|^2 = \frac{\hat{p}(1-\hat{p})}{n} \times \frac{1}{(1-4\hat{p}/3)^2} \tag{1.7}$$

(Kimura and Ohta 1972).

Example. Consider the sequences of human and orangutan 12s rRNA genes from the mitochondrial genome, summarized in Table 1.2. From the table, $x = 90$ out of the $n = 948$ sites are different, so that $\hat{p} = x/n = 0.09494$. By equation (1.6), $\hat{d} = 0.1015$. Equation (1.7) gives the variance of $\hat{d}$ as 0.0001188 and standard error 0.0109. The approximate 95% confidence interval is thus $0.1015 \pm 1.96 \times 0.0109$ or $(0.0801, 0.1229)$. □

Table 1.2 Numbers and frequencies (in parentheses) of sites for the 16 site configurations (patterns) in human and orangutan mitochondrial 12s rRNA genes

		Human			Sum (π_i)
Orangutan	T	C	A	G	
T	179 (0.188819)	23 (0.024262)	1 (0.001055)	0 (0)	0.2141
C	30 (0.031646)	219 (0.231013)	2 (0.002110)	0 (0)	0.2648
A	2 (0.002110)	1 (0.001055)	291 (0.306962)	10 (0.010549)	0.3207
G	0 (0)	0 (0)	21 (0.022152)	169 (0.178270)	0.2004
Sum (π_j)	0.2226	0.2563	0.3323	0.1888	1

GenBank accession numbers for the human and orangutan sequences are D38112 and NC_001646, respectively (Horai et al. 1995). There are 954 sites in the alignment, but six sites involve alignment gaps and are removed, leaving 948 sites in each sequence. The average base frequencies in the two sequences are 0.2184 (T), 0.2605 (C), 0.3265 (A), and 0.1946 (G).

1.2.2 The K80 model

Substitutions between the two pyrimidines (T ↔ C) or between the two purines (A ↔ G) are called transitions, while those between a pyrimidine and a purine (T, C ↔ A, G) are called transversions. In real data, transitions often occur at higher rates than transversions. Thus Kimura (1980) proposed a model that accounts for different transition and transversion rates. Note that the biologist's use of the term transition (as opposed to transversion) has nothing to do with the probabilist's use of the same term (as in transition probability). Typically the usage is clear from the context with little risk of confusion.

Let the substitution rates be α for transitions and β for transversions. This model is referred to as K80, also known as Kimura's two-parameter model. The rate matrix is as follows (see also Fig. 1.2)

$$Q = \begin{bmatrix} -(\alpha+2\beta) & \alpha & \beta & \beta \\ \alpha & -(\alpha+2\beta) & \beta & \beta \\ \beta & \beta & -(\alpha+2\beta) & \alpha \\ \beta & \beta & \alpha & -(\alpha+2\beta) \end{bmatrix}. \quad (1.8)$$

The total substitution rate for any nucleotide is $\alpha + 2\beta$, and the distance between two sequences separated by time t is $d = (\alpha + 2\beta)t$. Note that αt is the expected number of transitions per site and $2\beta t$ is the expected number of transversions per site. One can use αt and βt as the two parameters in the model, but it is often more convenient to use the distance d and the transition/transversion rate ratio $\kappa = \alpha/\beta$. The matrix of transition probabilities is obtained as

$$P(t) = e^{Qt} = \begin{bmatrix} p_0(t) & p_1(t) & p_2(t) & p_2(t) \\ p_1(t) & p_0(t) & p_2(t) & p_2(t) \\ p_2(t) & p_2(t) & p_0(t) & p_1(t) \\ p_2(t) & p_2(t) & p_1(t) & p_0(t) \end{bmatrix}, \quad (1.9)$$

where the three distinct elements of the matrix are

$$p_0(t) = \tfrac{1}{4} + \tfrac{1}{4}e^{-4\beta t} + \tfrac{1}{2}e^{-2(\alpha+\beta)t} = \tfrac{1}{4} + \tfrac{1}{4}e^{-4d/(\kappa+2)} + \tfrac{1}{2}e^{-2d(\kappa+1)/(\kappa+2)},$$

$$p_1(t) = \tfrac{1}{4} + \tfrac{1}{4}e^{-4\beta t} - \tfrac{1}{2}e^{-2(\alpha+\beta)t} = \tfrac{1}{4} + \tfrac{1}{4}e^{-4d/(\kappa+2)} - \tfrac{1}{2}e^{-2d(\kappa+1)/(\kappa+2)}, \quad (1.10)$$

$$p_2(t) = \tfrac{1}{4} - \tfrac{1}{4}e^{-4\beta t} = \tfrac{1}{4} - \tfrac{1}{4}e^{-4d/(\kappa+2)}$$

(Kimura 1980; Li 1986). Note that $p_0(t) + p_1(t) + 2p_2(t) = 1$.

The sequence data can be summarized as the proportions of sites with transitional and transversional differences. Let these be S and V, respectively. Again by the symmetry of the model (equation 1.9), the probability that a site is occupied by nucleotides with a transitional difference is $E(S) = p_1(t)$. Similarly $E(V) = 2p_2(t)$. Equating these to the observed proportions S and V leads to two simultaneous equations in two unknowns, which are easily solved to give

$$\hat{d} = -\tfrac{1}{2}\log(1 - 2S - V) - \tfrac{1}{4}\log(1 - 2V),$$
$$\hat{\kappa} = \frac{2 \times \log(1 - 2S - V)}{\log(1 - 2V)} - 1 \quad (1.11)$$

(Kimura 1980; Jukes 1987). Equivalently the transition distance αt and the transversion distance $2\beta t$ are estimated as

$$\widehat{\alpha t} = -\tfrac{1}{2}\log(1 - 2S - V) + \tfrac{1}{4}\log(1 - 2V),$$
$$\widehat{2\beta t} = -\tfrac{1}{2}\log(1 - 2V). \quad (1.12)$$

The distance formula is applicable only if $1 - 2S - V > 0$ and $1 - 2V > 0$. As S and V are multinomial proportions with $\mathrm{var}(S) = S(1-S)/n$, $\mathrm{var}(V) = V(1-V)/n$, and $\mathrm{cov}(S, V) = -SV/n$, we can use the delta technique to derive the variance–covariance matrix of $\hat{d}$ and $\hat{\kappa}$ (see Appendix B). In particular, the variance of $\hat{d}$ is

$$\mathrm{var}(\hat{d}) = [a^2 S + b^2 V - (aS + bV)^2]/n, \quad (1.13)$$

where

$$a = (1 - 2S - V)^{-1},$$
$$b = \tfrac{1}{2}\left[(1 - 2S - V)^{-1} + (1 - 2V)^{-1}\right]. \quad (1.14)$$

Example. For the 12s rRNA data of Table 1.2, the proportions of transitional and transversional differences are $S = (23 + 30 + 10 + 21)/948 = 0.08861$ and $V = (1 + 0 + 2 + 0 + 2 + 1 + 0 + 0)/948 = 0.00633$. Thus equations (1.11) and (1.13) give the distance and standard error as 0.1046 ± 0.0116 (Table 1.3). The estimate $\hat{\kappa} = 30.836$ indicates that the transition rate is ~ 30 times higher than the transversion rate. □

1.2.3 HKY85, F84, TN93, etc.

1.2.3.1 TN93

The models of Jukes and Cantor (1969) and Kimura (1980) have symmetrical substitution rates, with $q_{ij} = q_{ji}$ for all i and j. Such Markov chains have $\pi_i = 1/4$ for all i as the stationary distribution; that is, when the substitution process reaches equilibrium, the sequence will have equal proportions of the four nucleotides. This assumption is unrealistic for virtually every real data set. Here we consider a few models that accommodate unequal base compositions. The model of Tamura and Nei (1993), referred to as TN93, has most of the commonly used models as special cases.

Table 1.3 Estimates of distance between the human and orangutan 12s rRNA genes

Model and method	$\hat{d} \pm$ S.E.	Estimates of other parameters
Distance formula		
JC69	0.1015 ± 0.0109	
K80	0.1046 ± 0.0116	$\hat{\kappa} = 30.83 \pm 13.12$
F81	0.1016	
F84	0.1050	$\hat{\kappa} = 15.548$
TN93	0.1078	$\hat{\kappa}_1 = 44.228, \hat{\kappa}_1 = 21.789$
Maximum likelihood		
JC69 and K80	As above	
F81	0.1017 ± 0.0109	$\hat{\pi} = (0.2251, 0.2648, 0.3188, 0.1913)$
F84	0.1048 ± 0.0117	$\hat{\kappa} = 15.640$,
		$\hat{\pi} = (0.2191, 0.2602, 0.3286, 0.1921)$
HKY85	0.1048 ± 0.0117	$\hat{\kappa} = 32.137$,
		$\hat{\pi} = (0.2248, 0.2668, 0.3209, 0.1875)$
TN93	0.1048 ± 0.0117	$\hat{\kappa}_1 = 44.229, \hat{\kappa}_1 = 21.781$,
		$\hat{\pi} = (0.2185, 0.2604, 0.3275, 0.1936)$
GTR (REV)	0.1057 ± 0.0119	$\hat{a} = 2.0431, \hat{b} = 0.0821, \hat{c} = 0.0000$,
		$\hat{d} = 0.0670, \hat{e} = 0.0000$,
		$\hat{\pi} = (0.2184, 0.2606, 0.3265, 0.1946)$
UNREST	0.1057 ± 0.0120	See equation (1.59) for the estimated Q;
		$\hat{\pi} = (0.2184, 0.2606, 0.3265, 0.1946)$

We present detailed results for this model, which also apply to its special cases. The substitution-rate matrix under the TN93 model is

$$Q = \begin{bmatrix} -(\alpha_1\pi_C + \beta\pi_R) & \alpha_1\pi_C & \beta\pi_A & \beta\pi_G \\ \alpha_1\pi_T & -(\alpha_1\pi_T + \beta\pi_R) & \beta\pi_A & \beta\pi_G \\ \beta\pi_T & \beta\pi_C & -(\alpha_2\pi_G + \beta\pi_Y) & \alpha_2\pi_G \\ \beta\pi_T & \beta\pi_C & \alpha_2\pi_A & -(\alpha_2\pi_A + \beta\pi_Y) \end{bmatrix}. \quad (1.15)$$

While parameters $\pi_T, \pi_C, \pi_A, \pi_G$ are used to specify the substitution rates, they also give the stationary (equilibrium) distribution, with $\pi_Y = \pi_T + \pi_C$ and $\pi_R = \pi_A + \pi_G$ to be the frequencies of pyrimidines and purines, respectively.

The matrix of transition probabilities over time t is $P(t) = \{p_{ij}(t)\} = e^{Qt}$. The standard way for calculating an algebraic function, such as the exponential, of a matrix Q, is to *diagonalize Q* (e.g. Schott 1997, Chapter 3). Suppose Q can be written in the form

$$Q = U\Lambda U^{-1}, \quad (1.16)$$

where U is a nonsingular matrix and U^{-1} is its inverse, and Λ is a diagonal matrix $\Lambda = \text{diag}\{\lambda_1, \lambda_2, \lambda_3, \lambda_4\}$. Then we have $Q^2 = (U\Lambda U^{-1}) \cdot (U\Lambda U^{-1}) = U\Lambda^2 U^{-1} = U \text{diag}\{\lambda_1^2, \lambda_2^2, \lambda_3^2, \lambda_4^2\}U^{-1}$. Similarly $Q^m = U \text{diag}\{\lambda_1^m, \lambda_2^m, \lambda_3^m, \lambda_4^m\}U^{-1}$ for any

1.2 Markov models of nucleotide substitution

integer m. In general, any algebraic function h of matrix Q can be calculated as $h(Q) = U \text{diag}\{h(\lambda_1), h(\lambda_2), h(\lambda_3), h(\lambda_4)\}U^{-1}$ as long as $h(Q)$ exists. Thus, given equation (1.16),

$$P(t) = e^{Qt} = U \text{diag}\{\exp(\lambda_1 t), \exp(\lambda_2 t), \exp(\lambda_3 t), \exp(\lambda_4 t)\}U^{-1}. \quad (1.17)$$

The λs are the eigenvalues (or latent roots) of Q, and columns of U and rows of U^{-1} are the corresponding right and left eigenvectors of Q, respectively. Equation (1.16) is also known as the spectral decomposition of Q. The reader should consult a textbook on linear algebra for calculation of eigenvalues and eigenvectors of a matrix (e.g. Schott 1997, Chapter 3). For the TN93 model, the solution is analytical. We have $\lambda_1 = 0$, $\lambda_2 = -\beta$, $\lambda_3 = -(\pi_R \alpha_2 + \pi_Y \beta)$, and $\lambda_4 = -(\pi_Y \alpha_1 + \pi_R \beta)$,

$$U = \begin{bmatrix} 1 & 1/\pi_Y & 0 & \pi_C/\pi_Y \\ 1 & 1/\pi_Y & 0 & -\pi_T/\pi_Y \\ 1 & -1/\pi_R & \pi_G/\pi_R & 0 \\ 1 & -1/\pi_R & -\pi_A/\pi_R & 0 \end{bmatrix}, \quad (1.18)$$

$$U^{-1} = \begin{bmatrix} \pi_T & \pi_C & \pi_A & \pi_G \\ \pi_T \pi_R & \pi_C \pi_R & -\pi_A \pi_Y & -\pi_G \pi_Y \\ 0 & 0 & 1 & -1 \\ 1 & -1 & 0 & 0 \end{bmatrix}. \quad (1.19)$$

Substituting Λ, U, and U^{-1} into equation (1.17) gives

$P(t) =$

$$\begin{bmatrix} \pi_T + \frac{\pi_T \pi_R}{\pi_Y} e_2 + \frac{\pi_C}{\pi_Y} e_4 & \pi_C + \frac{\pi_C \pi_R}{\pi_Y} e_2 - \frac{\pi_C}{\pi_Y} e_4 & \pi_A(1 - e_2) & \pi_G(1 - e_2) \\ \pi_T + \frac{\pi_T \pi_R}{\pi_Y} e_2 - \frac{\pi_T}{\pi_Y} e_4 & \pi_C + \frac{\pi_C \pi_R}{\pi_Y} e_2 + \frac{\pi_T}{\pi_Y} e_4 & \pi_A(1 - e_2) & \pi_G(1 - e_2) \\ \pi_T(1 - e_2) & \pi_C(1 - e_2) & \pi_A + \frac{\pi_A \pi_Y}{\pi_R} e_2 + \frac{\pi_G}{\pi_R} e_3 & \pi_G + \frac{\pi_G \pi_Y}{\pi_R} e_2 - \frac{\pi_G}{\pi_R} e_3 \\ \pi_T(1 - e_2) & \pi_C(1 - e_2) & \pi_A + \frac{\pi_A \pi_Y}{\pi_R} e_2 - \frac{\pi_A}{\pi_R} e_3 & \pi_G + \frac{\pi_G \pi_Y}{\pi_R} e_2 + \frac{\pi_A}{\pi_R} e_3 \end{bmatrix}$$
(1.20)

where $e_2 = \exp(\lambda_2 t) = \exp(-\beta t)$, $e_3 = \exp(\lambda_3 t) = \exp[-(\pi_R \alpha_2 + \pi_Y \beta)t]$, $e_4 = \exp(\lambda_4 t) = \exp[-(\pi_Y \alpha_1 + \pi_R \beta)t]$.

When t increases from 0 to ∞, the diagonal element $p_{jj}(t)$ decreases from 1 to π_j, while the off-diagonal element $p_{ij}(t)$ increases from 0 to π_j, with $p_{ij}(\infty) = \pi_j$, irrespective of the starting nucleotide i. The limiting distribution $(\pi_T, \pi_C, \pi_A, \pi_G)$ is also the stationary distribution.

We now consider estimation of the sequence distance under the model. First the definition of distance. The substitution rate of nucleotide i is $-q_{ii} = \sum_{j \neq i} q_{ij}$, and differs among the four nucleotides. When the substitution process is in equilibrium, the amount of time the Markov chain spends in the four states T, C, A, and G is proportional to the equilibrium frequencies π_T, π_C, π_A, and π_G, respectively. Similarly, if we consider a long DNA sequence in substitution equilibrium, the proportions of sites occupied by nucleotides T, C, A, and G are π_T, π_C, π_A, and π_G, respectively. The average substitution rate is thus

$$\lambda = -\sum_i \pi_i q_{ii} = 2\pi_T \pi_C \alpha_1 + 2\pi_A \pi_G \alpha_2 + 2\pi_Y \pi_R \beta. \tag{1.21}$$

The distance between two sequences separated by time t is $d = \lambda t$.

To derive a distance estimate, we use the same strategy as for the K80 model discussed above. Let S_1 be the proportion of sites occupied by two different pyramidines (i.e. sites occupied by TC or CT in the two sequences), S_2 the proportion of sites with two different purines (i.e. sites with AG or GA), and V the proportion of sites with a transversional difference. Next, we need to derive the expected probabilities of such sites: $E(S_1)$, $E(S_2)$, and $E(V)$. We cannot use the symmetry argument as for JC69 and K80 since Q is not symmetrical. However, Q satisfies the following condition

$$\pi_i q_{ij} = \pi_j q_{ji}, \quad \text{for all } i \neq j. \tag{1.22}$$

Equivalently, $\pi_i p_{ij}(t) = \pi_j p_{ji}(t)$, for all t and for all $i \neq j$. Markov chains satisfying such conditions are said to be *time-reversible*. Reversibility means that the process will look the same whether time runs forward or backward, that is, whether we view the substitution process from the present into the future or from the present back into the past. As a result, given two sequences, the probability of data at a site is the same whether one sequence is ancestral to the other or both are descendants of an ancestral sequence. Equivalently, equation (1.22) means that the expected amount of change from i to j is equal to the amount of change from j to i; note that the rates of change may be different in the two directions: $q_{ij} \neq q_{ji}$. Now consider sequence 1 to be the ancestor of sequence 2, separated by time t. Then

$$E(S_1) = \pi_T p_{TC}(t) + \pi_C p_{CT}(t) = 2\pi_T p_{TC}(t). \tag{1.23}$$

The first term in the sum is the probability that any site has nucleotide T in sequence 1 and C in sequence 2. This equals the probability of having T in sequence 1, given by π_T, times the transition probability $p_{TC}(t)$ that T will become C in sequence 2 time t later. We refer to the nucleotides across sequences at a site as a *site configuration* or *site pattern*. Thus $\pi_T p_{TC}(t)$ is the probability of observing site pattern TC. The second term in the sum, $\pi_C p_{CT}(t)$, is the probability for site pattern CT. Similarly $E(S_2) = 2\pi_A p_{AG}(t)$ and $E(V) = 2\pi_T p_{TA}(t) + 2\pi_T p_{TG}(t) + 2\pi_C p_{CA}(t) + 2\pi_C p_{CG}(t)$. Equating the observed proportions S_1, S_2, and V to their expected probabilities leads to three simultaneous equations in three unknowns: e_2, e_3, and e_4 in the transition-probability matrix (1.20) or equivalently, d, $\kappa_1 = \alpha_1/\beta$, and $\kappa_2 = \alpha_2/\beta$. Note that the nucleotide frequency parameters π_T, π_C, π_A, and π_G can be estimated using the average observed frequencies. Solving the system of equations gives the following estimates

$$\hat{d} = \frac{2\pi_T \pi_C}{\pi_Y}(a_1 - \pi_R b) + \frac{2\pi_A \pi_G}{\pi_R}(a_2 - \pi_Y b) + 2\pi_Y \pi_R b,$$

$$\hat{\kappa}_1 = \frac{a_1 - \pi_R b}{\pi_Y b},$$ (1.24)

$$\hat{\kappa}_2 = \frac{a_2 - \pi_Y b}{\pi_R b},$$

where

$$a_1 = -\log\left(1 - \frac{\pi_Y S_1}{2\pi_T \pi_C} - \frac{V}{2\pi_Y}\right),$$

$$a_2 = -\log\left(1 - \frac{\pi_R S_2}{2\pi_A \pi_G} - \frac{V}{2\pi_R}\right),$$ (1.25)

$$b = -\log\left(1 - \frac{V}{2\pi_Y \pi_R}\right)$$

(Tamura and Nei 1993).

The formulae are inapplicable whenever π_Y or π_R is 0 or any of the arguments to the logarithm functions are ≤ 0, as may happen when the sequences are divergent. The variance of the estimated distance $\hat{d}$ can be obtained by using the delta technique, ignoring errors in the estimates of nucleotide frequencies and noting that S_1, S_2, and V are multinomial proportions. This is similar to the calculation under the model of Kimura (1980) (see Tamura and Nei 1993).

Example. For the 12s rRNA data of Table 1.2, we have the observed proportions $S_1 = (23+30)/948 = 0.05591$, $S_2 = (10+21)/948 = 0.03270$, and $V = 6/948 = 0.00633$. Equation (1.24) gives the estimates as $\hat{d} = 0.1078$, $\hat{\kappa}_1 = 44.228$, and $\hat{\kappa}_1 = 21.789$. □

1.2.3.2 HKY85, F84, etc.

Two commonly used models are special cases of the TN93 model. The first is due to Hasegawa and colleagues (Hasegawa et al. 1984, 1985). This is now commonly known as HKY85, instead of HYK84, apparently due to my misnaming (Yang 1994b). The model is obtained by setting $\alpha_1 = \alpha_2 = \alpha$ or $\kappa_1 = \kappa_2 = \kappa$ in the TN93 model (Table 1.1). The transition-probability matrix is given by equation (1.20), with α_1 and α_2 replaced by α. It is not straightforward to derive a distance formula under this model (Yang 1994b), although Rzhetsky and Nei (1994) suggested a few possibilities.

16 · 1 Models of nucleotide substitution

The second special case of the TN93 model was implemented by Joseph Felsenstein in his DNAML program since Version 2.6 (1984) of the PHYLIP package. This is now known as the F84 model. The rate matrix was first published by Hasegawa and Kishino (1989) and Kishino and Hasegawa (1989). It is obtained by setting $\alpha_1 = (1 + \kappa/\pi_Y)\beta$ and $\alpha_2 = (1 + \kappa/\pi_R)\beta$ in the TN93 model, requiring one fewer parameter (Table 1.1). Under this model, the eigenvalues of the Q matrix become $\lambda_1 = 0, \lambda_2 = -\beta, \lambda_3 = \lambda_4 = -(1+\kappa)\beta$. There are only three distinct eigenvalues, as for the K80 model, and thus it is possible to derive a distance formula.

From equation (1.21), the sequence distance is $d = \lambda t = 2(\pi_T\pi_C + \pi_A\pi_G + \pi_Y\pi_R)\beta t + 2(\pi_T\pi_C/\pi_Y + \pi_A\pi_G/\pi_R)\kappa\beta t$. The expected probabilities of sites with transitional and transversional differences are

$$E(S) = 2(\pi_T\pi_C + \pi_A\pi_G) + 2\left(\frac{\pi_T\pi_C\pi_R}{\pi_Y} + \frac{\pi_A\pi_G\pi_Y}{\pi_R}\right)e^{-\beta t}$$

$$- 2\left(\frac{\pi_T\pi_C}{\pi_Y} + \frac{\pi_A\pi_G}{\pi_R}\right)e^{-(\kappa+1)\beta t}, \quad (1.26)$$

$$E(V) = 2\pi_Y\pi_R(1 - e^{-\beta t}).$$

By equating the observed proportions S and V to their expectations, one can obtain a system of two equations in two unknowns, which can be solved to give

$$\hat{d} = 2\left(\frac{\pi_T\pi_C}{\pi_Y} + \frac{\pi_A\pi_G}{\pi_R}\right)a - 2\left(\frac{\pi_T\pi_C\pi_R}{\pi_Y} + \frac{\pi_A\pi_G\pi_Y}{\pi_R} - \pi_Y\pi_R\right)b,$$

$$\hat{\kappa} = a/b - 1, \quad (1.27)$$

where

$$a = \overline{(\kappa+1)\beta t} = -\log\left(1 - \frac{S}{2[(\pi_T\pi_C/\pi_Y) + (\pi_A\pi_G/\pi_R)]} - \frac{[(\pi_T\pi_C\pi_R/\pi_Y) + (\pi_A\pi_G\pi_Y/\pi_R)]V}{2(\pi_T\pi_C\pi_R + \pi_A\pi_G\pi_Y)}\right), \quad (1.28)$$

$$b = \overline{\beta t} = -\log\left(1 - \frac{V}{2\pi_Y\pi_R}\right)$$

(Tateno et al. 1994; Yang 1994a). The approximate variance of $\hat{d}$ can be obtained in a similar way to that under K80 (Tateno et al. 1994). The estimated distance under F84 for the 12s rRNA genes is shown in Table 1.3.

If we assume $\alpha_1 = \alpha_2 = \beta$ in the TN93 model, we obtain the F81 model (Felsenstein 1981) (Table 1.1). A distance formula was derived by Tajima and Nei (1982). Estimates under this and some other models for the 12s rRNA data set of Table 1.2 are listed in Table 1.3. It may be mentioned that the matrices Λ, U, U^{-1} and $P(t)$ derived for the TN93 model hold for its special cases, such as JC69 (Jukes and Cantor 1969), K80 (Kimura 1980), F81 (Felsenstein 1981), HKY85 (Hasegawa et al. 1984, 1985), and F84. Under some of those simpler models, simplifications are possible (see Exercise 1.2).

1.2.4 The transition/transversion rate ratio

Unfortunately three definitions of the 'transition/transversion rate ratio' are in use in the literature. The first is the ratio of the numbers (or proportions) of transitional and transversional differences between the two sequences, without correcting for multiple hits (e.g. Wakeley 1994). This is $E(S)/E(V) = p_1(t)/(2p_2(t))$ under the K80 model (see equation 1.10). For infinitely long sequences, this is close to $\alpha/(2\beta)$ under K80 when the sequences are very similar. At intermediate levels of divergence, $E(S)/E(V)$ increases with $\alpha/(2\beta)$, but the pattern is complex. When the sequences are very different, $E(S)/E(V)$ approaches $1/2$ irrespective of $\alpha/(2\beta)$. Figure 1.5 plots the $E(S)/E(V)$ ratio against the sequence divergence. Thus the ratio is meaningful only for closely related sequences. In real data sets, however, highly similar sequences may not contain much information and the estimate may involve large sampling errors. In general, the $E(S)/E(V)$ ratio is a poor measure of the transition/transversion rate difference and should be avoided.

The second measure is $\kappa = \alpha/\beta$ in the models of Kimura (1980) and Hasegawa et al. (1985), with $\kappa = 1$ meaning no rate difference between transitions and transversions. A third measure may be called the average transition/transversion ratio, and is the ratio of the expected numbers of transitional and transversional substitutions between the two sequences. This is the same measure as the first one, except that it corrects for multiple hits. For a general substitution-rate matrix (the UNREST model in Table 1.1), this is

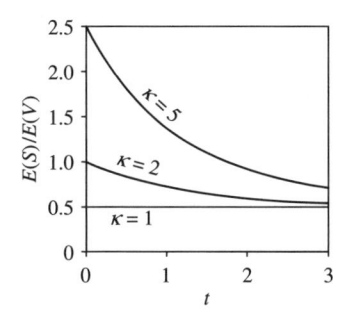

$$R = \frac{\pi_T q_{TC} + \pi_C q_{CT} + \pi_A q_{AG} + \pi_G q_{GA}}{\pi_T q_{TA} + \pi_T q_{TG} + \pi_C q_{CA} + \pi_C q_{CG} + \pi_A q_{AT} + \pi_A q_{AC} + \pi_G q_{GT} + \pi_G q_{GC}}. \tag{1.29}$$

Fig. 1.5 The transition/transversion ratio $E(S)/E(V)$ under the K80 model (Kimura 1980) plotted against sequence divergence t. This is $p_1/(2p_2)$ in equation (1.10) and corresponds to infinitely long sequences.

Table 1.4 Average transition/transversion ratio R

Model	Average transition/transversion ratio (R)
JC69	$1/2$
K80	$\kappa/2$
F81	$\dfrac{\pi_T\pi_C + \pi_A\pi_G}{\pi_Y\pi_R}$
F84	$\dfrac{\pi_T\pi_C(1+\kappa/\pi_Y) + \pi_A\pi_G(1+\kappa/\pi_R)}{\pi_Y\pi_R}$
HKY85	$\dfrac{(\pi_T\pi_C + \pi_A\pi_G)\kappa}{\pi_Y\pi_R}$
TN93	$\dfrac{\pi_T\pi_C\kappa_1 + \pi_A\pi_G\kappa_2}{\pi_Y\pi_R}$
REV (GTR)	$\dfrac{\pi_T\pi_C a + \pi_A\pi_G f}{\pi_T\pi_A b + \pi_T\pi_G c + \pi_C\pi_A d + \pi_C\pi_G e}$
UNREST	See equation (1.29) in text

Note that the Markov chain spends a proportion π_T of time in state T, while q_{TC} is the rate that T changes to C. Thus $\pi_T q_{TC}$ is the amount of 'flow' from T to C. The numerator in (1.29) is thus the average amount of transitional change while the denominator is the amount of transversional change. Table 1.4 gives R for commonly used simple models. Under the model of Kimura (1980), $R = \alpha/(2\beta)$ and equals $1/2$ when there is no transition/transversion rate difference. As from each nucleotide one change is a transition and two changes are transversions, we expect to see twice as many transversions as transitions, hence the ratio $1/2$.

The parameter κ has different definitions under the F84 and HKY85 models (Table 1.1). Without transition-transversion rate difference, $\kappa_{F84} = 0$ and $\kappa_{HKY85} = 1$. Roughly, $\kappa_{HKY85} \simeq 1 + 2\kappa_{F84}$. By forcing the average ratio R to be identical under the two models (Table 1.4), one can derive a more accurate approximation (Goldman 1993)

$$\kappa_{HKY} \simeq 1 + \frac{(\pi_T\pi_C/\pi_Y) + (\pi_A\pi_G/\pi_R)}{\pi_T\pi_C + \pi_A\pi_G}\kappa_{F84}. \tag{1.30}$$

Overall, R is more convenient to use for comparing estimates under different models, while κ is more suitable for formulating the null hypothesis of no transition/transversion rate difference.

1.3 Variable substitution rates across sites

All models discussed in Section 1.2 assume that different sites in the sequence evolve in the same way and at the same rate. This assumption may be unrealistic in real data.

1.3 Variable substitution rates across sites • 19

First, the mutation rate may vary among sites. Second, mutations at different sites may be fixed at different rates due to their different roles in the structure and function of the gene and thus different selective pressures acting on them. When the rates vary, the substitutional hotspots may accumulate many changes, while the conserved sites remain unchanged. Thus, for the same amount of evolutionary change or sequence distance, we will observe fewer differences than if the rate is constant. In other words, ignoring variable rates among sites leads to underestimation of the sequence distance.

One can accommodate the rate variation by assuming that rate r for any site is a random variable drawn from a statistical distribution. The most commonly used distribution is the gamma. The resulting models are represented by a suffix '+Γ', such as JC69+Γ, K80+Γ, etc., and the distances are sometimes called *gamma distances*. The density function of the gamma distribution is

$$g(r; \alpha, \beta) = \frac{\beta^\alpha}{\Gamma(\alpha)} e^{-\beta r} r^{\alpha-1}, \quad \alpha > 0, \ \beta > 0, \ r > 0, \tag{1.31}$$

where α and β are the shape and scale parameters. The mean and variance are $E(r) = \alpha/\beta$ and $\text{var}(r) = \alpha/\beta^2$. To avoid using too many parameters, we set $\beta = \alpha$ so that the mean of the distribution is 1, with variance $1/\alpha$. The shape parameter α is then inversely related to the extent of rate variation at sites (Fig. 1.6). If $\alpha > 1$, the distribution is bell-shaped, meaning that most sites have intermediate rates around 1, while few sites have either very low or very high rates. In particular, when $\alpha \to \infty$, the distribution degenerates into the model of a single rate for all sites. If $\alpha \leq 1$, the distribution has a highly skewed L-shape, meaning that most sites have very low rates of substitution or are nearly 'invariable', but there are some substitution hotspots

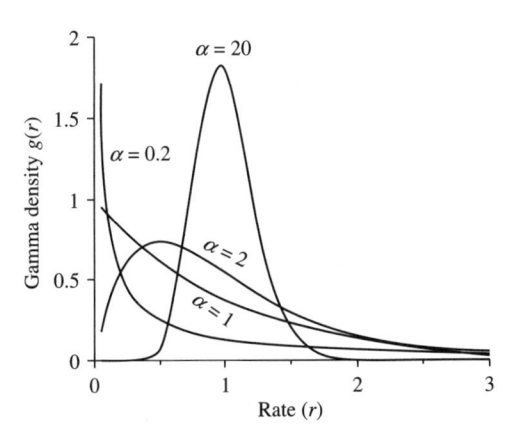

Fig. 1.6 Probability density function of the gamma distribution for variable rates among sites. The scale parameter of the distribution is fixed so that the mean is 1; as a result, the density involves only the shape parameters α. The x-axis is the substitution rate, while the y-axis is proportional to the number of sites with that rate.

with high rates. Estimation of α from real data requires joint comparison of multiple sequences as it is virtually impossible to do so using only two sequences. We will discuss the estimation in Section 4.3 in Chapter 4. Here we assume that α is given.

With variable rates among sites, the sequence distance is defined as the expected number of substitutions per site, averaged over all sites. Here we will derive a gamma distance under the K80 model, and comment on similar derivations under other models. To avoid confusion about the notation, we use d and κ as parameters under the K80 model and α and β as parameters of the gamma distribution. Since the mean rate is 1, the distance averaged across all sites is still d. If a site has rate r, both transition and transversion rates at the site are multiplied by r, so that the distance between the sequences at that site is dr. The transition/transversion rate ratio κ remains constant among sites. As in the case of one rate for all sites, we use the observed proportions, S and V, of sites with transitional and transversional differences and equate them to their expected probabilities under the model. If the rate r for a site is given, the probability that the site has a transitional difference will be $p_1(dr)$, with p_1 given in equation (1.10). However, r is an unknown random variable, so we have to consider contributions from sites with different rates. In other words, we average over the distribution of r to calculate the unconditional probability

$$E(S) = \int_0^\infty p_1(d \cdot r) g(r) \, dr$$

$$= \int_0^\infty \left[\tfrac{1}{4} + \tfrac{1}{4} \exp\left(\frac{-4d \cdot r}{\kappa + 2} \right) - \tfrac{1}{2} \exp\left(\frac{-2(\kappa + 1)d \cdot r}{\kappa + 2} \right) \right] g(r) \, dr \quad (1.32)$$

$$= \tfrac{1}{4} + \tfrac{1}{4} \left(1 + \frac{4d}{(\kappa+2)\alpha} \right)^{-\alpha} - \tfrac{1}{2} \left(1 + \frac{2(\kappa+1)d}{(\kappa+2)\alpha} \right)^{-\alpha}.$$

Similarly the probability that we observe a transversional difference is

$$E(V) = \int_0^\infty 2 p_2(d \cdot r) g(r) \, dr = \tfrac{1}{2} - \tfrac{1}{2} \left(1 + \frac{4d}{(\kappa+2)\alpha} \right)^{-\alpha}. \quad (1.33)$$

Equating the above to the observed proportions S and V leads to

$$\hat{d} = \tfrac{1}{2}\alpha\left[(1 - 2S - V)^{-1/\alpha} - 1\right] + \tfrac{1}{4}\alpha\left[(1 - 2V)^{-1/\alpha} - 1\right],$$

$$\hat{\kappa} = \frac{2\left[(1 - 2S - V)^{-1/\alpha} - 1\right]}{\left[(1 - 2V)^{-1/\alpha} - 1\right]} - 1 \quad (1.34)$$

(Jin and Nei 1990). Compared with equation (1.11) for the one-rate model, the only change is that the logarithm function $\log(y)$ becomes $-\alpha(y^{-1/\alpha} - 1)$. This is a general

1.3 Variable substitution rates across sites

feature of gamma distances. The large-sample variance of $\hat{d}$ is given by equation (1.13) except that now

$$a = (1 - 2S - V)^{-1/\alpha - 1},$$
$$b = \tfrac{1}{2}[(1 - 2S - V)^{-1/\alpha - 1} + (1 - 2V)^{-1/\alpha - 1}]. \tag{1.35}$$

Similarly, the expected proportion of different sites under the JC69+Γ model is

$$p = \int_0^\infty \left(\tfrac{3}{4} - \tfrac{3}{4} e^{-4d \cdot r/3}\right) g(r)\, dr = \tfrac{3}{4} - \tfrac{3}{4}\left(1 + \frac{4d}{3\alpha}\right)^{-\alpha}. \tag{1.36}$$

The JC69+Γ distance is thus

$$\hat{d} = \tfrac{3}{4}\alpha\left[(1 - \tfrac{4}{3}\hat{p})^{-1/\alpha} - 1\right] \tag{1.37}$$

(Golding 1983), with variance

$$\operatorname{var}(\hat{d}) = \operatorname{var}(\hat{p})\left|\frac{dd}{dp}\right|^2 = \frac{\hat{p}(1-\hat{p})}{n}\left(1 - \tfrac{4}{3}\hat{p}\right)^{-2/\alpha - 2}. \tag{1.38}$$

In general, note that equation (1.17) can be written equivalently as

$$p_{ij}(t) = \sum_{k=1}^{4} c_{ijk} e^{\lambda_k t} = \sum_{k=1}^{4} u_{ik} u_{kj}^{-1} e^{\lambda_k t}, \tag{1.39}$$

where λ_k is the kth eigenvalue of the rate matrix Q, u_{ik} is the ikth element of U, and u_{kj}^{-1} is the kjth element of U^{-1} in equation (1.17). Thus the probability of observing nucleotides i and j in the two sequences at a site is

$$f_{ij}(t) = \int_0^\infty \pi_i p_{ij}(t \cdot r) g(r)\, dr = \pi_i \sum_{k=1}^{4} c_{ijk} \left(1 - \frac{\lambda_k t}{\alpha}\right)^{-\alpha}. \tag{1.40}$$

The exponential functions under the one-rate model are replaced by the power functions under the gamma model. Under the one-rate model, we can view the exponential functions as unknowns to solve the equations, and now we can view those power functions as unknowns. Thus, one can derive a gamma distance under virtually every model for which a one-rate distance formula is available. Those include the F84 model (Yang 1994a) and the TN93 model (Tamura and Nei 1993), among others.

Example. We calculate the sequence distance between the two mitochondrial 12s rRNA genes under the K80+Γ model, assuming a fixed $\alpha = 0.5$. The estimates of the distance and the transition/transversion rate ratio κ are $\hat{d} \pm \text{SE} = 0.1283 \pm 0.01726$ and $\hat{\kappa} \pm \text{SE} = 37.76 \pm 16.34$. Both estimates are much larger than under the one-rate model (Table 1.3). It is well known that ignoring rate variation among sites leads to underestimation of both the sequence distance and the transition/transversion rate ratio (Wakeley 1994; Yang 1996a). The underestimation is more serious at larger distances and with more variable rates (that is, smaller α). □

1.4 Maximum likelihood estimation

In this section, we discuss the maximum likelihood (ML) method for estimating sequence distances. ML is a general methodology for estimating parameters in a model and for testing hypotheses concerning the parameters. It plays a central role in statistics and is widely used in molecular phylogenetics. It forms the basis of much material covered later in this book. We will focus mainly on the models of Jukes and Cantor (1969) and Kimura (1980), re-deriving the distance formulae discussed earlier. While discovering what we already know may not be very exciting, it may be effective in helping us understand the workings of the likelihood method. Note that ML is an 'automatic' method, as it tells us how to proceed even when the estimation problem is difficult and the model is complex when no analytical solution is available and our intuition fails. Interested readers should consult a statistics textbook, for example DeGroot and Schervish (2002), Kalbfleisch (1985), Edwards (1992) at elementary levels, or Cox and Hinkley (1974) or Stuart et al. (1999) at more advanced levels.

1.4.1 The JC69 model

Let X be the data and θ the parameter we hope to estimate. The probability of observing data X, when viewed as a function of the unknown parameter θ with the data given, is called the *likelihood function*: $L(\theta; X) = f(\theta|X)$. According to the *likelihood principle*, the likelihood function contains all information in the data about θ. The value of θ that maximizes the likelihood, say $\hat{\theta}$, is our best point estimate, called the *maximum likelihood estimate* (MLE). Furthermore, the likelihood curve around $\hat{\theta}$ provides information about the uncertainty in the point estimate. The theory applies to problems with a single parameter as well as problems involving multiple parameters, in which case θ is a vector.

Here we apply the theory to estimation of the distance between two sequences under the JC69 model (Jukes and Cantor 1969). The single parameter is the distance d. The data are two aligned sequences, each n sites long, with x differences. From equation (1.5), the probability that a site has different nucleotides between two sequences separated by distance d is

$$p = 3p_1 = \tfrac{3}{4} - \tfrac{3}{4}e^{-4d/3}. \tag{1.41}$$

1.4 Maximum likelihood estimation

Thus, the probability of observing the data, that is, x differences out of n sites, is given by the binomial probability

$$L(d;x) = f(x|d) = Cp^x(1-p)^{n-x} = C\left(\tfrac{3}{4} - \tfrac{3}{4}e^{-4d/3}\right)^x \left(\tfrac{1}{4} + \tfrac{3}{4}e^{-4d/3}\right)^{n-x}. \quad (1.42)$$

As the data x are observed, this probability is now considered a function of the parameter d. Values of d with higher L are better supported by the data than values of d with lower L. As multiplying the likelihood by any function of the data that is independent of the parameter θ will not change our inference about θ, the likelihood is defined up to a proportionality constant. We will use this property to introduce two changes to the likelihood of equation (1.42). First, the binomial coefficient $C = [n!/x!(n-x)!]$ is a constant and will be dropped. Second, to use the same definition of likelihood for all substitution models, we will distinguish 16 possible data outcomes at a site (the 16 possible site patterns) instead of just two outcomes (that is, difference with probability p and identity with probability $1-p$) as in equation (1.42). Under JC69, the four constant patterns (TT, CC, AA, GG) have the same probability of occurrence, as do the 12 variable site patterns (TC, TA, TG etc.). This will not be the case for other models. Thus the redefined likelihood is given by the multinomial probability with 16 cells

$$L(d;x) = \left(\tfrac{1}{4}p_1\right)^x \left(\tfrac{1}{4}p_0\right)^{n-x} = \left(\tfrac{1}{16} - \tfrac{1}{16}e^{-4d/3}\right)^x \left(\tfrac{1}{16} + \tfrac{3}{16}e^{-4d/3}\right)^{n-x}, \quad (1.43)$$

where p_0 and p_1 are from equation (1.3). Each of the 12 variable site patterns has probability $p_1/4$ or $p/12$. For example, the probability for site pattern TC is equal to $1/4$, the probability that the starting nucleotide is T, times the transition probability $p_{TC}(t) = p_1$ from equation (1.3). Similarly, each of the four constant site patterns (TT, CC, AA, GG) has probability $p_0/4$ or $(1-p)/4$. The reader can verify that equations (1.42) and (1.43) differ only by a proportionality constant (Exercise 1.4).

Furthermore, the likelihood L is typically extremely small and awkward to work with. Thus its logarithm $\ell(d) = \log\{L(d)\}$ is commonly used instead. As the logarithm function is monotonic, we achieve the same result; that is, $L(d_1) > L(d_2)$ if and only if $\ell(d_1) > \ell(d_2)$. The *log likelihood function* is thus

$$\ell(d;x) = \log\{L(d;x)\} = x\log\left(\tfrac{1}{16} - \tfrac{1}{16}e^{-4d/3}\right) + (n-x)\log\left(\tfrac{1}{16} + \tfrac{3}{16}e^{-4d/3}\right). \quad (1.44)$$

To estimate d, we maximize L or equivalently its logarithm ℓ. By setting $d\ell/dd = 0$, we can determine that ℓ is maximized at the MLE

$$\hat{d} = -\tfrac{3}{4}\log\left(1 - \tfrac{4}{3}\tfrac{x}{n}\right). \quad (1.45)$$

This is the distance formula (1.6) under JC69 that we derived earlier.

We now discuss some statistical properties of MLEs. Under quite mild regularity conditions which we will not go into, the MLEs have nice asymptotic (large-sample) properties (see, e.g., Stuart et al. 1999, pp. 46–116). For example, they are asymptotically unbiased, consistent, and efficient. Unbiasedness means that the expectation of the estimate equals the true parameter value: $E(\hat{\theta}) = \theta$. Consistency means that the estimate $\hat{\theta}$ converges to the true value θ when the sample size $n \to \infty$. Efficiency means that no other estimate can have a smaller variance than the MLE. Furthermore, the MLEs are asymptotically normally distributed. These properties are known to hold in large samples. How large the sample size has to be for the approximation to be reliable depends on the particular problem.

Another important property of MLEs is that they are invariant to transformations of parameters or reparametrizations. The MLE of a function of parameters is the same function of the MLEs of the parameters: $\hat{h}(\theta) = h(\hat{\theta})$. Thus if the same model can be formulated using either parameters θ_1 or θ_2, with θ_1 and θ_2 constituting a one-to-one mapping, use of either parameter leads to the same inference. For example, we can use the probability of a difference between the two sequences p as the parameter in the JC69 model instead of the distance d. The two form a one-to-one mapping through equation (1.41). The log likelihood function for p corresponding to equation (1.43) is $L(p; x) = (p/12)^x [(1-p)/4]^{n-x}$, from which we get the MLE of p: $\hat{p} = x/n$. We can then view d as a function of p and obtain its MLE $\hat{d}$, as given by equation (1.45). Whether we use p or d as the parameter, the same inference is made, and the same log likelihood is achieved: $\ell(\hat{d}) = \ell(\hat{p}) = x \log(x/(12n)) + (n - x) \log[(n - x)/(4n)]$.

Two approaches can be used to calculate a confidence interval for the MLE. The first relies on the theory that the MLE $\hat{\theta}$ is asymptotically normally distributed around the true value θ when the sample size $n \to \infty$. The asymptotic variance can be calculated using either the observed information $-d^2\ell/d\theta^2$ or the expected (Fisher) information $-E(d^2\ell/d\theta^2)$. While both are reliable in large samples, the observed information is preferred in real data analysis (e.g. Efron and Hinkley 1978). This is equivalent to using a quadratic polynomial to approximate the log likelihood around the MLE. Here we state the result for the multivariate case, with k parameters in the model:

$$\hat{\theta} \sim N_k(\theta, -H^{-1}), \quad \text{with } H = \{d^2\ell/d\theta_i d\theta_j\}. \tag{1.46}$$

In other words, the MLEs $\hat{\theta}$ have an asymptotic k-variate normal distribution, with the mean to be the true values θ, and the variance–covariance matrix to be $-H^{-1}$, where H is the matrix of second derivatives, also known as the Hessian matrix (Stuart et al. 1999, pp. 73–74).

In our example, the asymptotic variance for $\hat{d}$ is

$$\text{var}(\hat{d}) = -\left(\frac{d^2\ell}{dd^2}\right)^{-1} = \frac{\hat{p}(1-\hat{p})}{(1 - 4\hat{p}/3)^2 n}. \tag{1.47}$$

This is equation (1.7). An approximate 95% confidence interval for d can be constructed as $\hat{d} \pm 1.96\sqrt{\text{var}(\hat{d})}$.

The normal approximation has a few drawbacks. First, if the log likelihood curve is not symmetrical around the MLE, the normal approximation will be unreliable. For example, if the parameter is a probability, which ranges from 0 to 1, and the MLE is close to 0 or 1, the normal approximation may be very poor. Second, the confidence interval constructed in this way includes parameter values that have lower likelihood than values outside the interval. Third, even though the MLEs are invariant to reparametrizations, the confidence intervals constructed using the normal approximation are not.

These problems are circumvented by the second approach, which is based on the likelihood ratio. In large samples, the likelihood ratio test statistic, $2(\ell(\hat{\theta}) - \ell(\theta))$, where θ is the true parameter value and $\hat{\theta}$ is the MLE, has a χ_k^2 distribution with the degree of freedom k equal to the number of parameters. Thus one can lower the log likelihood from the peak $\ell(\hat{\theta})$ by $\chi_{k,5\%}^2/2$ to construct a 95% confidence (likelihood) region. Here $\chi_{k,5\%}^2$ is the 5% critical value of the χ^2 distribution with k degrees of freedom. The likelihood region contains parameter values with the highest likelihood, values that cannot be rejected by a likelihood ratio test at the 5% level when compared against $\hat{\theta}$. This likelihood ratio approach is known to give more reliable intervals than the normal approximation. The normal approximation works well for some parametrizations but not for others; the likelihood interval automatically uses the best parametrization.

Example. For the 12s rRNA data of Table 1.2, we have $\hat{p} = x/n = 90/948 = 0.09494$ and $\hat{d} = 0.1015$. The variance of $\hat{d}$ is 0.0001188, so that the 95% confidence interval based on the normal approximation is $(0.0801, 0.1229)$. If we use p as the parameter instead, we have $\text{var}(\hat{p}) = \hat{p}(1-\hat{p})/n = 0.00009064$, so that the 95% confidence interval for p is $(0.0763, 0.1136)$. These two intervals do not match. For example, if we use the lower bound for p to calculate the lower bound for d, the result will be different. The log-likelihood curves are shown in Fig. 1.7, with the peak at $\ell(\hat{d}) = \ell(\hat{p}) = -1710.577$. By lowering the log likelihood ℓ by $\chi_{1,5\%}^2/2 = 3.841/2 = 1.921$ from its peak, we obtain the 95% likelihood intervals $(0.0817, 0.1245)$ for d and $(0.0774, 0.1147)$ for p. Compared with the intervals based on the normal approximation, the likelihood intervals are asymmetrical and are shifted to the right, reflecting the steeper drop of log likelihood and thus more information on the left side of the MLE. Also the likelihood intervals for p and d match each other in that the lower bounds are related through their functional relationship (1.41), as are the upper bounds. □

1.4.2 The K80 model

The likelihood theory applies to models with multiple parameters. We apply the method to estimation of the sequence distance d and the transition/transversion rate ratio κ under the K80 model (Kimura 1980). The data are the number of sites with

26 • 1 Models of nucleotide substitution

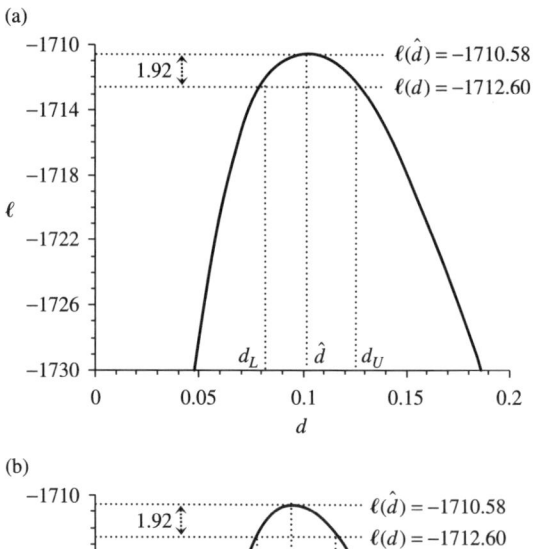

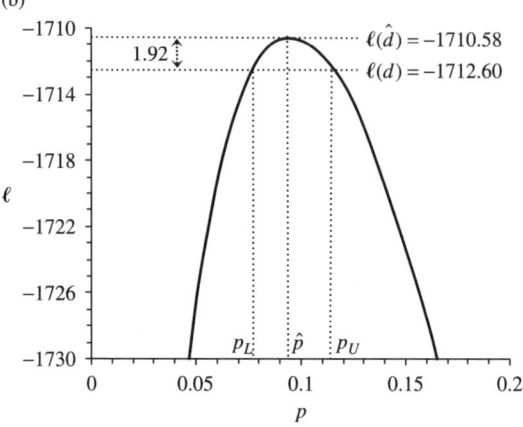

Fig. 1.7 Log likelihood curves and construction of confidence (likelihood) intervals under the JC69 substitution model. The parameter in the model is the sequence distance d in (a) and the probability of different sites p in (b). The mitochondrial 12s rRNA genes of Table 1.2 are analysed.

transitional (n_S) and transversional (n_V) differences, with the number of constant sites to be $n - n_S - n_V$. In deriving the probabilities of observing such sites, we again consider all 16 site patterns, as for the JC69 model. Thus the probability is $p_0/4$ for any constant site (e.g. TT), $p_1/4$ for any site with a transitional difference (e.g. TC), and $p_2/4$ for any site with a transversional difference (e.g. TA), with p_0, p_1, p_2 given in equation (1.10). The log likelihood is

$$\ell(d, \kappa | n_S, n_V) = \log\{f(n_S, n_V | d, \kappa)\} \\ = (n - n_S - n_V) \log(p_0/4) + n_S \log(p_1/4) + n_V \log(p_2/4). \tag{1.48}$$

MLEs of d and κ can be derived from the likelihood equation $\partial \ell/\partial d = 0$, $\partial \ell/\partial \kappa = 0$. The solution can be shown to be equation (1.11), with $S = n_S/n$ and $V = n_V/n$. A simpler argument relies on the invariance property of the MLEs. Suppose we consider the probabilities of transitional and transversional differences $E(S)$ and $E(V)$ as parameters in the model instead of d and κ. The log likelihood is equation (1.48) with $p_1 = E(S)$ and $p_2 = E(V)/2$. The MLEs of $E(S)$ and $E(V)$ are simply S and V. The MLEs of d and κ can be obtained through the one-to-one mapping between the two sets of parameters, which involves the same step taken when we derived equation (1.11) in Subsection 1.2.2 by equating the observed proportions S and V to their expected probabilities $E(S)$ and $E(V)$.

Example. For the 12s rRNA data of Table 1.2, we have $S = 0.08861$ and $V = 0.00633$. The MLEs are thus $\hat{d} = 0.1046$ for the sequence distance and $\hat{\kappa} = 30.83$ for the transition/transversion rate ratio. These are the same as calculated in Subsection 1.2.2. The maximized log likelihood is $\ell(\hat{d}, \hat{\kappa}) = -1637.905$. Application of equation (1.46) leads to the variance–covariance matrix (see Appendix B)

$$\operatorname{var}\begin{pmatrix} \hat{d} \\ \hat{\kappa} \end{pmatrix} = \begin{pmatrix} 0.0001345 & 0.007253 \\ 0.007253 & 172.096 \end{pmatrix} \quad (1.49)$$

From this, one can get the approximate SEs to be 0.0116 for d and 13.12 for κ. The log likelihood surface contour is shown in Fig. 1.8, which indicates that the data are much more informative about d than about κ. One can lower the log likelihood from its peak by $\chi^2_{2,5\%}/2 = 5.991/2 = 2.996$, to construct a 95% confidence (likelihood) region for the parameters (Fig. 1.8). □

*1.4.3 Profile and integrated likelihood methods

Suppose we are interested in the sequence distance d under the K80 model (Kimura 1980) but not in the transition/transversion rate ratio κ. However, we want to consider κ in the model as transition and transversion rates are known to differ and the rate difference may affect our estimation of d. Parameter κ is thus appropriately called a *nuisance parameter*, while d is our parameter of interest. Dealing with nuisance parameters is commonly considered a weakness of the likelihood method. The approach we described above, estimating both d and κ with ML and using $\hat{d}$ while ignoring $\hat{\kappa}$, is known variously as the *relative likelihood*, *pseudo likelihood*, or *estimated likelihood*, since the nuisance parameters are replaced by their estimates.

A more respected approach is the *profile likelihood*, which defines a log likelihood for the parameters of interest only, calculated by optimizing the nuisance parameters at fixed values of the parameters of interest. In other words, the profile log likelihood is $\ell(d) = \ell(d, \hat{\kappa}_d)$, where $\hat{\kappa}_d$ is the MLE of κ for the given d. This is a pragmatic approach that most often leads to reasonable answers. The likelihood interval for $\hat{d}$ is constructed from the profile likelihood in the usual way.

28 • 1 Models of nucleotide substitution

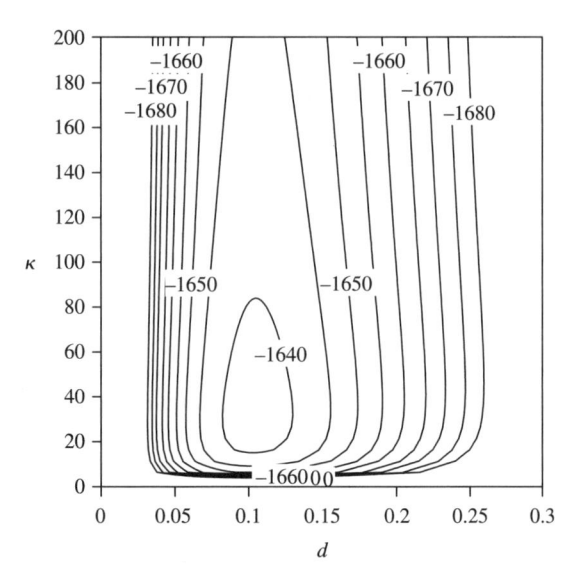

Fig. 1.8 Log-likelihood contours for the sequence distance d and the transition/transversion rate ratio κ under the K80 model. The mitochondrial 12s rRNA genes of Table 1.2 are analysed. The peak of the surface is at the MLEs $\hat{d} = 0.1046$, $\hat{\kappa} = 30.83$, with $\ell = -1637.905$. The 95% likelihood region is surrounded by the contour line at $\ell = -1637.905 - 2.996 = -1640.901$ (not shown).

Example. For the 12s rRNA genes, the highest likelihood $\ell(\hat{d}) = -1637.905$ is achieved at $\hat{d} = 0.1046$ and $\hat{\kappa} = 30.83$. Thus the point estimate of d is the same as before. We fix d at different values. For each fixed d, the log likelihood (1.48) is a function of the nuisance parameter κ, and is maximized to estimate κ. Let the estimate be $\hat{\kappa}_d$, with the subscript indicating it is a function of d. It does not seem possible to derive $\hat{\kappa}_d$ analytically, so we use a numerical algorithm instead (Section 4.5 discusses such algorithms). The optimized likelihood is the profile likelihood for d: $\ell(d) = \ell(d, \hat{\kappa}_d)$. This is plotted against d in Fig. 1.9(a), together with the estimate $\hat{\kappa}_d$. We lower the log likelihood by $\chi^2_{1,5\%}/2 = 1.921$ to construct the profile likelihood interval for d: $(0.0836, 0.1293)$. □

If the model involves many parameters, and in particular, if the number of parameters increases without bound with the increase of the sample size, the likelihood method may run into deep trouble, so deep that the MLEs may not even be consistent (e.g. Kalbfleisch and Sprott 1970; Kalbfleisch 1985, pp. 92–96). A useful strategy in this case is to assign a statistical distribution to describe the variation in the parameters, and integrate them out in the likelihood. Here we apply this approach to dealing with nuisance parameters, known as *integrated likelihood* or *marginal likelihood*. This has much of the flavour of a Bayesian approach. Let $f(\kappa)$ be the distribution assigned

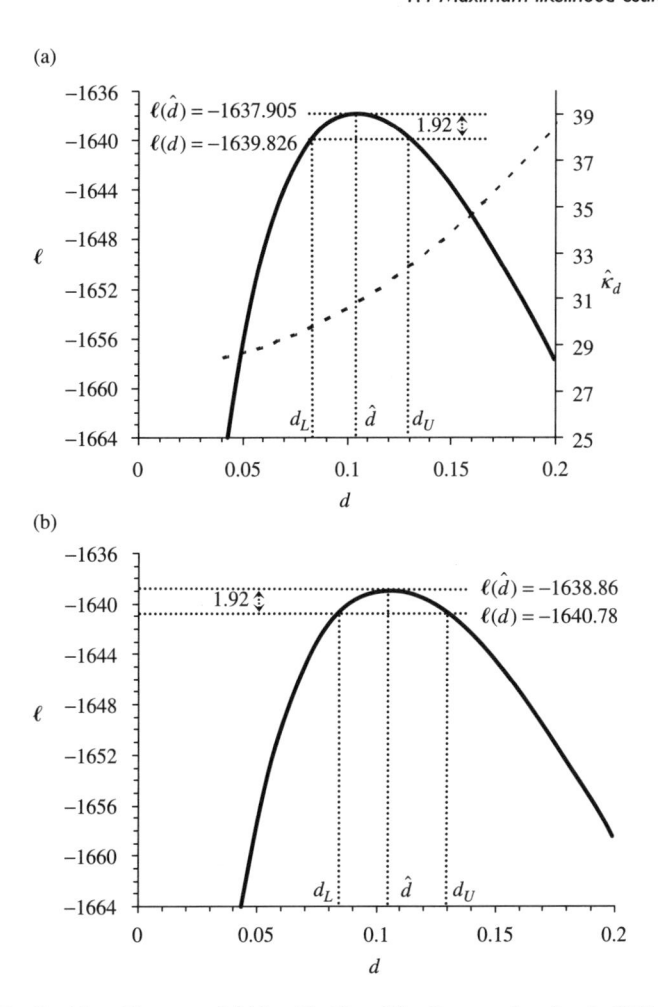

Fig. 1.9 Profile (a) and integrated (b) log likelihood for distance d under the K80 model. The mitochondrial 12s rRNA genes of Table 1.2 are analysed. (a) The profile likelihood $\ell(d) = \ell(d, \hat{\kappa}_d)$ is plotted against d. The estimated nuisance parameter $\hat{\kappa}_d$ at fixed d is also shown. The profile log likelihood is lowered from the peak by 1.921 to construct a likelihood interval for parameter d. (b) The likelihood for d is calculated by integrating over the nuisance parameter κ using equation (1.50), with a uniform prior $\kappa \sim U(0, 99)$.

to κ, also known as a prior. Then the integrated likelihood is

$$L(d) = \int_0^\infty f(\kappa) f(n_S, n_V | d, \kappa) \, d\kappa$$

$$= \int_0^\infty f(\kappa) \left(\frac{p_0}{4}\right)^{n-n_S-n_V} \left(\frac{p_1}{4}\right)^{n_S} \left(\frac{p_2}{4}\right)^{n_V} d\kappa, \tag{1.50}$$

where p_0, p_1, and p_2 are from equation (1.10). For the present problem it is possible to use an improper prior: $f(\kappa) = 1, 0 < \kappa < \infty$. As this does not integrate to 1, it is not a proper probability density and is thus called an *improper prior*. The integrated likelihood is then

$$L(d) = \int_0^\infty f(n_S, n_V | d, \kappa) \, d\kappa = \int_0^\infty \left(\frac{p_0}{4}\right)^{n-n_S-n_V} \left(\frac{p_1}{4}\right)^{n_S} \left(\frac{p_2}{4}\right)^{n_V} d\kappa. \quad (1.51)$$

Example. We apply the integrated likelihood approach under the K80 model (equation 1.50) to the 12s rRNA data of Table 1.2. We use a uniform prior $\kappa \sim U(0, c)$, with $c = 99$ so that $f(\kappa) = 1/c$ in equation (1.50). Analytical calculation of the integral appears awkward, so a numerical method is used instead. The log integrated likelihood $\ell(d) = \log\{L(d)\}$, with $L(d)$ given by equation (1.50), is plotted in Fig. 1.9(b). This is always lower than the profile log likelihood (Fig. 1.9a). The MLE of d is obtained numerically as $\hat{d} = 0.1048$, with the maximum log likelihood $\ell(\hat{d}) = -1638.86$. By lowering ℓ by 1.921, we construct the likelihood interval for d to be $(0.0837, 0.1295)$. For this example, the profile and integrated likelihood methods produced very similar MLEs and likelihood intervals. □

1.5 Markov chains and distance estimation under general models

We have discussed most of the important properties of continuous-time Markov chains that will be useful in this book. In this section we provide a more systematic overview, and also discuss two general Markov-chain models: the general time-reversible model and the general unconstrained model. The theory will be applied in a straightforward manner to model substitutions between amino acids and between codons in Chapter 2. Note that Markov chains (processes) are classified according to whether time and state are discrete or continuous. In the Markov chains we consider in this chapter, the states (the four nucleotides) are discrete while time is continuous. In Chapter 5, we will encounter Markov chains with discrete time and either discrete or continuous states. Interested readers should consult any of the many excellent textbooks on Markov chains and stochastic processes (e.g. Grimmett and Stirzaker 1992; Karlin and Taylor 1975; Norris 1997; Ross 1996). Note that some authors use the term Markov chains if time is discrete and Markov process if time is continuous.

1.5.1 General theory

Let the state of the chain at time t be $X(t)$. This is one of the four nucleotides T, C, A, or G. We assume that different sites in a DNA sequence evolve independently, and the Markov-chain model is used to describe nucleotide substitutions at any site. The Markov chain is characterized by its generator matrix or the substitution-rate matrix $Q = \{q_{ij}\}$, where q_{ij} is the instantaneous rate of change from i to j; that is,

$\Pr\{X(t + \Delta t) = j | X(t) = i\} = q_{ij}\Delta t$, for any $j \neq i$. If q_{ij} does not depend on time, as we assume here, the process is said to be *time-homogeneous*. The diagonals q_{ii} are specified by the requirement that each row of Q sums to zero, that is, $q_{ii} = -\sum_{j \neq i} q_{ij}$. Thus $-q_{ii}$ is the substitution rate of nucleotide i, that is, the rate at which the Markov chain leaves state i. The general model without any constraint on the structure of Q will have 12 free parameters. This was called the UNREST model by Yang (1994b).

The Q matrix fully determines the dynamics of the Markov chain. For example, it specifies the *transition-probability matrix* over any time $t > 0$: $P(t) = \{p_{ij}(t)\}$, where $p_{ij}(t) = \Pr\{X(t) = j | X(0) = i\}$. Indeed $P(t)$ is the solution to the following differential equation

$$dP(t)/dt = P(t)Q, \quad (1.52)$$

with the boundary condition $P(0) = I$, the identity matrix (e.g. Grimmett and Stirzaker 1992, p. 242). This has the solution

$$P(t) = e^{Qt} \quad (1.53)$$

(e.g. Chapter 8 in Lang 1987).

As Q and t occur only in the form of a product, it is conventional to multiple Q by a scale factor so that the average rate is 1. Time t will then be measured by distance, that is, the expected number of substitutions per site. Thus we use Q to define the relative substitution rates only.

If the Markov chain $X(t)$ has the initial distribution $\pi^{(0)} = (\pi_T^{(0)}, \pi_C^{(0)}, \pi_A^{(0)}, \pi_G^{(0)})$, then time t later the distribution $\pi^{(t)} = (\pi_T^{(t)}, \pi_C^{(t)}, \pi_A^{(t)}, \pi_G^{(t)})$ will be given by

$$\pi^{(t)} = \pi^{(0)} P(t). \quad (1.54)$$

If a long sequence initially has the four nucleotides in proportions $\pi_T^{(0)}$, $\pi_C^{(0)}$, $\pi_A^{(0)}$, $\pi_G^{(0)}$, then time t later the proportions will become $\pi^{(t)}$. For example, consider the frequency of nucleotide T in the target sequence: $\pi_T^{(t)}$. Such a T can result from any nucleotide in the source sequence at time 0. Thus $\pi_T^{(t)} = \pi_T^{(0)} p_{TT}(t) + \pi_C^{(0)} p_{CT}(t) + \pi_A^{(0)} p_{AT}(t) + \pi_G^{(0)} p_{GT}(t)$, which is equation (1.54).

If the initial and target distributions are the same, $\pi^{(0)} = \pi^{(t)}$, the chain will stay in that distribution forever. The chain is then said to be *stationary* or at equilibrium, and the distribution (let it be π) is called the *stationary* or *steady-state distribution*. Our Markov chain allows any state to change into any other state in finite time with positive probability. Such a chain is called *irreducible* and has a unique stationary distribution, which is also the *limiting distribution* when time $t \to \infty$. As indicated above, the stationary distribution is given by

$$\pi P(t) = \pi. \quad (1.55)$$

1 Models of nucleotide substitution

This is equivalent to

$$\pi Q = 0 \tag{1.56}$$

(e.g. Grimmett and Stirzaker 1992, p. 244). Note that the total amount of flow into any state j is $\sum_{i \ne j} \pi_i q_{ij}$, while the total amount of flow out of state j is $-\pi_j q_{jj}$. Equation (1.56) states that the two are equal when the chain is stationary; that is $\sum_i \pi_i q_{ij} = 0$ for any j. Equation (1.56), together with the obvious constraints $\pi_j \ge 0$ and $\sum_j \pi_j = 1$, allows us to determine the stationary distribution from Q for any Markov chain.

1.5.1.1 Estimation of sequence distance under the general model

The rate matrix Q without any constraint involves 12 parameters (Table 1.1). If Q defines the relative rates only, 11 free parameters are involved. This model, called UNREST in Yang (1994b), can in theory identify the root of the two-sequence tree (Fig. 1.10a), so that two branch lengths are needed in the model: t_1 and t_2. The likelihood is given by the multinomial probability with 16 cells, corresponding to the 16 possible site patterns. Let $f_{ij}(t_1, t_2)$ be the probability for the ijth cell, that is, the probability that any site has nucleotide i in sequence 1 and j in sequence 2. Since such a site can result from all four possible nucleotides in the ancestor, we have to average over them

$$f_{ij}(t_1, t_2) = \sum_k \pi_k p_{ki}(t_1) p_{kj}(t_2). \tag{1.57}$$

Let n_{ij} be the number of sites in the ijth cell. The log likelihood is then

$$\ell(t_1, t_2, Q) = \sum_{i,j} n_{ij} \log\{f_{ij}(t_1, t_2)\}. \tag{1.58}$$

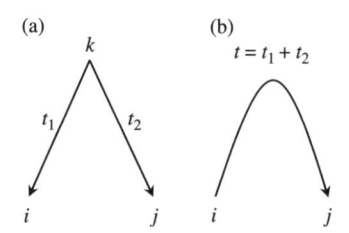

Fig. 1.10 A tree for two sequences, showing the observed nucleotides i and j at one site and the direction of evolution. (a) Two sequences diverged from a common ancestor (root of the tree) t_1 and t_2 time units ago; time is measured by the distance or the amount of sequence change. (b) Sequence 1 is ancestral to sequence 2. Under time-reversible models, we cannot identify the root of the tree, as the data will look the same whether both sequences were descendants of a common ancestor (as in a), or one sequence is ancestral to the other (as in b), or wherever the root of the tree is along the single branch connecting the two sequences.

The model involves 13 parameters: 11 relative rates in Q plus two branch lengths. Note that the frequency parameters π_T, π_C, π_A, π_G are determined from Q using equation (1.56) and are not free parameters. There are two problems with this unconstrained model. First, numerical methods are necessary to find the MLEs of parameters as no analytical solution seems possible. The eigenvalues of Q may be complex numbers. Second, and more importantly, typical data sets may not have enough information to estimate so many parameters. In particular, even though t_1 and t_2 are identifiable, their estimates are highly correlated. For this reason the model is not advisable for use in distance calculation.

Example. For the 12s rRNA data of Table 1.2, the log likelihood appears flat when t_1 and t_2 are estimated as separate parameters. We thus force $t_1 = t_2$ during the numerical maximization of the log likelihood. The estimate of the sequence distance $t = (t_1 + t_2)$ is 0.1057, very close to estimates under other models (Table 1.3). The MLE of rate matrix Q is

$$Q = \begin{pmatrix} -1.4651 & 1.3374 & 0.1277 & 0 \\ 1.2154 & -1.2220 & 0.0066 & 0 \\ 0.0099 & 0.0808 & -0.5993 & 0.5086 \\ 0 & 0 & 0.8530 & -0.8530 \end{pmatrix}, \quad (1.59)$$

scaled so that the average rate is $-\sum_i \pi_i q_{ii} = 1$. The steady-state distribution is calculated from equation (1.56) to be $\hat{\pi} = (0.2184, 0.2606, 0.3265, 0.1946)$, virtually identical to the observed frequencies (Table 1.2). □

1.5.2 The general time-reversible (GTR) model

A Markov chain is said to be time-reversible if and only if

$$\pi_i q_{ij} = \pi_j q_{ji}, \quad \text{for all } i \neq j. \quad (1.60)$$

Note that π_i is the proportion of time the Markov chain spends in state i, and $\pi_i q_{ij}$ is the amount of 'flow' from states i to j, while $\pi_j q_{ji}$ is the flow in the opposite direction. Equation (1.60) is known as the *detailed-balance* condition and means that the flow between any two states in the opposite direction is the same. There is no biological reason to expect the substitution process to be reversible, so reversibility is a mathematical convenience. Models discussed in this chapter, including JC69 (Jukes and Cantor 1969), K80 (Kimura 1980), F84, HKY85 (Hasegawa et al. 1985), and TN93 (Tamura and Nei 1993), are all time-reversible. Equation (1.60) is equivalent to

$$\pi_i p_{ij}(t) = \pi_j p_{ji}(t), \quad \text{for all } i \neq j \text{ and for any } t. \quad (1.61)$$

Another equivalent condition for reversibility is that the rate matrix can be written as a product of a symmetrical matrix multiplied by a diagonal matrix; the diagonals in the diagonal matrix will then specify the equilibrium frequencies. Thus the rate matrix for the general time-reversible model of nucleotide substitution is

$$Q = \{q_{ij}\} = \begin{bmatrix} \cdot & a\pi_C & b\pi_A & c\pi_G \\ a\pi_T & \cdot & d\pi_A & e\pi_G \\ b\pi_T & d\pi_C & \cdot & f\pi_G \\ c\pi_T & e\pi_C & f\pi_A & \cdot \end{bmatrix}$$

$$= \begin{bmatrix} \cdot & a & b & c \\ a & \cdot & d & e \\ b & d & \cdot & f \\ c & e & f & \cdot \end{bmatrix} \begin{bmatrix} \pi_T & 0 & 0 & 0 \\ 0 & \pi_C & 0 & 0 \\ 0 & 0 & \pi_A & 0 \\ 0 & 0 & 0 & \pi_G \end{bmatrix}, \quad (1.62)$$

with the diagonals of Q determined by the requirement that each row of Q sums to 0. This matrix involves nine free parameters: the rates a, b, c, d, e, and f and three frequency parameters. The model was first applied by Tavaré (1986) to sequence distance calculation and by Yang (1994b) to estimation of relative substitution rates (substitution pattern) between nucleotides using ML. It is commonly known as GTR or REV.

Keilson (1979) discussed a number of nice mathematical properties of reversible Markov chains. One of them is that all eigenvalues of the rate matrix Q are real (see Section 2.6). Thus efficient and stable numerical algorithms can be used to calculate the eigenvalues of Q for the GTR model. Alternatively, it appears possible to diagonalize Q of (1.62) analytically: one eigenvalue is 0, so that the characteristic equation that the eigenvalues should satisfy (e.g. Lang 1987, Chapter 8) is a cubic equation, which is solvable. Even so, analytical calculation appears awkward.

In phylogenetic analysis of sequence data, reversibility leads to an important simplification to the likelihood function. The probability of observing site pattern ij in equation (1.57) becomes

$$f_{ij}(t_1, t_2) = \sum_k \pi_k p_{ki}(t_1) p_{kj}(t_2)$$

$$= \sum_k \pi_i p_{ik}(t_1) p_{kj}(t_2) \quad (1.63)$$

$$= \pi_i p_{ij}(t_1 + t_2).$$

The second equality is because of the reversibility condition $\pi_k p_{ki}(t_1) = \pi_i p_{ik}(t_1)$, while the third equality is due to the Chapman–Kolmogorov theorem (equation 1.4).

Two remarks are in order. First, f_{ij} depends on $t_1 + t_2$ but not on t_1 and t_2 individually; thus we can estimate $t = t_1 + t_2$ but not t_1 and t_2. Equation (1.63) thus becomes

$$f_{ij}(t) = \pi_i p_{ij}(t). \quad (1.64)$$

Second, while we defined f_{ij} as the probability of a site when both sequences are descendants of a common ancestor (Fig. 1.10a), $\pi_i p_{ij}(t)$ is the probability of the site if sequence 1 is ancestral to sequence 2 (Fig. 1.10b). The probability is the same if we consider sequence 2 as the ancestor of sequence 1, or wherever we place the root along the single branch linking the two sequences. Thus under the model, the log likelihood (1.58) becomes

$$\ell(t, a, b, c, d, e, \pi_T, \pi_C, \pi_A) = \sum_i \sum_j n_{ij} \log\{f_{ij}(t)\} = \sum_i \sum_j n_{ij} \log\{\pi_i p_{ij}(t)\}. \tag{1.65}$$

We use Q to represent the relative rates, with $f = 1$ fixed, and multiply the whole matrix by a scale factor so that the average rate is $-\sum_i \pi_i q_{ii} = 1$. Time t is then the distance $d = -t \sum_i \pi_i q_{ii} = t$. The model thus involves nine parameters, which can be estimated numerically by solving a nine-dimensional optimization problem. Sometimes the base frequency parameters are estimated using the average observed frequencies, in which case the dimension is reduced to six.

Note that the log likelihood functions under the JC69 (Jukes and Cantor 1969) and K80 (Kimura 1980) models, that is, equations (1.44) and (1.48), are special cases of equation (1.65). Under these two models, the likelihood equation is analytically tractable, so that numerical optimization is not needed. Equation (1.65) also gives the log likelihood for other reversible models such as F81 (Felsenstein 1981), HKY85 (Hasegawa et al. 1985), F84, and TN93 (Tamura and Nei 1993). MLEs under those models were obtained through numerical optimization for the 12s rRNA genes of Table 1.2 and listed in Table 1.3. Note that the distance formulae under those models, discussed in Section 1.2, are not MLEs, despite claims to the contrary. First, the observed base frequencies are in general not MLEs of the base frequency parameters. Second, all 16 site patterns have distinct probabilities under those models and are not collapsed in the likelihood function (1.65), but the distance formulae collapsed some site patterns, such as the constant patterns TT, CC, AA, GG. Nevertheless, it is expected that the distance formulae will give estimates very close to the MLEs (see, e.g., Table 1.3).

Under the gamma model of variable rates among sites, the log likelihood is still given by equation (1.65) but with $f_{ij}(t)$ given by equation (1.40). This is the ML procedure described by Gu and Li (1996) and Yang and Kumar (1996).

Besides the ML estimation, a few distance formulae are suggested in the literature for the GTR and even the UNREST models. We consider the GTR model first. Note that in matrix notation, equation (1.64) becomes

$$F(t) = \{f_{ij}(t)\} = \Pi P(t), \tag{1.66}$$

where $\Pi = \text{diag}\{\pi_T, \pi_C, \pi_A, \pi_G\}$. Noting $P(t) = e^{Qt}$, we can estimate Qt by

$$\overline{Qt} = \log\{\hat{P}\} = \log\{\hat{\Pi}^{-1}\hat{F}\} \tag{1.67}$$

where we use the average observed frequencies to estimate Π and $\hat{f}_{ij} = \hat{f}_{ji} = (n_{ij} + n_{ji})/n$ to estimate the F matrix. The logarithm of $\hat{P}$ is computed by diagonalizing $\hat{P}$. When Q is defined as the relative substitution rates with the average rate to be 1, both t and Q can be recovered from the estimate of Qt; that is

$$\hat{t} = -\text{trace}\{\hat{\Pi} \log(\hat{\Pi}^{-1}\hat{F})\}, \qquad (1.68)$$

where trace$\{A\}$ is the sum of the diagonal elements of matrix A. Note that $-\text{trace}\{\hat{\Pi}\overline{Qt}\} = -\sum_i \hat{\pi}_i \hat{q}_{ii} \hat{t}$, the sequence distance. This approach was first suggested by Tavaré (1986, equation 3.12), although Rodriguez et al. (1990) were the first to publish equation (1.68). A number of authors (e.g. Gu and Li 1996; Yang and Kumar 1996; Waddell and Steel 1997) apparently rediscovered the distance formula, and also extended the distance to the case of gamma-distributed rates among sites, using the same idea for deriving gamma distances under JC69 and K80 (see Section 1.3).

The distance (1.68) is inapplicable when any of the eigenvalues of $\hat{P}$ is ≤ 0, which can occur often, especially at high sequence divergences. This is similar to the inapplicability of the JC69 distance when more than 75% of sites are different. As there are nine free parameters in the model and also nine free observables in the symmetrical matrix $\hat{F}$, the invariance property of MLEs suggests that equation (1.68), if applicable, should give the MLEs.

Next we describe a distance suggested by Barry and Hartigan (1987a), which works without the reversibility assumption and even without assuming a stationary model:

$$\hat{d} = -\tfrac{1}{4} \log\{\text{Det}(\hat{\Pi}^{-1}\hat{F})\}, \qquad (1.69)$$

where Det(A) is the determinant of matrix A, which is equal to the product of the eigenvalues of A. The distance is inapplicable when the determinant is ≤ 0 or when any of the eigenvalues of $\hat{\Pi}^{-1}\hat{F}$ is ≤ 0. Barry and Hartigan (1987a) referred to equation (1.69) as the *asynchronous distance*. It is now commonly known as the *Log-Det distance*.

Let us consider the behaviour of the distance under simpler stationary models in very long sequences. In such a case, $\hat{\Pi}^{-1}\hat{F}$ will approach the transition-probability matrix $P(t)$, and its determinant will approach $\exp(\sum_k \lambda_k t)$, where the λ_ks are the eigenvalue of the rate matrix Q (see equation 1.17). Thus $\hat{d}$ (equation 1.69) will approach $-(1/4)\sum_k \lambda_k t$. For the K80 model (Kimura 1980), the eigenvalues of the rate matrix (1.8) are $\lambda_1 = 0$, $\lambda_2 = -4\beta$, $\lambda_3 = \lambda_4 = -2(\alpha + \beta)$, so that $\hat{d}$ approaches $(\alpha + 2\beta)t$, which is the correct sequence distance. Obviously this will hold true for the simpler JC69 model as well. However, for more complex models with unequal base frequencies, $\hat{d}$ of equation (1.69) does not estimate the correct distance even though it grows linearly with time. For example, under the TN93 model (Tamura and Nei 1993), $\hat{d}$ approaches $(1/4)(\pi_Y \alpha_1 + \pi_R \alpha_2 + 2\beta)t$.

Barry and Hartigan (1987a) defined $\hat{f}_{ij} = n_{ij}/n$, so that $\hat{F}$ is not symmetrical, and interpreted $\hat{\Pi}^{-1}\hat{F}$ as an estimate of $P_{12}(t)$, the matrix of transition probabilities from sequences 1 to 2. The authors argued that the distance should work even if the substitution process is not homogeneous or stationary, that is, if there is systematic drift in base compositions during the evolutionary process. Evidence for the performance of the distance when different sequences have different base compositions is mixed. The distance appears to have acquired a paranormal status when it was rediscovered or modified by Lake (1994), Steel (1994b), and Zharkikh (1994), among others.

1.6 Discussions

1.6.1 Distance estimation under different substitution models

One might expect more complex models to be more realistic and to produce more reliable distance estimates. However, the situation is more complex. At small distances, the different assumptions about the structure of the Q matrix do not make much difference, and simple models such as JC69 and K80 produce very similar estimates to those under more complex models. The two 12s rRNA genes analysed in this chapter are different at about 10% of the sites. The different distance formulae produced virtually identical estimates, all between 0.10 and 0.11 (Table 1.3). This is despite the fact that simple models like JC69 are grossly wrong, judged by the log likelihood values achieved by the models. The rate variation among sites has much more impact, as seen in Section 1.3.

At intermediate distances, for example, when the sequences are about 20% or 30% different, different model assumptions become more important. It may be favourable to use realistic models for distance estimation, especially if the sequences are not short. At large distances, for example, when the sequences are >40% different, the different methods often produce very different estimates, and furthermore, the estimates, especially those under more complex models, involve large sampling errors. Sometimes the distance estimates become infinite or the distance formulae become inapplicable. This happens far more often under more complex models than under simpler models. In such cases, a useful approach is to add more sequences to break down the long distances and to use a likelihood-based approach to compare all sequences jointly on a phylogeny.

In this regard, the considerable interest in distance formulae under general models and their rediscoveries reflect more the mathematical tractability of two-sequence analysis than the biological utility of such distances. Unfortunately the mathematical tractability ends as soon as we move on to compare three sequences.

1.6.2 Limitations of pairwise comparison

If there are only two sequences in the whole data set, pairwise comparison is all we can do. If we have multiple sequences, however, pairwise comparison may be hampered as it ignores the other sequences, which should also provide information about the relatedness of the two compared sequences. Here I briefly comment on two obvious limitations of the pairwise approach. The first is lack of internal consistency. Suppose we use the K80 model for pairwise comparison of three sequences: a, b, and c.

Let $\hat{\kappa}_{ab}$, $\hat{\kappa}_{bc}$, and $\hat{\kappa}_{ca}$ be the estimates of the transition/transversion rate ratio κ in the three comparisons. Considering that the three sequences are related by a phylogenetic tree, we see that we estimated κ for the branch leading to sequence a as $\hat{\kappa}_{ab}$ in one comparison but as $\hat{\kappa}_{ca}$ in another. This inconsistency is problematic when complex models involving unknown parameters are used, and when information about model parameters is visible only when one compares multiple sequences simultaneously. An example is the variation of evolutionary rates among sites. With only two sequences, it is virtually impossible to decide whether a site has a difference because the rate at the site is high or because the overall divergence between the two sequences is high. Even if the parameters in the rate distribution (such as the shape parameter α of the gamma distribution) are fixed, the pairwise approach does not guarantee that a high-rate site in one comparison is also a high-rate site in another.

A second limitation is important in analysis of highly divergent sequences, in which substitutions have nearly reached *saturation*. The distance between two sequences is the sum of branch lengths on the phylogeny along the path linking the two sequences. By adding branch lengths along the tree, the pairwise distance can become large even if all branch lengths on the tree are small or moderate. As discussed above, large distances involve large sampling errors in the estimates or even cause the distance formulae to be inapplicable. By summing up branch lengths, the pairwise approach exacerbates the problem of saturation and may be expected to be less tolerant of high sequence divergences than likelihood or Bayesian methods, which compare all sequences simultaneously.

1.7 Exercises

1.1 Use the transition probabilities under the JC69 model (equation 1.3) to confirm the Chapman–Kolmogorov theorem (equation 1.4). It is sufficient to consider two cases: (a) $i = T, j = T$; and (b) $i = T, j = C$. For example, in case (a), confirm that $p_{TT}(t_1 + t_2) = p_{TT}(t_1)p_{TT}(t_2) + p_{TC}(t_1)p_{CT}(t_2) + p_{TA}(t_1)p_{AT}(t_2) + p_{TG}(t_1)p_{GT}(t_2)$.

1.2 Derive the transition-probability matrix $P(t) = e^{Qt}$ for the JC69 model (Jukes and Cantor 1969). Set $\pi_T = \pi_C = \pi_A = \pi_G = 1/4$ and $\alpha_1 = \alpha_2 = \beta$ in the rate matrix (1.15) for the TN93 model to obtain the eigenvalues and eigenvectors of Q under JC69, using results of Subsection 1.2.3. Alternatively you can derive the eigenvalues and eigenvectors from equation (1.1) directly. Then apply equation (1.17).

1.3 Derive the transition-probability matrix $P(t)$ for the Markov chain with two states 0 and 1 and generator matrix $Q = \begin{pmatrix} -u & u \\ v & -v \end{pmatrix}$. Confirm that the spectral decomposition of Q is given as

$$Q = U\Lambda U^{-1} = \begin{pmatrix} 1 & -u \\ 1 & v \end{pmatrix} \begin{pmatrix} 0 & 0 \\ 0 & -u-v \end{pmatrix} \begin{pmatrix} v/(u+v) & u/(u+v) \\ -1/(u+v) & 1/(u+v) \end{pmatrix}, \quad (1.70)$$

so that

$$P(t) = e^{Qt} = \frac{1}{u+v}\begin{pmatrix} v + ue^{-(u+v)t} & u - ue^{-(u+v)t} \\ v - ve^{-(u+v)t} & u + ve^{-(u+v)t} \end{pmatrix}. \quad (1.71)$$

Note that the stationary distribution of the chain is given by the first row of U^{-1}, as $[v/(u+v), u/(u+v)]$, which can also be obtained from $P(t)$ by letting $t \to \infty$. A special case is $u = v = 1$, when we have

$$P(t) = \begin{pmatrix} \frac{1}{2} + \frac{1}{2}e^{-2t} & \frac{1}{2} - \frac{1}{2}e^{-2t} \\ \frac{1}{2} - \frac{1}{2}e^{-2t} & \frac{1}{2} + \frac{1}{2}e^{-2t} \end{pmatrix}. \quad (1.72)$$

This is the binary equivalent of the JC69 model.

1.4 Confirm that the two likelihood functions for the JC69 model, equations (1.42) and (1.43), are proportional and the proportionality factor is a function of n and x but not of d. Confirm that the likelihood equation, $d\ell/dd = d\log\{L(d)\}/dd = 0$, is the same whichever of the two likelihood functions is used.

*1.5 Suppose $x = 9$ heads and $r = 3$ tails are observed in $n = 12$ independent tosses of a coin. Derive the MLE of the probability of heads (θ). Consider two mechanisms by which the data are generated.

(a) *Binomial*. The number $n = 12$ tosses was fixed beforehand. In $n = 12$ tosses, $x = 9$ heads were observed. Then the number of heads x has a binomial distribution, with probability

$$f(x|\theta) = \binom{n}{x} \theta^x (1-\theta)^{n-x}. \quad (1.73)$$

(b) *Negative binomial*. The number of tails $r = 3$ was fixed beforehand, and the coin was tossed until $r = 3$ tails were observed, at which point it was noted that $x = 9$ heads were observed. Then x has a negative binomial distribution, with probability

$$f(x|\theta) = \binom{r+x-1}{x} \theta^x (1-\theta)^{n-x}. \quad (1.74)$$

Confirm that under both models, the MLE of θ is x/n.

2

Models of amino acid and codon substitution

2.1 Introduction

In Chapter 1 we discussed continuous-time Markov chain models of nucleotide substitution and their application to estimate the distance between two nucleotide sequences. This chapter discusses similar Markov chain models to describe substitutions between amino acids in proteins or between codons in protein-coding genes. We make a straightforward use of the Markov chain theory introduced in Chapter 1, except that the states of the chain are now the 20 amino acids or the 61 sense codons (in the universal genetic code), instead of the four nucleotides.

With protein-coding genes, we have the advantage of being able to distinguish the *synonymous* or *silent* substitutions (nucleotide substitutions that do not change the encoded amino acid) from the *nonsynonymous* or *replacement* substitutions (those that do change the amino acid). As natural selection operates mainly at the protein level, synonymous and nonsynonymous mutations are under very different selective pressures and are fixed at very different rates. Thus comparison of synonymous and nonsynonymous substitution rates provides a means to understanding the effect of natural selection on the protein, as pointed out by pioneers of molecular evolution as soon as DNA sequencing techniques became available (e.g. Kafatos *et al.* 1977; Kimura 1977; Jukes and King 1979; Miyata and Yasunaga 1980). This comparison does not require estimation of absolute substitution rates or knowledge of the divergence times. Chapter 8 provides a detailed discussion of models developed to detect selection through phylogenetic comparison of multiple sequences. In this chapter, we consider comparison of two sequences only, to calculate two distances: one for synonymous substitutions and another for nonsynonymous substitutions.

2.2 Models of amino acid replacement

2.2.1 Empirical models

A distinction can be made between empirical and mechanistic models of amino acid substitution. *Empirical* models attempt to describe the relative rates of substitution between amino acids without considering explicitly factors that influence the evolutionary process. They are often constructed by analysing large quantities of sequence data, as compiled from databases. *Mechanistic models*, on the other hand, consider the biological process involved in amino acid substitution, such as mutational biases in the DNA, translation of the codons into amino acids, and acceptance or rejection of the resulting amino acid after filtering by natural selection. Mechanistic models have more interpretative power and are particularly useful for studying the forces and mechanisms of gene sequence evolution. For reconstruction of phylogenetic trees, empirical models appear at least equally efficient.

Empirical models of amino acid substitution are all constructed by estimating relative substitution rates between amino acids under the general time-reversible model. The rate q_{ij} from amino acids i to j is assumed to satisfy the detailed balance condition

$$\pi_i q_{ij} = \pi_j q_{ji}, \text{ for any } i \neq j. \tag{2.1}$$

This is equivalent to the requirement that the rate matrix can be written as the product of a symmetrical matrix and a diagonal matrix:

$$Q = S\Pi, \tag{2.2}$$

where $S = \{s_{ij}\}$ with $s_{ij} = s_{ji}$ for all $i = j$, and $\Pi = \text{diag}\{\pi_1, \pi_2, \ldots, \pi_{20}\}$, with π_j to be the equilibrium frequency of amino acid j (see Subsection 1.5.2). Whelan and Goldman (2001) referred to the s_{ij}s as the *amino acid exchangeabilities*.

The first empirical amino acid substitution matrix was constructed by Dayhoff and colleagues (Dayhoff *et al.* 1978). They compiled and analysed protein sequences available at the time, using a parsimony argument to reconstruct ancestral protein sequences and tabulating amino acid changes along branches on the phylogeny. To reduce the impact of multiple hits, the authors used only similar sequences that were different from one another at < 15% of sites. Inferred changes were merged across all branches without regard for their different lengths. Dayhoff *et al.* (1978) approximated the transition-probability matrix for an expected distance of 0.01 changes per site, called 1 PAM (for point-accepted mutations). This is $P(0.01)$ in our notation, from which the instantaneous rate matrix Q can be constructed. (See Kosiol and Goldman (2005) for a discussion of this construction.) The resulting rate matrix is known as the DAYHOFF matrix.

The DAYHOFF matrix was updated by Jones *et al.* (1992), who analysed a much larger collection of protein sequences, using the same approach as did Dayhoff *et al.* (1978). The updated matrix is known as the JTT matrix.

A variation to these empirical models is to replace the equilibrium amino acid frequencies (the π_js) in the empirical matrices by the frequencies observed in the data being analysed, while using the amino acid exchangeabilities from the empirical model (Cao *et al.* 1994). This strategy adds 19 free frequency parameters and is often found to improve the fit of the model considerably. The models are then known as DAYHOFF-F, JTT-F, etc., with a suffix '-F'.

It is also straightforward to estimate the rate matrix Q under the general time-reversible model from the data set by using maximum likelihood (ML) (Adachi and Hasegawa 1996*a*). This is the same procedure as used to estimate the pattern of nucleotide substitution (Yang 1994b). We will describe the details of likelihood calculation on a phylogeny in Chapter 4. The reversible rate matrix involves $19 \times 20/2 - 1 = 189$ relative rate parameters in the symmetrical matrix S of exchangeabilities as well as 19 free amino acid frequency parameters, with a total of 208 parameters (see equation 2.2). Usually the amino acid frequency parameters are estimated using the observed frequencies, reducing the dimension of the optimization problem by 19. The data set should be relatively large to allow estimation of so many parameters; 50 or 100 reasonably divergent protein sequences appear sufficient to provide good estimates. Examples of such matrices are the mtREV or MTMAM models for vertebrate or mammalian mitochondrial proteins (Adachi and Hasegawa 1996a; Yang et al. 1998) and the cpREV model for chloroplast proteins (Adachi et al. 2000). Whelan and Goldman (2001) used this approach to estimate a combined substitution-rate matrix from 182 alignments of nuclear proteins. This is known as the WAG matrix, and is an update to the DAYHOFF and JTT matrices.

Figure 2.1 presents the symmetrical matrix S of exchangeabilities under a few empirical models. Several features of these matrices are worth noting. First, amino acids with similar physico-chemical properties tend to interchange with each other at higher rates than dissimilar amino acids (Zuckerkandl and Pauling 1965; Clark 1970; Dayhoff et al. 1972, 1978; Grantham 1974). This pattern is particularly conspicuous when we compare rates between amino acids separated by a difference of only one codon position (Miyata et al. 1979). For example, aspartic acid (D) and glutamic acid (E) have high rates of exchange, as do isoleucine (I) and valine (V); in each pair the amino acids are similar. Cysteine, however, has low rates of exchange with all other amino acids. Second, the 'mutational distance' between amino acids determined by the structure of the genetic code has a major impact on the interchange rates, with amino acids separated by differences of two or three codon positions having lower rates than amino acids separated by a difference of one codon position. For example, empirical models for nuclear proteins (DAYHOFF, JTT, and WAG) and for mitochondrial proteins (MTMAM) have very different interchange rates between arginine (R) and lysine (K). These two amino acids are chemically similar and interchange frequently in nuclear proteins as they can reach each other through a change of one codon position. However, they rarely interchange in mitochondrial proteins as the codons differ at two or three positions in the mitochondrial code (Fig. 2.1) (Adachi and Hasegawa 1996a). Similarly, arginine (R) interchanges, albeit at low rates, with methionine (M), isoleucine (I), and threonine (T) in nuclear proteins but such changes are virtually absent in mitochondrial proteins. Furthermore, the two factors may be operating at the same time. When codons were assigned to amino acids during the origin and evolution of the genetic code, error minimization appears to have played a role so that amino acids with similar chemical properties tend to be assigned codons close to one another in the code (e.g. Osawa and Jukes 1989; Freeland and Hurst 1998).

Empirical amino acid substitution matrices are also used in alignment of multiple protein sequences. Cost (weight) matrices are used to penalize mismatches, with heavier penalties applied to rarer changes. The penalty for a mismatch between amino acids i and j is usually defined as $-\log(p_{ij}(t))$, where $p_{ij}(t)$ is the transition probability

2.2 Models of amino acid replacement • 43

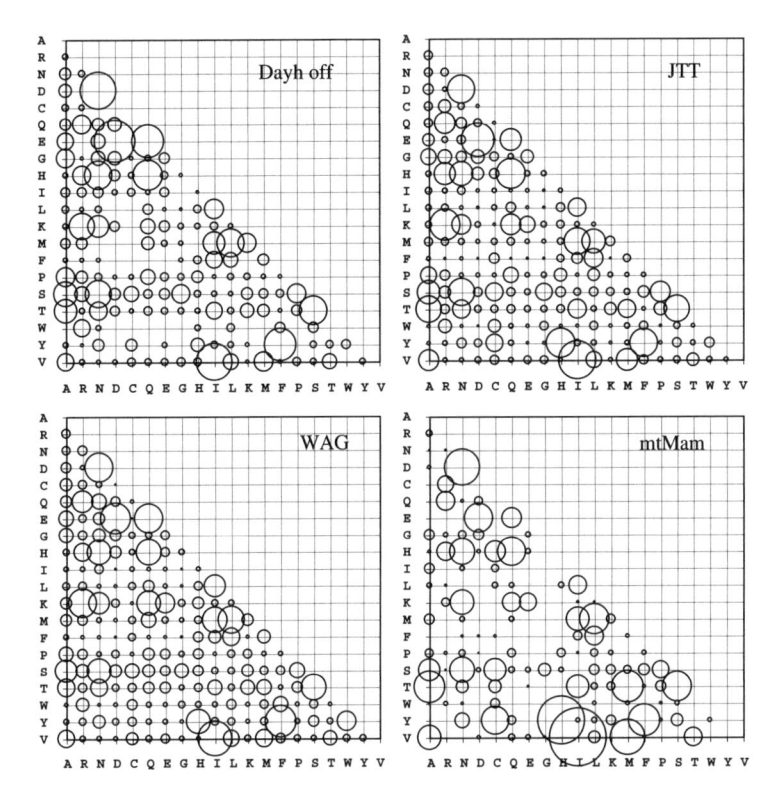

Fig. 2.1 Amino acid exchangeabilities under different empirical models: DAYHOFF (Dayhoff *et al.* 1978), JTT (Jones *et al.* 1992), WAG (Whelan and Goldman 2001), and MTMAM (Yang *et al.* 1998). The size (volume) of the bubble represents the exchangeability between the two amino acids (s_{ij} in equation 2.2). Note that the first three matrices are for nuclear genes, while MTMAM was estimated from the 12 protein-coding genes in the mitochondrial genomes of 20 mammalian species (Yang *et al.* 1998). The latter model is very similar to the mtREV24 model of Adachi and Hasegawa (1996*a*), estimated from 24 vertebrate species.

between i and j, and t measures the sequence divergence. Such matrices include Dayhoff *et al.*'s PAM matrices at different sequence distances, such as PAM$_{100}$ for $t = 1$ substitution per site, and PAM$_{250}$ for $t = 2.5$, as well as similar BLOSSUM matrices (Henikoff and Henikoff 1992) and the Gonnet matrix (Gonnet *et al.* 1992). Such matrices are in general too crude for phylogenetic analysis.

2.2.2 Mechanistic models

Yang *et al.* (1998) implemented a few mechanistic models of amino acid substitution. They are formulated at the level of codons and explicitly model the biological processes involved, i.e. different mutation rates between nucleotides, translation of the codon triplet into an amino acid, and acceptance or rejection of the amino acid due to selective pressure on the protein. We will discuss models of codon substitution later in this chapter. Such codon-based models of amino acid substitution may be called *mechanistic*. Yang et al. (1998) presented an approach for constructing a Markov-process model of amino acid replacement from a model of codon substitution by aggregating synonymous codons into one state (the encoded amino acid). Analysis of the mitochondrial genomes of 20 mammalian species suggests that the mechanistic models fit the data better than the empirical models such as DAYHOFF (Dayhoff et al. 1978) and JTT (Jones et al. 1992). Some of the mechanistic models implemented incorporate physico-chemical properties of amino acids (such as size and polarity), by assuming that dissimilar amino acids have lower exchange rates. While the use of such chemical properties improved the fit of the models, the improvement was not extraordinary, perhaps reflecting our poor understanding of which of the many chemical properties are most important and how they affect amino acid substitution rates. As Zuckerkandl and Pauling (1965) remarked, 'apparently chemists and protein molecules do not share the same opinions regarding the definition of the most prominent properties of a residue'. More research is needed to improve models of codon substitution to generate more realistic mechanistic models of amino acid substitution.

2.2.3 Among-site heterogeneity

The evolutionary process must vary considerably across sites or regions of a protein as they perform different roles in the structure and function of the protein and are thus under very different selective pressures. The simplest example of such among-site heterogeneity is variable substitution rates. Empirical models of amino acid substitution can be combined with the gamma model of rates among sites (Yang 1993, 1994a), leading to models such as DAYHOFF+Γ, JTT+Γ, etc., with a suffix '+Γ'. This is the same gamma model as discussed in Section 1.3, and assumes that the *pattern* of amino acid substitution is the same among all sites in the protein, but the rate is variable. The rate matrix for a site with rate r is rQ, with the same Q shared for all sites. The shape parameter α of the gamma distribution measures how variable the rates are, with small αs indicating strong rate variation. Table 2.1 provides a small sample of estimates of α from real data. It seems that in every functional protein the rates are significantly variable among sites, and in most proteins the shape parameter α is less than 1, indicating strong rate variation.

Besides the rates, the pattern of amino acid substitution represented by the matrix of relative substitution rates may also vary among sites. For example, different sites in the protein may prefer different amino acids. Bruno (1996) described an amino acid model in which a set of amino acid frequency parameters are used for each site in the sequence. This model involves too many parameters. Thorne et al. (1996) and Goldman et al. (1998) described models that allow for a few classes of sites that evolve under different Markov-chain models with different rate matrices. Such site classes may correspond to secondary structural categories in the protein. Similar models

Table 2.1 Maximum likelihood estimates of the gamma shape parameter α from a few data sets

Data	$\hat{\alpha}$	Ref.
DNA sequences		
1063 human and chimpanzee mitochondrial D-loop HVI sequences	0.42–0.45	Excoffier and Yang (1999)
SSU rRNA from 40 species (archaea, bacteria, and eukaryotes)	0.60	Galtier (2001)
LSU rRNA from 40 species archaea, bacteria, and eukaryotes)	0.65	Galtier (2001)
13 hepatitis B viral genomes	0.26	Yang et al. (1995b)
Protein sequences		
6 nuclear proteins (APO3, ATP7, BDNF, CNR1, EDG1, ZFY) from 46 mammalian species	0.12-0.93	Pupko et al. (2002b)
4 nuclear proteins (A2AB, BRCA1, IRBP, vmF) from 28 mammalian species	0.29-3.0	Pupko et al. (2002b)
12 mitochondrial proteins concatenated from 18 mammalian species	0.29	Cao et al. (1999)
45 chloroplast proteins concatenated from 10 plants and cynobacteria species	0.56	Adachi et al. (2000)

For nucleotide sequences, the substitution models HKY85+Γ or GTR+Γ were assumed, which account for different transition and transversion rates and different nucleotide frequencies. For amino acid models, JTT+Γ, mtREV+Γ, or cpREV+Γ were assumed. Estimates under the 'G+I' models, which incorporates a proportion of invariable sites in addition to the gamma distribution of rates for sites, are not included in this table, as parameter α has different interpretations in the '+Γ' and '+I + Γ' models.

were described by Koshi and Goldstein (1996b) and Koshi et al. (1999). Fitting such models require joint analysis of multiple sequences on a phylogeny.

2.3 Estimation of distance between two protein sequences

2.3.1 The Poisson model

If every amino acid has the same rate λ of changing into any other amino acid, the number of substitutions over time t will be a Poisson-distributed variable. This is the amino acid equivalent of the Jukes and Cantor (1969) model for nucleotide substitution. The sequence distance, defined as the expected number of amino acid substitutions per site, is then $d = 19\lambda t$, where t is the total time separating the two sequences (that is, twice the time of divergence). We let $\lambda = 1/19$ so that the substitution rate of each amino acid is 1 and then time t is measured by the distance. We thus use t and d interchangeably. Suppose x out of n sites are different between two protein sequences, with the proportion $\hat{p} = x/n$. The MLE of distance t is then

$$\hat{t} = -\frac{19}{20}\log\left(1 - \frac{20}{19}\hat{p}\right). \tag{2.3}$$

2.3.2 Empirical models

Under empirical models such as DAYHOFF (Dayhoff et al. 1978), JTT (Jones et al. 1992), or WAG (Whelan and Goldman 2001), different amino acids have different substitution rates. The Q matrix representing the relative substitution rates is assumed to be known, so that the only parameter to be estimated is the sequence distance d. It is a common practice to multiply Q by a scale constant so that the average rate is $-\sum_i \pi_i q_{ii} = 1$. Then $t = d$. Under empirical models DAYHOFF-F, JTT-F, etc., amino acid frequencies in the observed data are used to replace the equilibrium frequencies in the empirical model. Then again only distance t needs to be estimated.

It is straightforward to use ML to estimate t, as explained in Subsection 1.5.2 for nucleotides. Let n_{ij} be the number of sites occupied by amino acids i and j in the two sequences. The log likelihood function is

$$\ell(t) = \sum_i \sum_j n_{ij} \log\{f_{ij}(t)\} = \sum_i \sum_j n_{ij} \log\{\pi_i p_{ij}(t)\}, \quad (2.4)$$

where $f_{ij}(t)$ is the probability of observing a site with amino acids i and j in the two sequences, π_i is the equilibrium frequency of amino acid i, and $p_{ij}(t)$ is the transition probability. This is equation (1.65). The one-dimensional optimization problem can be easily managed numerically. Calculation of the transition-probability matrix $P(t) = \{p_{ij}(t)\}$ is discussed in Section 2.6.

An alternative approach is to estimate t by matching the observed proportion of different sites between the two sequences with the expected proportion under the model

$$p = \sum_{i \neq j} f_{ij}(t) = \sum_{i \neq j} \pi_i p_{ij}(t) = \sum_i \pi_i (1 - p_{ii}(t)). \quad (2.5)$$

Figure 2.2 shows the relationship between t and p for several commonly used models. As $f_{ij}(t)$ is not the same for all $i \neq j$, the two approaches produce different estimates. They should be close if the empirical model is not too wrong.

2.3.3 Gamma distances

If the rates vary according to the gamma distribution with given shape parameter α, and the relative rates are the same between any two amino acids, the sequence distance becomes

$$\hat{t} = \frac{19}{20}\alpha \left[\left(1 - \frac{20}{19}\hat{p}\right)^{-1/\alpha} - 1 \right], \quad (2.6)$$

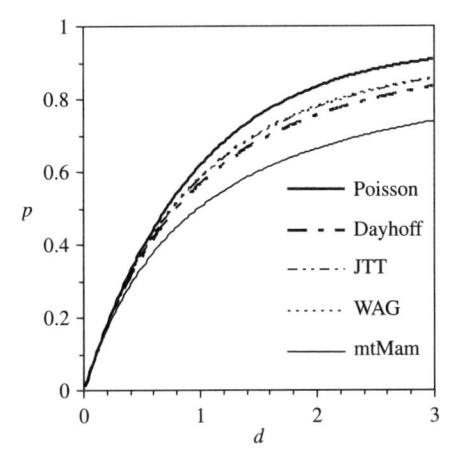

Fig. 2.2 The expected proportion of different sites (p) between two sequences separated by time or distance d under different models. The models are, from top to bottom, Poisson, WAG (Whelan and Goldman 2001), JTT (Jones et al. 1992), DAYHOFF (Dayhoff et al. 1978), and MTMAM (Yang et al. 1998). Note that the results for WAG, JTT, and DAYHOFF are almost identical.

where $\hat{p}$ is the proportion of different sites. This gamma distance under the Poisson model is very similar to the gamma distance under the JC69 model for nucleotides (equation 1.37).

For empirical models such as DAYHOFF, one can use ML to estimate the sequence distance under the gamma model. The theory discussed in Section 1.3 for nucleotides can be implemented for amino acids in a straightforward manner.

2.3.4 Example: distance between cat and rabbit p53 genes

We calculate the distance between the amino acid sequences of the tumour suppressor protein p53 from the cat (*Felis catus*, GenBank accession number D26608) and the rabbit (*Oryctolagus cuniculus*, X90592). There are 386 and 391 amino acids in the cat and rabbit sequences, respectively. We delete alignment gaps, with 382 sites left in the sequence. Out of them, 66 sites ($\hat{p} = 17.3\%$ of all sites) are different. Application of the Poisson correction (equation 2.3) gives the distance as 0.191 amino acid replacements per site. If we assume that the rates are gamma distributed with the shape parameter $\alpha = 0.5$, the distance becomes 0.235 by equation (2.6). If we use the WAG model (Whelan and Goldman 2001), the MLE of the distance (from equation 2.4) is 0.195 if the rate is assumed to be constant across sites, and 0.237 if the rate is assumed to be gamma distributed with shape parameter $\alpha = 0.5$. As is commonly the case, the rate variation among sites had a much greater impact on the distance estimates than the assumed substitution model.

2.4 Models of codon substitution

Markov-chain models of codon substitution were proposed by Goldman and Yang (1994) and Muse and Gaut (1994). In these models, the codon triplet is considered the unit of evolution, and a Markov chain is used to describe substitutions from one codon to another codon. The state space of the chain is the sense codons in the genetic code (that is, 61 sense codons in the universal code or 60 in the vertebrate mitochondrial code). Stop codons are not allowed inside a functional protein and are not considered in the chain. As before, we construct the Markov model by specifying the substitution-rate matrix, $Q = \{q_{ij}\}$, where q_{ij} is the instantaneous rate from codons i to j ($i \neq j$). The model in common use is a simplified version of the model of Goldman and Yang (1994), which specifies the substitution rate as

$$q_{ij} = \begin{cases} 0, & \text{if } i \text{ and } j \text{ differ at two or three codon positions,} \\ \pi_j, & \text{if } i \text{ and } j \text{ differ by a synonymous transversion,} \\ \kappa \pi_j, & \text{if } i \text{ and } j \text{ differ by a synonymous transition,} \\ \omega \pi_j, & \text{if } i \text{ and } j \text{ differ by a nonsynonymous transversion,} \\ \omega \kappa \pi_j, & \text{if } i \text{ and } j \text{ differ by a nonsynonymous transition.} \end{cases} \quad (2.7)$$

Here κ is the transition/transversion rate ratio, ω is the nonsynonymous/synonymous rate ratio or the acceptance rate of Miyata and Yasunaga (1980), and π_j is the equilibrium frequency of codon j. Parameters κ and π_j characterize processes at the DNA level, while parameter ω characterizes selection on nonsynonymous mutations. Mutations are assumed to occur independently at the three codon positions so that simultaneous changes at two or three positions are considered negligible. Different assumptions can be made concerning the equilibrium codon frequency π_j. One can assume that each codon has the same frequency, or codon frequencies are expected from the frequencies of four nucleotides (with three free parameters) or from three sets of nucleotide frequencies for the three codon positions (with nine free parameters). The most parameter-rich model uses all codon frequencies as parameters with the constraint that their sum is 1. Those models are referred to as Fequal, F1 × 4, F3 × 4, and F61.

The substitution rates to the target codon j = CTA (Leu) are given in Table 2.2 and illustrated in Fig. 2.3. Note that the model specified by equation (2.7) is very similar to the HKY85 model of nucleotide substitution (Hasegawa et al. 1985). A GTR-type model can also be implemented, which accommodates six different substitution types instead of just two (transitions and transversions); see equation (1.62).

It is easy to confirm that the Markov chain specified by the rate matrix Q of equation (2.7) satisfies the detailed-balance conditions (equation 2.1) for time-reversible chains. To relate the model to real observed data, we have to calculate the transition-probability matrix $P(t) = \{p_{ij}(t)\}$. Numerical algorithms for this calculation are discussed in Section 2.6.

Table 2.2 Substitution rates to target codon CTA (Leu)

Substitution	Relative rate	Substitution type
TTA (Leu) → CTA (Leu)	$q_{TTA, CTA} = \kappa \pi_{CTA}$	Synonymous transition
ATA (Ile) → CTA (Leu)	$q_{ATA, CTA} = \omega \pi_{CTA}$	Nonsynonymous transversion
GTA (Val) → CTA (Leu)	$q_{GTA, CTA} = \omega \pi_{CTA}$	Nonsynonymous transversion
CTT (Leu) → CTA (Leu)	$q_{CTT, CTA} = \pi_{CTA}$	Synonymous transversion
CTC (Leu) → CTA (Leu)	$q_{CTC, CTA} = \pi_{CTA}$	Synonymous transversion
CTG (Leu) → CTA (Leu)	$q_{CTG, CTA} = \kappa \pi_{CTA}$	Synonymous transition
CCA (Pro) → CTA (Leu)	$q_{CCA, CTA} = \kappa \omega \pi_{CTA}$	Nonsynonymous transition
CAA (Gln) → CTA (Leu)	$q_{CAA, CTA} = \omega \pi_{CTA}$	Nonsynonymous transversion
CGA (Arg) → CTA (Leu)	$q_{CGA, CTA} = \omega \pi_{CTA}$	Nonsynonymous transversion

Instantaneous rates from all other codons to CTA are 0.

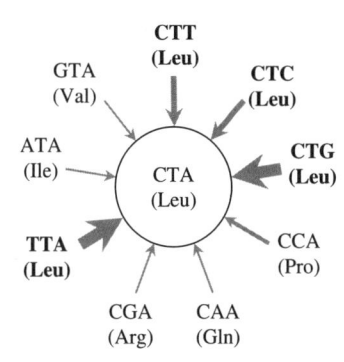

Fig. 2.3 Substitution rates to the same target codon CTA from its nine neighbours. Neighbours are codons that differ from the concerned codon at only one position. Some codons have fewer than nine neighbours as changes to and from stop codons are disallowed. The thickness of the lines represent different rates. The diagram is drawn using $\kappa = 2$ and $\omega = 1/3$, so that there are four different rates to codon CTA, in proportions 1:2:3:6, for nonsynonymous transversion, nonsynonymous transition, synonymous transversion, and synonymous transition, respectively (see equation 2.7).

2.5 Estimation of synonymous and nonsynonymous substitution rates

Two distances are usually calculated between protein-coding DNA sequences, for synonymous and nonsynonymous substitutions, respectively. They are defined as the number of synonymous substitutions per synonymous site (d_S or K_S) and the number of nonsynonymous substitutions per nonsynonymous site (d_N or K_A). Two classes of methods have been developed for estimating d_S and d_N: the heuristic counting methods and the ML method.

2.5.1 Counting methods

The counting methods proceed similarly to distance calculation under nucleotide-substitution models such as JC69 (Jukes and Cantor 1969). They involve three steps:

(i) Count synonymous and nonsynonymous *sites*.
(ii) Count synonymous and nonsynonymous *differences*.
(iii) Calculate the proportions of differences and correct for multiple hits.

The difference is that now a distinction is made between the synonymous and nonsynonymous types when sites and differences are counted.

The first counting methods were developed in the early 1980s shortly after DNA sequencing techniques were invented (Miyata and Yasunaga 1980; Perler *et al.* 1980). Miyata and Yasunaga (1980) assumed a simple mutation model with equal rates between nucleotides (as in JC69) when counting sites and differences, and used amino acid chemical distances developed by Miyata *et al.* (1979) to weight evolutionary pathways when counting differences between codons that differ at two or three positions (see below). The method was simplified by Nei and Gojobori (1986), who abandoned the weighting scheme and used equal weighting instead. Li *et al.* (1985) pointed out the importance of transition and transversion rate difference and dealt with it by partitioning codon positions into different degeneracy classes. Their procedure was improved by Li (1993), Pamilo and Bianchi (1993), Comeron (1995), and Ina (1995). Moriyama and Powell (1997) discussed the influence of unequal codon usage, which implies that the substitution rates are not symmetrical between codons, as assumed by the methods mentioned above. Yang and Nielsen (2000) accommodated both the transition/transversion rate difference and unequal codon frequencies in an iterative algorithm.

Here we describe the method of Nei and Gojobori (1986), referred to as NG86, to illustrate the basic concepts. This is a simplified version of the method of Miyata and Yasunaga (1980) and is similar to the JC69 model of nucleotide substitution. We then discuss the effects of complicating factors such as the unequal transition and transversion rates and unequal codon usage. It is worth noting that estimation of d_S and d_N is a complicated exercise, as manifested by the fact that even the simple transition/transversion rate difference is nontrivial to deal with. Unfortunately, different methods can often produce very different estimates.

2.5.1.1 Counting sites

Each codon has three nucleotide sites, which are divided into synonymous and nonsynonymous categories. Take the codon TTT (Phe) as an example. As each of the three codon positions can change into three other nucleotides, the codon has nine immediate neighbours: TTC (Phe), TTA (Leu), TTG (Leu), TCT (Ser), TAT (Tyr), TGT (Cys), CTT (Leu), ATT (Ile), and GTT (Val). Out of these, codon TTC codes for the same amino acid as the original codon (TTT). Thus there are $3 \times 1/9 = 1/3$ synonymous sites and $3 \times 8/9 = 8/3$ nonsynonymous sites in codon TTT (Table 2.3).

2.5 Synonymous and nonsynonymous substitution rates • 51

Table 2.3 Counting sites in codon TTT (Phe)

Target codon	Mutation type	Rate ($\kappa = 1$)	Rate ($\kappa = 2$)
TTC (Phe)	Synonymous	1	2
TTA (Leu)	Nonsynonymous	1	1
TTG (Leu)	Nonsynonymous	1	1
TCT (Ser)	Nonsynonymous	1	2
TAT (Tyr)	Nonsynonymous	1	1
TGT (Cys)	Nonsynonymous	1	1
CTT (Leu)	Nonsynonymous	1	2
ATT (Ile)	Nonsynonymous	1	1
GTT (Val)	Nonsynonymous	1	1
Sum		9	12
No. of syn. sites		1/3	1/2
No. of nonsyn. sites		8/3	5/2

κ is the transition/transversion rate ratio.

Mutations to stop codons are disallowed during the counting. We apply the procedure to all codons in sequence 1 and sum up the counts to obtain the total numbers of synonymous and nonsynonymous sites in the whole sequence. We then repeat the process for sequence 2, and average the numbers of sites between the two sequences. Let these be S and N, with $S + N = 3 \times L_c$, where L_c is the number of codons in the sequence.

2.5.1.2 Counting differences

The second step is to count the numbers of synonymous and nonsynonymous *differences* between the two sequences. In other words, the observed differences between the two sequences are partitioned between the synonymous and nonsynonymous categories. We proceed again codon by codon. This is straightforward if the two compared codons are identical (e.g. TTT vs. TTT), in which case the numbers of synonymous and nonsynonymous differences are 0, or if they differ at one codon position only (e.g. TTC vs. TTA), in which case whether the single difference is synonymous or nonsynonymous is obvious. However, when the two codons differ at two or three positions (e.g. CCT vs. CAG or GTC vs. ACT), there exist four or six evolutionary pathways from one codon to the other. The multiple pathways may involve different numbers of synonymous and nonsynonymous differences. Most counting methods give equal weights to the different pathways.

For example, there are two pathways between codons CCT and CAG (Table 2.4). The first goes through the intermediate codon CAT and involves two nonsynonymous differences while the second goes through CCG and involves one synonymous and one nonsynonymous difference. If we apply equal weights to the two pathways, there are 0.5 synonymous differences and 1.5 nonsynonymous differences between the two

Table 2.4 Two pathways between codons CCT and CAG

Pathway	Differences	
	Synonymous	Nonsynonymous
CCT (Pro) ↔ CAT (His) ↔ CAG (Gln)	0	2
CCT (Pro) ↔ CCG (Pro) ↔ CAG (Gln)	1	1
Average	0.5	1.5

codons. If the synonymous rate is higher than the nonsynonymous rate, as in almost all genes, the second pathway should be more likely than the first. Without knowing the d_N/d_S ratio and the sequence divergence beforehand, it is difficult to weight the pathways appropriately. Nevertheless, Nei and Gojobori's (1986) computer simulation demonstrated that weighting had minimal effects on the estimates, especially if the sequences are not very divergent.

The counting is done codon by codon across the sequence, and the numbers of differences are summed to produce the total numbers of synonymous and non-synonymous differences between the two sequences. Let these be S_d and N_d, respectively.

2.5.1.3 Correcting for multiple hits

We now have

$$p_S = S_d/S,$$
$$p_N = N_d/N \tag{2.8}$$

to be the proportions of differences at the synonymous and nonsynonymous sites, respectively. These are equivalent to the proportion of differences under the JC69 model for nucleotides. Thus we apply the JC69 correction for multiple hits

$$d_S = -\tfrac{3}{4} \log\left(1 - \tfrac{4}{3} p_S\right),$$
$$d_N = -\tfrac{3}{4} \log\left(1 - \tfrac{4}{3} p_N\right). \tag{2.9}$$

As pointed out by Lewontin (1989), this step is logically flawed. The JC69 formula is suitable for noncoding regions and assumes that any nucleotide can change into *three* other nucleotides with equal rates. When we focus on synonymous sites and differences only, each nucleotide does not have *three* other nucleotides to change into. In practice, the effect of multiple-hit correction is minor, at least at low sequence divergences, so that the bias introduced by the correction formula is not very important.

2.5.1.4 Application to the rbcL genes

We apply the NG86 method to estimate d_S and d_N between the cucumber and tobacco genes for the plastid protein ribulose-1,5-bisphosphate carboxylase/oxygenase large subunit (*rbcL*). The GenBank accession numbers are NC_007144 for the cucumber (*Cucumis sativus*) and Z00044 for the tobacco (*Nicotiana tabacum*). There are 476 and 477 codons in the cucumber and tobacco genes, respectively, with 481 codons in the alignment. We delete codons that are alignment gaps in either species, leaving 472 codons in the sequence.

A few basic statistics from the data are listed in Table 2.5, obtained by applying the nucleotide-based model HKY85 (Hasegawa *et al.* 1985) to analyse the three codon positions separately. The base compositions are unequal, and the third codon positions are AT-rich. The estimates of the transition/transversion rate ratio for the three codon positions are in the order $\hat{\kappa}_3 > \hat{\kappa}_1 > \hat{\kappa}_2$. Estimates of the sequence distance are in the same order $\hat{d}_3 > \hat{d}_1 > \hat{d}_2$. Such patterns are common for protein-coding genes, and reflect the structure of the genetic code and the fact that essentially all proteins are under selective constraint, with higher synonymous than nonsynonymous substitution rates. When the genes are examined codon by codon, 345 codons are identical between the two species, while 115 differ at one position, 95 of which are synonymous and 20 nonsynonymous. Ten codons differ at two positions, and two codons differ at all three positions.

We then apply the NG86 method. The 1416 nucleotide sites are partitioned into $S = 343.5$ synonymous sites and $N = 1072.5$ nonsynonymous sites. There are 141 differences observed between the two sequences, which are partitioned into $S_d = 103.0$ synonymous differences and $N_d = 38.0$ nonsynonymous differences. The proportions of differences at the synonymous and nonsynonymous sites are thus $p_S = S_d/S = 0.300$ and $p_N = N_d/N = 0.035$. Application of the JC69 correction gives $d_S = 0.383$ and $d_N = 0.036$, with the ratio $\hat{\omega} = d_N/d_S = 0.095$. According to this estimate, the protein is under strong selective constraint, and a nonsynonymous mutation has only 9.5% of the chance of a synonymous mutation of spreading through the population.

Table 2.5 Basic statistics for the cucumber and tobacco *rbcL* genes

Position	Sites	π_T	π_C	π_A	π_G	$\hat{\kappa}$	$\hat{d}$
1	472	0.179	0.196	0.239	0.386	2.202	0.057
2	472	0.270	0.226	0.299	0.206	2.063	0.026
3	472	0.423	0.145	0.293	0.139	6.901	0.282
All	1416	0.291	0.189	0.277	0.243	3.973	0.108

Base frequencies are observed frequencies at the three codon positions, averaged over the two sequences. These are very close to MLEs under the HKY85 model (Hasegawa *et al.* 1985). The transition/transversion rate ratio κ and sequence distance d are estimated under the HKY85 model.

2.5.1.5 Transition/transversion rate difference and codon usage

Due to the structure of the genetic code, transitions at the third codon positions are more likely to be synonymous than transversions are. Thus one may observe many synonymous substitutions, relative to nonsynonymous substitutions, not because natural selection has removed nonsynonymous mutations but because transitions occur at higher rates than transversions, producing many synonymous mutations. Ignoring the transition/transversion rate difference thus leads to underestimation of the number of synonymous sites (S) and overestimation of the number of nonsynonymous sites (N), resulting in overestimation of d_S and underestimation of d_N (Li et al. 1985).

To accommodate the different transition and transversion rates, Li et al. (1985) classified each nucleotide site, that is, each position in a codon, into *nondegenerate*, *two-fold degenerate*, and *four-fold degenerate* classes. The degeneracy of a codon position is determined by how many of the three possible mutations are synonymous. At a four-fold degenerate site, every change is synonymous, while at the nondegenerate site, every change is nonsynonymous. At a two-fold site, only one change is synonymous. For example, the third position of codon TTT (Phe) is two-fold degenerate. In the universal code the third position of ATT (Ile) is three-fold degenerate but is often grouped into the two-fold degenerate class. Let the total number of sites in the three degeneracy classes, averaged over the two sequences, be L_0, L_2, and L_4. Similarly the numbers of transitional and transversional differences within each degeneracy class are counted. Li et al. (1985) then used the K80 model (Kimura 1980) to estimate the number of transitions and the number of transversions per site within each degeneracy class. Let these be A_i and B_i, $i = 0, 2, 4$, with $d_i = A_i + B_i$ to be the total distance. A_i and B_i are estimates of αt and $2\beta t$ in the K80 model, given by equation (1.12) in Chapter 1. Thus $L_2 A_2 + L_4 d_4$ and $L_2 B_2 + L_0 d_0$ are the total numbers of synonymous and nonsynonymous substitutions between the two sequences, respectively. To estimate d_S and d_N, we also need the numbers of synonymous and nonsynonymous sites. Each four-fold degenerate site is a synonymous site and each non-degenerate site is a nonsynonymous site. The case with two-fold degenerate sites is less clear. Li et al. (1985) counted each two-fold site as 1/3 synonymous and 2/3 nonsynonymous, based on the assumption of equal mutation rates. The numbers of synonymous and nonsynonymous sites are thus $L_2/3 + L_4$ and $2L_2/3 + L_0$, respectively. The distances then become

$$d_S = \frac{L_2 A_2 + L_4 d_4}{L_2/3 + L_4},$$

$$d_N = \frac{L_2 B_2 + L_0 d_0}{2L_2/3 + L_0}$$

(2.10)

(Li et al. 1985). This method is referred to as LWL85.

Li (1993) and Pamilo and Bianchi (1993) pointed out that the rule of counting a two-fold site as 1/3 synonymous and 2/3 nonsynonymous ignores the transition/transversion rate difference, and causes underestimation of the number of synonymous sites and overestimation of d_S (and underestimation of d_N). Instead, these authors suggested the following formulae

$$d_S = \frac{L_2A_2 + L_4A_4}{L_2 + L_4} + B_4,$$
$$d_N = A_0 + \frac{L_0B_0 + L_2B_2}{L_0 + L_2}. \quad (2.11)$$

This method, known as LPB93, effectively uses the distance at the four-fold sites, $A_4 + B_4$, to estimate the synonymous distance d_S, but replaces the transition distance A_4 with an average between two-fold and four-fold sites $(L_2A_2 + L_4A_4)/(L_2 + L_4)$, assuming that the transition rate is the same at the two-fold and four-fold sites. Similarly d_N may be considered an estimate of the distance at the nondegenerate site, $(A_0 + B_0)$, but with the transversion distance B_0 replaced by the average over the two-fold degenerate and nondegenerate sites, $(L_0B_0 + L_2B_2)/(L_0 + L_2)$.

An alternative approach is to calculate each of d_S and d_N as the number of substitutions divided by the number of sites, as in the LWL85 method, but replace 1/3 with an estimate of the proportion ρ of two-fold sites that are synonymous:

$$d_S = \frac{L_2A_2 + L_4d_4}{\rho L_2 + L_4},$$
$$d_N = \frac{L_2B_2 + L_0d_0}{(1-\rho)L_2 + L_0}. \quad (2.12)$$

We can use the distances A_4 and B_4 at the four-fold sites to estimate the transition/transversion rate ratio and to partition two-fold sites into synonymous and nonsynonymous categories. In other words, $\hat{\kappa} = 2A_4/B_4$ is an estimate of κ in the K80 model and

$$\hat{\rho} = \hat{\kappa}/(\hat{\kappa} + 2) = A_4/(A_4 + B_4). \quad (2.13)$$

We will refer to equations (2.12) and (2.13) as the LWL85m method. Similar but slightly more complicated distances were defined by Tzeng et al. (2004) in their modification of the LWL85 method.

It is not essential to partition sites according to codon degeneracy. One can take into account the transition/transversion rate difference by counting synonymous and nonsynonymous sites in proportion to synonymous and nonsynonymous 'mutation' rates. This is the approach of Ina (1995) and it has the advantage of not having to deal with the irregularities of the genetic code, such as the existence of three-fold sites and the fact that not all transversions at two-fold sites are synonymous. Table 2.3 explains the counting of sites in codon TTT when the transition/transversion rate ratio $\kappa = 2$ is given. One simply uses the mutation rates to the nine neighbouring codons as weights when partitioning sites, with each transitional change (to codons TTC, TCT, and CTT) having twice the rate of each transversional change (to the six other codons). Thus

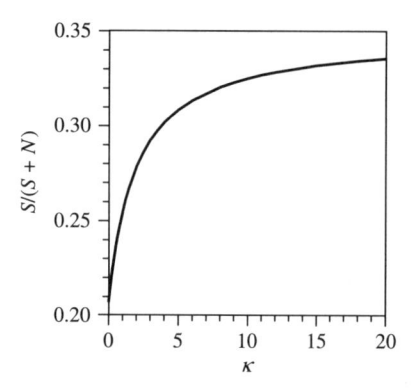

Fig. 2.4 Proportion of synonymous sites $S/(S+N)$ as a function of the transition/transversion rate ratio κ. Codon frequencies are assumed to be the same (1/61). Redrawn from Yang and Nielsen (1998).

there are $3 \times 2/12 = 1/2$ synonymous sites and $3 \times 10/12 = 5/2$ nonsynonymous sites in codon TTT, compared with 1/3 and 8/3 when $\kappa = 1$. The percentage of synonymous sites as a function of κ under the universal code is plotted in Fig. 2.4. Ina (1995) described two methods for estimating κ. The first uses the third codon positions and tends to overestimate κ. The second uses an iterative algorithm. Another commonly used approach is to use the four-fold degenerate sites, as mentioned above (Pamilo and Bianchi 1993; Yang and Nielsen 2000).

Besides the transition–transversion rate difference, another major complicating factor to estimation of d_S and d_N is the unequal codon frequencies. Recall that the JC69 and K80 models of nucleotide substitution assume symmetrical rates and predict uniform base frequencies. Similarly symmetrical codon substitution rates predict equal frequencies of all sense codons, and the fact that the observed codon frequencies are not equal means that the rates are not symmetrical, as assumed by the methods discussed above. Overall, rates to common codons are expected to be higher than rates to rare codons. Such rate differences affect our counting of sites and differences. Yang and Nielsen (2000) accommodated both the transition/transversion rate difference and unequal codon frequencies in an iterative procedure for estimating d_S and d_N.

Example. We apply the methods discussed above to the cucumber and tobacco *rbcL* genes (Table 2.6). The numbers of non degenerate, two-fold degenerate, and four-fold degenerate sites are $L_0 = 916.5$, $L_2 = 267.5$, and $L_4 = 232.0$. The 141 differences between the two sequences are partitioned into 15.0 transitions and 18.0 transversions at the nondegenerate sites, 44.0 transitions and 8.5 transversions at the two-fold degenerate sites, and 32.0 transitions and 23.5 transversions at the four-fold degenerate sites. Application of the K80 correction formula leads to $A_0 = 0.0169$ and $B_0 = 0.0200$ as the transitional and transversional distances at the nondegenerate sites. Similarly $A_2 = 0.2073$ and $B_2 = 0.0328$ at the two-fold sites and $A_4 = 0.1801$ and $B_4 = 0.1132$

Table 2.6 Estimates of d_S and d_N between the cucumber and tobacco *rbcL* genes

Model	$\hat{\kappa}$	$\hat{S}$	$\hat{N}$	$\hat{d}_S$	$\hat{d}_N$	$(\hat{\omega})$	$\hat{t}$	ℓ
Counting methods								
NG86 (Nei and Gojobori 1986)	1	343.5	1072.5	0.383	0.036	0.095		
LWL85 (Li et al. 1985)	N/A	321.2	1094.8	0.385	0.039	0.101		
LWL85m (equation 2.12)	3.18	396.3	1019.7	0.312	0.042	0.134		
LPB93 (Li 1993; Pamilo and Bianchi 1993)	N/A	N/A	N/A	0.308	0.040	0.129		
Ina95 (Ina 1995)	5.16	418.9	951.3	0.313	0.041	0.131		
YN00 (Yang and Nielsen 2000)	2.48	308.4	1107.6	0.498	0.035	0.071		
ML method (Goldman and Yang 1994)								
(A) Fequal, $\kappa = 1$	1	360.7	1055.3	0.371	0.037	0.096	0.363	−2466.33
(B) Fequal, κ estimated	2.59	407.1	1008.9	0.322	0.037	0.117	0.358	−2454.26
(C) F1 × 4, $\kappa = 1$ fixed	1	318.9	1097.1	0.417	0.034	0.081	0.361	−2436.17
(D) F1 × 4, κ estimated	2.53	375.8	1040.2	0.362	0.036	0.099	0.367	−2424.98
(E) F3 × 4, $\kappa = 1$ fixed	1	296.6	1119.4	0.515	0.034	0.066	0.405	−2388.35
(F) F3 × 4, κ estimated	3.13	331.0	1085.0	0.455	0.036	0.078	0.401	−2371.86
(G) F61, $\kappa = 1$ fixed	1	263.3	1152.7	0.551	0.034	0.061	0.389	−2317.76
(H) F61, κ estimated	2.86	307.4	1108.6	0.473	0.035	0.074	0.390	−2304.47

NG86, LWL85, and LPB93 were implemented in MEGA3.1 (Kumar et al. 2005a) and PAML (Yang 1997a). Both programs use equal weighting of pathways when the two compared codons differ at two or three positions. Slightly different results were obtained for LPB93 from BAMBE4.2 (Xia and Xie 2001), which implements the weighting scheme described by Li et al. (1985). Ina's program was used for Ina's method (Ina 1995), and YN00 (Yang 1997a) was used for YN00 (Yang and Nielsen 2000). The likelihood method was implemented in the CODEML program in PAML (Yang 1997a). The models assumed are: Fequal, with equal codon frequencies ($\pi_j = 1/61$ for all j); F1 × 4, with four nucleotide frequencies used to calculate the expected codon frequencies (using three free parameters); F3 × 4, with nucleotide frequencies at three codon positions used to calculate codon frequencies (9 free parameters); F61, with all 61 codon frequencies used as free parameters (60 free parameters because the sum is 1). '$\kappa = 1$ fixed' means that transition and transversion rates are assumed to be equal, while 'κ estimated' accommodates different transition and transversion rates (see equation 2.7). ℓ is the loglikelihood value under the model.

at the four-fold sites. The LWL85 method calculates the numbers of synonymous and nonsynonymous sites as $S = L_2/3 + L_4 = 321.2$ and $N = L_2 \times 2/3 + L_0 = 1094.8$, so that $d_S = 0.385$ and $d_N = 0.039$, with $d_N/d_S = 0.101$.

The LPB93 method gives $d_S = 0.308$ and $d_N = 0.040$, with $d_N/d_S = 0.129$. If we use $\hat{\rho} = A_4/(A_4 + B_4) = 0.6141$ to estimate the percentage of two-fold sites that are synonymous, and apply equation (2.12), we get $S = 396.3$ and $N = 1019.7$ for the LWL85m method. The distances are then $d_S = 0.312$ and $d_N = 0.042$, with $d_N/d_S = 0.134$. Note that S is much larger than S calculated for the NG86 method, which ignores the transition/transversion rate difference. The YN00 method incorporates unequal codon frequencies as well as the transition/transversion rate difference. This gives $S = 308.4$, even smaller than S calculated by the NG86 method.

The biased codon usage had the opposite effect to the transition/transversion rate difference, and its effect counter-balanced the effect of the transition/transversion rate difference. The distance estimates then become $d_S = 0.498$ and $d_N = 0.035$, with $d_N/d_S = 0.071$.

2.5.2 Maximum likelihood method

2.5.2.1 Likelihood estimation of d_S and d_N

The ML method (Goldman and Yang 1994) fits a Markov model of codon substitution, such as equation (2.7), to data of two sequences to estimate parameters in the model, including t, κ, ω, and π_j. The likelihood function is given by equation (2.4), except that i and j now refer to the 61 sense codons (in the universal genetic code) rather than the 20 amino acids. The codon frequencies (if they are not fixed to 1/61) are usually estimated by using the observed frequencies in the data, while parameters t, κ, and ω will be estimated by numerical maximization of the log likelihood ℓ. Then d_S and d_N are calculated from the estimates of t, κ, ω, and π_j, according to their definitions. Here we describe the definition of d_S and d_N when parameters in the codon model are given, and the rate matrix Q and sequence distance t are known. In real data, the same calculation, with the parameters replaced by their MLEs, will produce the MLEs of d_S and d_N, according to the invariance property of MLEs.

We define the numbers of sites and substitutions on a per codon basis. First, the expected number of substitutions per codon from codons i to j, $i \neq j$, over any time t is $\pi_i q_{ij} t$. Thus the numbers of synonymous and nonsynonymous substitutions per codon between two sequences separated by time (or distance) t are

$$S_d = t\rho_S = \sum_{i \neq j,\, aa_i = aa_j} \pi_i q_{ij} t,$$
$$N_d = t\rho_N = \sum_{i \neq j,\, aa_i \neq aa_j} \pi_i q_{ij} t. \tag{2.14}$$

That is, S_d sums over codon pairs that differ by a synonymous difference and N_d sums over codon pairs that differ by a nonsynonymous difference, with aa_i to be the amino acid encoded by codon i. As the rate matrix is scaled so that one nucleotide substitution is expected to occur in one time unit, we have $S_d + N_d = t$, and ρ_S and ρ_N are the proportions of synonymous and nonsynonymous substitutions, respectively.

Next, we count sites. The proportions of synonymous and nonsynonymous 'mutations', ρ_S^1 and ρ_N^1, are calculated in the same way as ρ_S and ρ_N in equation (2.14) except that $\omega = 1$ is fixed (Goldman and Yang 1994). The numbers of synonymous and nonsynonymous sites per codon are then

$$S = 3\rho_S^1,$$
$$N = 3\rho_N^1. \tag{2.15}$$

2.5 Synonymous and nonsynonymous substitution rates • 59

The definition of sites by equation (2.15) is equivalent to Ina's (1995, Table 1) if codon frequencies are equal ($\pi_j = 1/61$) and the mutation rate is the same at the three codon positions. However, if the three codon positions have different base compositions and different rates, Ina's method counts each codon position as one site and will lead to nonsensical results. In equation (2.15), each codon is counted as three sites, but each codon position will be counted as more or less than one site depending on whether the mutation rate at the position is higher or lower than the average rate at all three positions. Another point worth mentioning is that the 'mutation rate' referred to here may have been influenced by selection acting on the DNA (but not selection on the protein). In other words, the proportions of sites represent what we would expect to observe if DNA-level processes, which cause the transition/transversion rate difference and unequal codon usage, are still operating but if there is no selection at the protein level (i.e. $\omega = 1$). The use of the term 'mutation rate' here may be confusing as the rate is affected by selection on the DNA. A discussion of definitions of sites is provided in Subsection 2.5.4.

The distances are then given by

$$d_S = S_d/S = t\rho_S/(3\rho_S^1),$$
$$d_N = N_d/N = t\rho_N/(3\rho_N^1).$$
(2.16)

Note that $\omega = d_N/d_S = (\rho_N/\rho_S)/(\rho_N^1/\rho_S^1)$ is a ratio of two ratios; the numerator ρ_N/ρ_S is the ratio of the numbers of synonymous and nonsynonymous substitutions that are inferred to have occurred, while the denominator ρ_N^1/ρ_S^1 is the corresponding ratio expected if there had been no selection on the protein (so that $\omega = 1$). Thus ω measures the perturbation in the proportions of synonymous and nonsynonymous substitutions caused by natural selection on the protein.

2.5.2.2 Estimation of d_S and d_N between the cucumber and tobacco rbcL genes

We use the ML method under different models to estimate d_S and d_N between the cucumber and tobacco *rbcL* genes, in comparison with the counting methods. The results are summarized in Table 2.6.

First, we assume equal transition and transversion rates ($\kappa = 1$) and equal codon frequencies ($\pi_j = 1/61$) (model A in Table 2.6). This is the model underlying the NG86 method. There are two parameters in the model: t and ω (see equation 2.7). The log likelihood contour is shown in Fig. 2.5. The MLEs are found numerically to be $\hat{t} = 0.363$ and $\hat{\omega} = 0.096$. The estimated sequence divergence corresponds to $472 \times 0.363 = 171.4$ nucleotide substitutions over the whole sequence, after correction for multiple hits, compared with 141 raw observed differences. Application of equation (2.14) gives 133.8 synonymous substitutions and 37.6 nonsynonymous substitutions for the whole sequence. We then calculate the proportions of synonymous and nonsynonymous sites to be $\rho_S^1 = 0.243$ and $\rho_N^1 = 0.757$. The numbers of sites for the whole sequence are $L_c \times 3\rho_S^1 = 360.7$ and $N = L_c \times 3\rho_N^1 = 1055.3$. Counting sites according to equation (2.15) under this model is equivalent to counting sites in

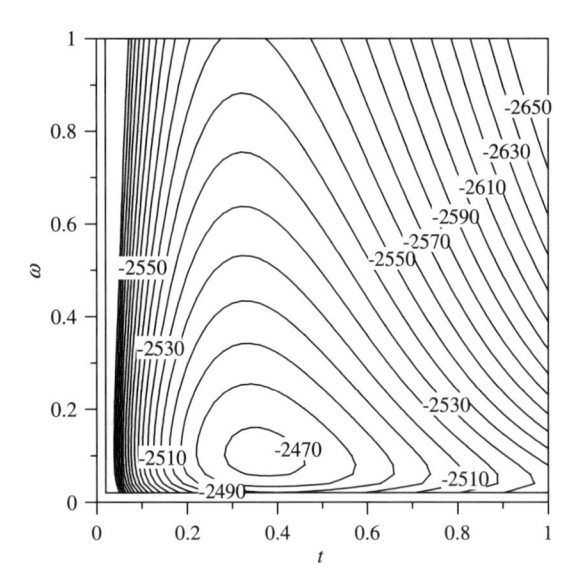

Fig. 2.5 Log likelihood contour as a function of the sequence distance t and the rate ratio ω for the cucumber and tobacco *rbcL* genes. The model assumes equal transition and transversion rates ($\kappa = 1$) and equal codon frequencies ($\pi_j = 1/61$).

the NG86 method explained in Subsection 2.5.1. The slight differences in the counts are due to the fact that the ML method averages over the codon frequencies (1/61) assumed in the model, while the NG86 method averages over the observed codons in the two sequences. The distances then become $d_S = 133.8/360.7 = 0.371$ and $d_N = 37.6/1055.3 = 0.037$, very similar to estimates from NG86.

Next, we assume equal codon frequencies but different transition and transversion rates (model B in Table 2.6: Fequal, κ estimated). This is the model underlying the LWL85 (Li *et al.* 1985), LPB93 (Li 1993; Pamilo and Bianchi 1993), and Ina's (1995) methods. The model involves three parameters: t, κ, and ω. Their MLEs are found numerically to be $\hat{t} = 0.358$, $\hat{\kappa} = 2.59$, and $\hat{\omega} = 0.117$. The ω ratio is 22% larger than the estimate obtained when the transition/transversion rate difference is ignored ($\kappa = 1$). The numbers of synonymous and nonsynonymous substitutions at those parameter values are 131.0 and 37.8, respectively. Counting sites by fixing $\omega = 1$ and using $\hat{\kappa} = 2.59$ gives $\rho_S^1 = 0.288$ and $\rho_N^1 = 0.719$, and $S = 407.1$ and $N = 1008.9$ for the whole sequence. Thus we have $d_S = 131.0/407.1 = 0.322$ and $d_N = 37.8/1008.9 = 0.037$. Note that incorporating the transition/transversion rate difference has had a much greater effect on the numbers of sites than on the numbers of substitutions. Also as $S + N$ is fixed, an increase in S means a decrease in N at the same time, leading to even greater differences in the ω ratio. As S is much smaller than N, d_S is affected more seriously than d_N.

Models that accommodate both the transition/transversion rate difference and unequal codon frequencies are also used to estimate d_S and d_N. The results are shown in Table 2.6. The counting method of Yang and Nielsen (2000) approximates the likelihood method under the F3 × 4 model (model F in Table 2.6), and indeed produced very similar results to the likelihood method.

2.5.3 Comparison of methods

The results of Table 2.6 suggest considerable differences in estimates of d_S and d_N and their ratio ω among the methods. Similar differences have been observed in analyses of many real and simulated data sets (e.g. Bielawski *et al.* 2000; Yang and Nielsen 2000). Based on those results, the following observations can be made:

(1) Ignoring the transition and transversion rate difference leads to underestimation of S, overestimation of d_S, and underestimation of the ω ratio (Li *et al.* 1985).
(2) Biased codon usage often has the opposite effect to the transition/transversion bias; ignoring codon usage bias leads to overestimation of S, underestimation of d_S, and overestimation of ω (Yang and Nielsen 2000). Extreme base or codon frequencies can overwhelm the effect of the transition/transversion rate difference, leading to the ironical result that NG86, which ignores both the transition/transversion rate difference and the codon usage bias, can produce more reliable estimates of d_S and d_N than LPB93, LWL85m, or Ina's method, which accommodates the transition/transversion rate difference but not the codon usage bias.
(3) Different methods or model assumptions can produce very different estimates even when the two sequences compared are highly similar. This is in contrast to distance calculation under nucleotide models, where the difference between methods is minimal at low sequence divergences (see, e.g., Table 1.3). The reason for this unfortunate lack of robustness to model assumptions is that different methods give different counts of sites.
(4) Assumptions appear to matter more than methods. The likelihood method can produce very different estimates under different assumptions. For example, the MLEs of ω in Table 2.6 are almost two-fold different. However, the counting and likelihood methods often produce similar estimates under the same assumptions. The NG86 method is seen to produce results similar to likelihood under the same model. Extensive simulations by Muse (1996) also confirmed the similarity of the two methods, at least if sequence divergence is not great. Similarly, LPB93 and Ina's method produced results similar to likelihood accommodating the transition/transversion rate difference but not codon usage.

Compared with the counting methods, the likelihood method has two clear advantages. The first is conceptual simplicity. In the counting method, dealing with features of DNA sequence evolution such as different transition and transversion rates and unequal codon frequencies is challenging. For example, some methods take into account the transition/transversion rate ratio κ, but a reliable estimate of κ is hard to obtain. All counting methods use nucleotide-based correction formulae to correct for multiple hits, which are logically flawed. In the likelihood method, we have to specify the instantaneous substitution rates only and leave it to the probability calculus to produce a sensible inference. At the level of instantaneous rates, there are no multiple changes in a codon and it is straightforward to decide whether each change is synonymous or nonsynonymous. The difficult tasks of estimating the transition/transversion rate ratio, weighting evolutionary pathways between codons, correcting for multiple hits at the same site, and dealing with irregularities of the genetic code are achieved automatically in the likelihood calculation. Second, it is much simpler to accommodate more realistic models of codon substitution in the likelihood method than in the counting method. For example, it is straightforward to use a GTR-style mutation model instead of the HKY85-style model in equation (2.7).

*2.5.4 Interpretation and a plethora of distances

2.5.4.1 More distances based on codon models

The expected number of substitutions from codon i to codon j over any time t is given by $\pi_i q_{ij} t$. Thus one can use the rate matrix Q and divergence time t to define various measures to characterize the evolutionary process of a protein-coding gene. For example, one can consider whether the $i \rightarrow j$ change is synonymous or nonsynonymous and contrast synonymous and nonsynonymous substitution rates. Similarly one can contrast the transitional and transversional substitutions, substitutions at the three codon positions, substitutions causing conservative or radical amino acid changes, and so on. Such measures are functions of parameters in the codon substitution model, such as κ, ω, and codon frequency π_j. As soon as the MLEs of model parameters are obtained, the rate matrix Q and divergence time t are known, so that the MLEs of such measures can be generated in a straightforward manner according to their definitions. Here we introduce a few more distance measures.

First the distances at the three codon positions can be calculated separately under the codon model: d_{1A}, d_{2A}, d_{3A}. The suffix 'A' in the subscript means 'after selection on the protein', as will become clear later. The expected number of nucleotide substitutions at the first codon position over any time t is the sum of $\pi_i q_{ij} t$ over all pairs of codons i and j that differ at the first position only (see equation 2.7). As there is one site at the first position in each codon, this is also the number of nucleotide substitutions per site at the first position

$$d_{1A} = \sum_{\{i,j\} \in A_1} \pi_i q_{ij} t, \qquad (2.17)$$

where the summation is over set A_1, which includes all codon pairs i and j that differ at position 1 only. Distances d_{2A} and d_{3A} for the second and third positions can be defined similarly. Here equation (2.17) serves two purposes: (i) to define the distance d_{1A}, as a function of parameters in the codon model $(t, \kappa, \omega, \pi_j)$ and (ii) to present the ML method for estimating the distance; that is, the MLE of d_{1A} is given by equation

(2.17), with parameters t, κ, ω, and π_j replaced by their MLEs. All distance measures discussed below in this section should be understood in this way.

Example. For the cucumber and tobacco *rbcL* genes, the distances calculated under the F3 × 4 codon model (model F in Table 2.6) are $d_{1A} = 0.046$, $d_{2A} = 0.041$, $d_{3A} = 0.314$, with the average to be 0.134. These are comparable with the distances calculated by applying the nucleotide-substitution model HKY85 (Hasegawa et al. 1985) to each codon position (Table 2.5): $d_1 = 0.057$, $d_2 = 0.026$, $d_3 = 0.282$. In the nucleotide-based analysis, five parameters (d, κ, and three nucleotide frequency parameters) are estimated at each of the three positions, and the substitution rate (or sequence distance) is allowed to vary freely among codon positions. In the codon-based analysis, 12 parameters (t, κ, ω, and nine nucleotide frequency parameters) are estimated for the whole data set, and the use of parameter ω to accommodate different synonymous and nonsynonymous rates leads to different substitution rates at the three codon positions, even though the codon model does not explicitly incorporate such rate differences. Whether the codon or nucleotide models produce more reliable distance estimates will depend on which models are more realistic biologically. □

Next we define three distances d_{1B}, d_{2B}, and d_{3B} for the three codon positions, where the suffix 'B' stands for 'before selection on the protein'. If there is no selection at the DNA level, these will be mutational distances at the three positions. Otherwise, they measure distances at the three positions before natural selection at the protein level has affected nucleotide substitution rates. For the first position we have

$$d_{1B} = \sum_{\{i,j\} \in A_1} \pi_i q_{ij}^1 t, \qquad (2.18)$$

where q_{ij}^1 is q_{ij} but calculated with $\omega = 1$ fixed, as we did when defining the proportion of synonymous sites (see equation 2.15). Distances d_{2B} and d_{3B} are defined similarly for the second and third positions.

Another distance is d_4, the expected number of substitutions per site at the four-fold degenerate sites of the third codon position, often used as an approximation to the neutral mutation rate. A commonly used heuristic approach to estimating d_4 applies a nucleotide model to data of four-fold degenerate sites at the third codon position. This approach counts a third codon position as a four-fold degenerate site only if the first two positions are identical across all sequences compared and if the encoded amino acid does not depend on the third position (e.g. Perna and Kocher 1995; Adachi and Hasegawa 1996a; Duret 2002; Kumar and Subramanian 2002; Waterston et al. 2002). This is a conservative definition of four-fold degenerate sites, based on comparison among sequences, and it differs from the definition used in the LWL85 method, which counts four-fold sites along each sequence and then takes an average between the two sequences. For example, if the codons in the two sequences are ACT and GCC, the LWL85 method will count the third position as four-fold degenerate, and the T–C difference at the third position as a four-fold degenerate difference. The heuristic approach does not use such a third codon position, as its status might have changed during the evolutionary history. This approach, however, has the drawback that the number of usable four-fold sites decreases with the increase of sequence divergence and with inclusion of more divergent sequences. Here we describe an ML method for estimating d_4, which overcomes this drawback and which uses a codon model instead of a nucleotide model to correct for multiple hits.

We define the expected number of four-fold degenerate sites per codon as the sum of frequencies (π_j) of all four-fold degenerate codons. This number does not decrease with the increase of sequence divergence. The expected number of four-fold degenerate nucleotide substitutions per codon is given by summing $\pi_i q_{ij} t$ over all codon pairs i and j that represent a four-fold degenerate substitution; that is, codons i and j are both four-fold degenerate and they have a difference at the third position. Then d_4 is given by the number of four-fold degenerate substitutions per codon divided by the number of four-fold degenerate sites per codon. Replacing parameters κ, ω, and π_j in this definition by their estimates gives the MLE of d_4. It is easy to see that d_4 defined this way converges to that in the heuristic approach when sequence divergence t approaches 0, but the ML method should work for divergent sequences. Also note that four-fold degenerate sites are counted in a similar way to the LWL85 method, although the number of four-fold degenerate substitutions is calculated differently.

Example. For the cucumber and tobacco *rbcL* genes, application of equation (2.18) under the F3 × 4 model (model F in Table 2.6) gives $d_{1B} = 0.463$, $d_{2B} = 0.518$, $d_{3B} = 0.383$, with the average over the three codon positions to be 0.455, which is d_S, as will be explained below. The ratios d_{1A}/d_{1B}, d_{2A}/d_{2B}, and d_{3A}/d_{3B} are the proportions of mutations at the three codon positions that are accepted after filtering by selection on the protein, and may be called the acceptance rates (Miyata *et al.* 1979). These are calculated to be 0.100, 0.078, and 0.820 for the three codon positions, respectively. At the second position, all mutations are nonsynonymous, so that the acceptance rate is exactly ω. At the first and third positions, some changes are synonymous, so that the acceptance rate is $> \omega$. In the heuristic method to estimate d_4, 215 four-fold degenerate sites (that is, 15.2% of all sites in the sequence) can be used, and application of the HKY85 model (Hasegawa *et al.* 1985) to data at these sites produces the estimate $d_4 = 0.344$. Using the ML method, 245.4 four-fold degenerate sites (17.3% of all sites) are used, with the MLE of d_4 to be 0.386. This is somewhat larger than the estimate from the heuristic method, but is very close to d_{3B}, which is calculated using 472 third-position sites. □

2.5.4.2 Mutational-opportunity distances d_S and d_N and physical-site distances d_S^* and d_N^*

Methods for estimating d_S and d_N have historically been developed to quantify the influence of natural selection on the protein on nucleotide substitution rates (Miyata and Yasunaga 1980; Gojobori 1983; Li *et al.* 1985). The proportions of synonymous and nonsynonymous sites are defined as the expected proportions of synonymous and nonsynonymous mutations. Essentially every method that has been developed is of this kind, including the counting methods LWL85 (Li *et al.* 1985),

LPB93 (Li 1993; Pamilo and Bianchi 1993), Ina's (1995) method, and YN00 (Yang and Nielsen 2000), as well as the ML method (Goldman and Yang 1994). Bierne and Eyre-Walker (2003) pointed out the existence of an alternative definition based on *physical sites* and argued that such a definition may be more appropriate under some situations. Here I provide an interpretation of d_S and d_N, define new distance measures d_S^* and d_N^* using the physical-site definition, and explore a few idealized models to illustrate their differences. I then briefly discuss the uses of this large collection of distance measures. I will focus on the likelihood method as it is conceptually simpler. However, the discussion should apply to the counting methods as well. The issue is the conceptual definition of parameters of biological interest, and sampling errors or estimation inefficiency are ignored in the discussion.

Interpretation of d_S and d_N. It is mathematically straightforward to show that d_S and d_N, defined in equation (2.16), satisfy the following relationships

$$d_S = (d_{1B} + d_{2B} + d_{3B})/3,$$
$$d_N = d_S \times \omega, \quad (2.19)$$

where d_{kB} ($k = 1, 2, 3$) is the mutation distance at the kth codon position, defined in equation (2.18). We will demonstrate this result below in a few simple models, but the reader may wish to confirm equation (2.19). Thus d_S is the average 'mutation' rate over the three codon positions. If silent site evolution is not neutral, d_S will be affected by selection on the DNA as well, and will measure the substitution rate before selection on the protein, averaged over the three codon positions. It would appear more appropriate to use symbols d_B (distance *before* selection on the protein) and d_A (distance *after* selection) in place of d_S and d_N.

Definition of d_S^* and d_N^*. Using a physical-site definition, we count synonymous and nonsynonymous sites S^* and N^* as follows. A nondegenerate site is counted as one nonsynonymous site, a two-fold degenerate site is counted as 1/3 synonymous and 2/3 nonsynonymous, a three-fold degenerate site as 2/3 synonymous and 1/3 nonsynonymous, and a four-fold degenerate site is counted as one synonymous site. The mutation/substitution model is ignored during the counting, as are differences in transition and transversion rates or in codon frequencies, even if such factors are considered in the model. The numbers of synonymous and nonsynonymous sites per codon are averaged over the codon frequencies (π_j) expected in the model. The numbers of synonymous and nonsynonymous substitutions per codon are calculated as before (equation 2.14). The distances are then defined as

$$d_S^* = S_d/S^*,$$
$$d_N^* = N_d/N^*. \quad (2.20)$$

The distances d_S and d_N produced by the methods of Miyata and Yasunaga (1980) and Nei and Gojobori (1986) can be interpreted using either the mutational-opportunity or the physical-site definitions. As these methods assume that the mutation rate is the same between any two nucleotides, there is no difference between the two definitions.

Example. We apply equations (2.20) to the cucumber and tobacco *rbcL* genes. The F3 × 4 model (model F in table 2.6) is assumed. The counts of (physical) sites are $S^* = 353.8$ and $N^* = 1062.2$. The distance estimates are thus $d_S^* = 0.425$ and $d_N^* = 0.036$.

We now consider a few idealized examples to contrast the different distance measures.

Example of the two-fold regular code (Bierne and Eyre-Walker 2003). Imagine a genetic code in which there are no stop codons and every codon is two-fold degenerate. Every first or second codon position is nondegenerate, and every third position is two-fold degenerate, with transitions to be synonymous and transversions nonsynonymous. This code would encode 32 amino acids. Suppose that the 64 codons have equal frequencies, and mutations occur according to the K80 model (Kimura 1980), with transition rate α and transversion rate β, with $\kappa = \alpha/\beta$. Suppose there is no selection on silent sites, and the proportion of nonsynonymous mutations that are neutral is ω; all other nonsynonymous mutations are lethal and removed by purifying selection. Thus over any time interval t, the expected number of synonymous substitutions per codon is αt and the expected number of nonsynonymous substitutions per codon is $2(\alpha + 2\beta)\omega t + 2\beta\omega t$.

Using a physical-site definition, we see that every codon has $S^* = 1/3$ synonymous sites and $N^* = 1 + 1 + 2/3 = 8/3$ nonsynonymous sites. The number of synonymous substitutions per (physical) synonymous site is then $d_S^* = \alpha t/(1/3) = 3\alpha t$. The synonymous rate is as high as the nucleotide substitution rate under the JC69 model in which both the transition and transversion rates are α. The number of nonsynonymous substitutions per (physical) nonsynonymous site is $d_N^* = [2(\alpha + 2\beta)\omega t + 2\beta\omega t]/(8/3) = 3(\alpha + 3\beta)\omega t/4$. The ratio $d_N^*/d_S^* = (1 + 3/\kappa)\omega/4$ differs from ω if $\kappa \neq 1$. Suppose the transversion rate β is the same in two genes, but the transition rate is $\alpha = \beta$ in the first gene and $\alpha = 5\beta$ in the second gene. Then we see that d_S^* is five times as high in the second gene as in the first, as one expects from a physical definition of sites. The ratio $d_N^*/d_S^* = \omega$ in the first gene, but $= 0.4\omega$ in the second. In the second gene, the nonsynonymous rate appears to be reduced relative to the synonymous rate, not because selective constraint on the protein has removed nonsynonymous mutations but because transitions occur at higher rates than transversions.

Using a mutational-opportunity definition of sites, we note that synonymous and nonsynonymous mutations occur in proportions $\alpha : 2\beta$ at the third codon position. Thus the number of synonymous sites in a codon is $S = \alpha/(\alpha + 2\beta)$. The number of nonsynonymous sites in a codon is $N = 1 + 1 + 2\beta/(\alpha + 2\beta)$, with $S + N = 3$. This is the counting method explained in Table 2.3 for LWL85m and Ina's methods (equation 2.12) and in Subsection 2.5.2 for the ML method. Then $d_S = \alpha t/S = (\alpha + 2\beta)t$, and $d_N = [2(\alpha + 2\beta)\omega t + 2\beta\omega t]/N = (\alpha + 2\beta)\omega t = d_S\omega$. Note that d_S is the mutation rate at each codon position. It is not the synonymous rate in the usual sense of the word. For the two genes with different transition rates discussed above ($\alpha = \beta$ versus $\alpha = 5\beta$), $d_S = 7\beta t$ in the second gene is only 7/3 times as high as in the first gene ($d_S = 3\beta t$), which would seem strange if one incorrectly uses a physical-site definition to interpret d_S. In both genes, $d_N/d_S = \omega$.

Example of the four-fold regular code. Imagine a genetic code in which all codons are four-fold degenerate, and the 64 sense codons encode 16 amino acids. Again suppose there is no selection on the DNA. Then d_S^* will be the mutation rate at the third position and will be equal to d_4 and d_{3B}, while d_S will be the average mutation rate over the three codon positions. Consider two cases. In the first, suppose mutations occur according to the K80 model, and the 64 codons have the same frequency. Then we have $d_S = d_S^*$, since the mutation rate is the same at the three codon positions, and $d_N = d_N^*$. In the second case, suppose the codon frequencies are unequal and the three codon positions have different base compositions. Then the mutation rate will differ among the three positions and d_S will be different from d_S^*.

Utilities of the distance measures. Both sets of distances (d_S and d_N versus d_S^* and d_N^*) are valid distance measures. They increase linearly with time and can be used to calculate species divergence times or to reconstruct phylogenetic trees. Both can be used to compare different genes with different codon usage patterns or to test models of molecular evolution. For testing adaptive protein evolution, d_S and d_N can be used while d_S^* and d_N^* are inappropriate. In likelihood-based methods, such tests are usually conducted using parameter ω in the codon model (see Chapter 8), so that estimation of rates of synonymous and nonsynonymous substitutions per site is not necessary. In counting methods, the ω ratio is calculated by estimating d_S and d_N first, and the sensitivity of d_S and d_N to model assumptions is a source of concern.

Distance measures d_{1A}, d_{2A}, d_{3A}, d_{1B}, d_{2B}, d_{3B}, as well as d_4, discussed in this subsection, all use physical-site definitions of sites. As estimates of the mutation rate at the third codon position, d_{3B} should be preferred to d_4, which should in turn be preferred to d_S^*, based on the numbers of sites used by these distances. For example, the synonymous substitution rate has been used to examine the correlation between the silent rate and codon usage bias (e.g. Bielawski et al. 2000; Bierne and Eyre-Walker 2003). For this purpose, d_{3B}, d_4, or d_S^* appear more appropriate than d_S if the codon usage bias is measured by GC$_3$, the GC content at the third codon position. It does not make much sense to correlate GC$_3$ with the average rate over all three codon positions (d_S). If the codon usage bias is measured by the *effective number of codons* or ENC (Wright 1990), the situation is less clear, since ENC depends on all positions and codon usage bias is more complex than base composition differences at the third codon position.

It is noted that distances based on physical sites are not so sensitive to model assumptions. In contrast, distances d_S and d_N, which define sites based on mutational opportunities, are much more sensitive to model assumptions; with extreme codon usage or base compositions, different methods or model assumptions can lead to estimates that are several-fold different. In such genes, d_S can be substantially different from d_S^*, d_4, or d_{3B}, and use of different distances can lead to different conclusions.

*2.6 Numerical calculation of the transition-probability matrix

Fitting Markov-chain models to real sequence data requires calculation of the transition-probability matrix $P(t)$, which is given as a matrix exponential through the Taylor expansion

$$P(t) = e^{Qt} = I + Qt + \tfrac{1}{2!}(Qt)^2 + \tfrac{1}{3!}(Qt)^3 + \tfrac{1}{4!}(Qt)^4 + \cdots . \tag{2.21}$$

For nucleotide models, the calculation is either analytically tractable or otherwise inexpensive due to the small size of the matrix. However, for amino acid and codon models, this calculation can be costly and unstable, so that use of a reliable algorithm is more important.

Moler and van Loan (1978) provide an excellent review of algorithms for calculating matrix exponentials. Here we focus on the case where Q is the rate matrix of a Markov chain. Direct use of equation (2.21) for numerical computation of $P(t)$ is not advisable. First, the diagonal elements of the matrix Qt are negative while the off-diagonals are positive. Thus cancellations during matrix multiplication lead to loss of significance digits. Second, many terms in the sum may be needed to achieve convergence, especially at large distances t. Here we describe two approaches that are more effective.

The first makes use of the relationship

$$e^{Qt} = (e^{Qt/m})^m \approx (I + Qt/m)^m. \tag{2.22}$$

For reasonably large m, the matrix $(I + Qt/m)$ will have positive elements only. If we choose $m = 2^k$ for some integer k, the mth power of the matrix $(I + Qt/m)$ can be calculated by squaring the matrix repetitively for k times. Also $(I + Qt/m)$ can be replaced by $I + Qt/m + 1/2(Qt/m)^2$, using the first three instead of two terms in the Taylor expansion of $e^{Qt/m}$. For small or intermediate distances, say with $t < 1$ change per site, $k = 5$ or 10 appears sufficient. For larger distances, say with $t > 3$, even larger ks may be necessary. This algorithm works for general rate matrices and is used in the PAML package (Yang 1997a) to calculate $P(t)$ for the UNREST model of nucleotide substitution (see Table 1.1).

A second approach is numerical computation of the eigenvalues and eigenvectors of the rate matrix Q, which is most effective for reversible Markov chains:

$$Q = U \Lambda U^{-1}, \tag{2.23}$$

where $\Lambda = \text{diag}\{\lambda_1, \lambda_2,..., \lambda_c\}$ has the eigenvalues of Q on the diagonal, while columns of U are the corresponding right eigenvectors and rows of $V = U^{-1}$ are the left

2.6 Numerical calculation of the transition-probability matrix • 69

eigenvectors. All these matrices are of size 20 × 20 for amino acids or 61 × 61 for codons (under the universal code); that is, $c = 20$ or 61. Then

$$P(t) = \exp(Qt) = U \exp(\Lambda t) U^{-1} = U \text{diag}\{\exp(\lambda_1 t), \exp(\lambda_2 t), \ldots, \exp(\lambda_c t)\} U^{-1}. \tag{2.24}$$

A general real matrix can have complex eigenvalues and eigenvectors and their numerical computation may be unstable. However, a real symmetrical matrix has only real eigenvalues and eigenvectors and their numerical computation is both fast and stable (Golub and Van Loan 1996). It can be proven that the rate matrix Q for a time-reversible Markov process is *similar* to a real symmetrical matrix and thus has only real eigenvalues and eigenvectors (Kelly 1979). Two matrices A and B are said to be *similar* if there exists a nonsingular matrix T so that

$$A = TBT^{-1} \tag{2.25}$$

Similar matrices have identical eigenvalues. Note that Q is similar to

$$B = \Pi^{1/2} Q \Pi^{-1/2}, \tag{2.26}$$

where $\Pi^{1/2} = \text{diag}\{\pi_T^{1/2}, \pi_C^{1/2}, \pi_A^{1/2}, \pi_G^{1/2}\}$ and $\Pi^{-1/2}$ is its inverse. From equation (2.2), B is symmetrical if Q is the rate matrix for a reversible Markov chain. Thus both B and Q have identical eigenvalues, all of which are real.

Thus we can calculate the eigenvalues and eigenvectors of the rate matrix Q for a reversible Markov chain by constructing the symmetrical matrix B and diagonalizing it.

$$B = R \Lambda R^{-1} \tag{2.27}$$

Then

$$Q = \Pi^{-1/2} B \Pi^{1/2} = (\Pi^{-1/2} R) \Lambda (R^{-1} \Pi^{1/2}). \tag{2.28}$$

Comparing equations (2.28) with (2.23), we get $U = (\Pi^{-1/2} R)$ and $U^{-1} = R^{-1} \Pi^{1/2}$.

For reversible Markov chains, the algorithm of matrix diagonalization (equation 2.28) is both faster and more accurate than the algorithm of repeated matrix squaring (equation 2.22). The former is particularly efficient when applied to likelihood calculation for multiple sequences on a phylogeny. In that case, we often need to calculate e^{Qt} for fixed Q and different ts. For example, when we fit empirical amino acid substitution models, the rate matrix Q is fixed while the branch length t varies. Then one has to diagonalize Q only once, after which calculation of $P(t)$ for each t involves only two matrix multiplications. The algorithm of repeated squaring can be useful for any substitution-rate matrix, including those for irreversible models.

2.7 Exercises

2.1 Obtain two sequences from GenBank, align the sequences and then apply the methods discussed in this chapter to estimate d_S and d_N and discuss their differences. One way of aligning protein-coding DNA sequences is to use CLUSTAL (Thompson et al. 1994) to align the protein sequences first and then construct the DNA alignment based on the protein alignment, using, for example, MEGA3.1 (Kumar et al. 2005a) or BAMBE (Xia and Xie 2001), followed by manual adjustments.

*2.2 Are there really three nucleotide sites in a codon? How many synonymous and nonsynonymous sites are in the codon TAT (use the universal code)?

2.3 Behaviour of LWL85 and related methods under the *two-fold and four-fold mixture regular code*. Imagine a genetic code in which a proportion γ of codons are four-fold degenerate while all other codons are two-fold degenerate. (If $\gamma = 48/64$, the code would encode exactly 20 amino acids.) Suppose that neutral mutations occur according to the K80 model, with transition rate α and transversion rate β, with $\alpha/\beta = \kappa$. The proportion of nonsynonymous mutations that are neutral is ω. The numbers of nondegenerate, two-fold, and four-fold degenerate sites in a codon are $L_0 = 2$, $L_2 = 1 - \gamma$, and $L_4 = \gamma$. Over time interval t, the numbers of transitional and transversional substitutions at the three degeneracy classes are thus $A_0 = \alpha t \omega$, $B_0 = 2\beta t \omega$, $A_2 = \alpha t$, $B_2 = 2\beta t \omega$, $A_4 = \alpha t$, $B_4 = 2\beta t$. (a) Show that the LWL85 method (equation 2.10) gives

$$d_S = \frac{3(\kappa + 2\gamma)\beta t}{1 + 2\gamma},$$
$$d_N = \frac{3(\kappa + 3 - \gamma)\beta t \omega}{4 - \gamma},$$
(2.29)

with the ratio $d_N/d_S = \omega[(\kappa + 3 - \gamma)(1 + 2\gamma)]/[(4 - \gamma)(\kappa + 2\gamma)]$, which becomes $\omega(\kappa + 3)/(4\kappa)$ if $\gamma = 0$ (so that the code is the two-fold regular code) and ω if $\gamma = 1$ (so that the code is the four-fold regular code). (b) Show that both LPB93 (equation 2.11) and LWL85m (equation 2.12) give $d_S = (\alpha + 2\beta)t$ and $d_N = d_S\omega$. (Comment: under this model, LWL85 gives d_S^* and d_N^*, distances using the physical-site definition.)

Part II

Phylogeny reconstruction

3

Phylogeny reconstruction: overview

This chapter provides an overview of phylogeny reconstruction methods. We introduce some basic concepts used to describe trees and discuss general features of tree-reconstruction methods. We will describe distance and parsimony methods as well, while likelihood and Bayesian methods are discussed in Chapters 4 and 5.

3.1 Tree concepts

3.1.1 Terminology

3.1.1.1 Trees, nodes (vertexes), and branches (edges)

A phylogeny or phylogenetic tree is a representation of the genealogical relationships among species, among genes, among populations, or even among individuals. Mathematicians define a graph as a set of *vertexes* and a set of *edges* connecting them, and a tree as a connected graph without loops (see, e.g., p. 1 in Tucker 1995). Biologists instead use *nodes* for vertexes and *branches* for edges. Here we consider species trees, but the description also applies to trees of genes or individuals. The *tips*, *leaves*, or *external nodes* represent present-day species, while the *internal nodes* usually represent extinct ancestors for which no sequence data are available. The ancestor of all sequences is the *root* of the tree. Trees are drawn equivalently with the root on the top, at the bottom, or on the side.

3.1.1.2 Root of the tree and rooting the tree

A tree with the root specified is called a *rooted tree* (Fig. 3.1a), while a tree in which the root is unknown or unspecified is called an *unrooted tree* (Fig. 3.1b). If the evolutionary rate is constant over time, an assumption known as the *molecular clock*, distance-matrix and maximum likelihood methods can identify the root and produce rooted trees. Such use of the clock assumption to determine the root of the tree is known as *molecular-clock rooting*. However, the clock assumption is most often violated, except for closely related species. Without the clock, most tree-reconstruction methods are unable to identify the root of the tree and produce unrooted trees. Then a commonly used strategy is the so-called *outgroup rooting*. Distantly related species, called the *outgroups*, are included in tree reconstruction, while in the reconstructed unrooted tree for all species the root is placed on the branch leading to the outgroups, so that the subtree for the *ingroups* is rooted. In the example of Fig. 3.1, the orangutan is used as the outgroup to root the tree for the ingroup species: human, chimpanzee, and

74 • *3 Phylogeny reconstruction: overview*

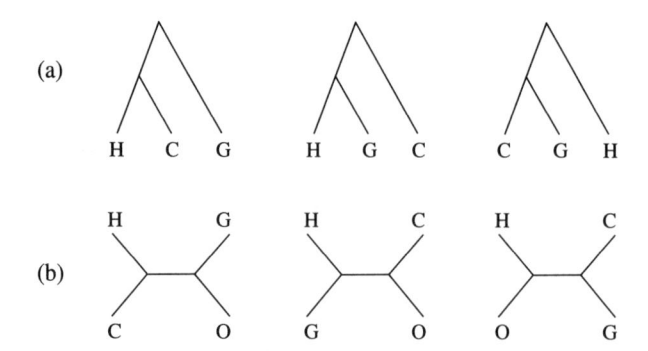

(a)

     H C G   H G C   C G H

(b)

  H   G   H   C   H   C

  C   O   G   O   O   G

Fig. 3.1 Outgroup rooting. To infer the relationships among human (H), chimpanzee (C), gorilla (G), represented in the three rooted trees in (a), we use orangutan (O) as the outgroup. Tree-reconstruction methods allow us to estimate an unrooted tree, that is, one of the trees in (b). As the root is along the branch leading to the outgroup, these three unrooted trees for all species correspond to the three rooted trees for the ingroup species H, C, and G.

gorilla. In general, outgroups closely related to the ingroup species are better than distantly related outgroups. In the universal tree of life, no outgroup species exist. Then a strategy is to root the tree using ancient gene duplications that occurred prior to the divergences of all existing life forms (Gogarten *et al.* 1989; Iwabe *et al.* 1989). The subunits of ATPase arose through a gene duplication before the divergence of eubacteria, eukaryotes, and archaebacteria. Protein sequences from both paralogues were used to construct a composite unrooted tree, and the root was placed on the branch separating the two duplicates. Elongation factors Tu and G constitute another ancient duplication, used in rooting the universal tree of life.

3.1.1.3 Tree topology, branch lengths, and the parenthesis notation

The branching pattern of a tree is called the *topology* of the tree. The length of a branch may represent the amount of sequence divergence or the time period covered by the branch. A tree showing only the tree topology without information about branch lengths is sometimes called a *cladogram* (Fig. 3.2a), while a tree showing both the topology and branch lengths is called a *phylogram* (Fig. 3.2b). For use in computer programs, trees are often represented using the parenthesis notation, also known as the Newick format. This uses a pair of parentheses to group sister taxa into one clade, with a semicolon marking the end of the tree. Branch lengths, if any, are prefixed by colons. For example, the trees in Fig. 3.2 are represented as

    a and b: ((((A, B), C), D), E);
    b: ((((A: 0.1, B: 0.2): 0.12, C: 0.3): 0.123, D: 0.4): 0.1234, E: 0.5);
    c: (((A, B), C), D, E);
    c: (((A: 0.1, B: 0.2): 0.12, C: 0.3): 0.123, D: 0.4, E: 0.6234);

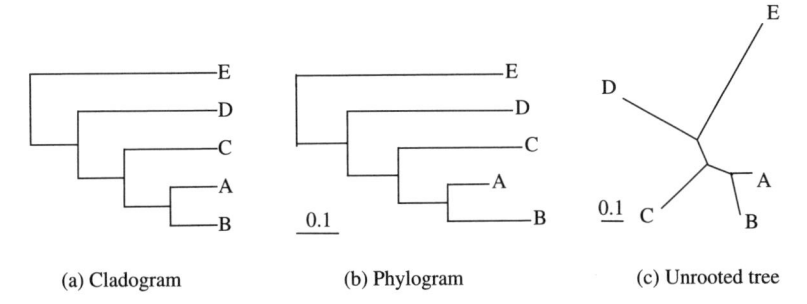

Fig. 3.2 The same tree shown in different styles. (a) The cladogram shows the tree topology without branch lengths or with branch lengths ignored. (b) In a phylogram, branches are drawn in proportion to their lengths. (c) In an unrooted tree, the location of the root is unknown or ignored.

Branch lengths here are measured by the expected number of nucleotide substitutions per site, like the sequence distances discussed in Chapter 1. This format is natural for representing rooted trees. Unrooted trees are represented as rooted and the representation is not unique since the root can be placed anywhere on the tree. For example, the tree in Fig. 3.2(c) can also be represented as '(A, B, (C, (D, E)));'.

3.1.1.4 Bifurcating and multifurcating trees

The number of branches connected to a node is called the *degree* of the node. Leaves have a degree of 1. If the root node has a degree greater than 2 or a nonroot node has a degree greater than 3, the node represents a *polytomy* or *multifurcation*. A tree with no polytomies is called a *binary tree*, *bifurcating tree*, or *fully resolved tree*. The most extreme unresolved tree is the *star* or *big-bang* tree, in which the root is the only internal node (see Fig. 3.3 for examples). A polytomy representing truly simultaneous species divergences is sometimes called a *hard polytomy*. It would seem extremely unlikely for one species to diverge into several at exactly the same time, and it may be argued that hard polytomies do not exist. Most often the polytomy represents lack of information in the data to resolve the relationship within a clade (a group of species). Such polytomies are called *soft polytomies*.

3.1.1.5 The number of trees

We can work out the total number of unrooted trees by the following *stepwise addition algorithm* (Cavalli-Sforza and Edwards 1967) (Fig. 3.4). We start with the single tree for the first three species. This tree has three branches to which the fourth species can be added. Thus there are three possible trees for the first four species. Each four-species tree has five branches, to which the fifth species can be added, resulting in five different five-species trees for each four-species tree. In general, a tree of the first $n - 1$ species has $(2n - 5)$ branches, to which the nth species can be added, so that

76 • *3 Phylogeny reconstruction: overview*

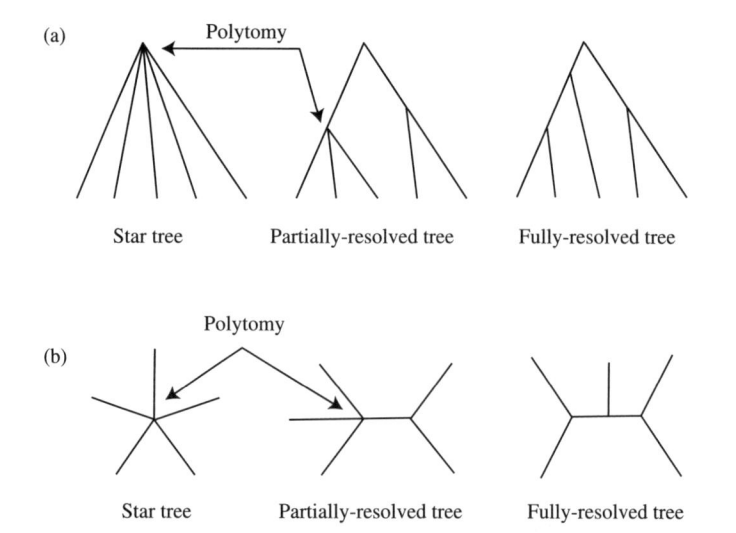

Fig. 3.3 Unresolved and resolved phylogenetic trees: (a) rooted trees, (b) unrooted trees.

Fig. 3.4 Enumeration of all trees for five taxa A, B, C, D, and E using the stepwise addition algorithm.

each of the $(n-1)$-species tree generates $(2n-5)$ new n-species trees. Thus the total number of unrooted bifurcating trees for n species is

$$T_n = T_{n-1} \times (2n-5) = 3 \times 5 \times 7 \times \cdots \times (2n-5). \tag{3.1}$$

To work out the number of rooted trees for n species, note that each unrooted tree has $(2n-3)$ branches, and the root can be placed on any of those branches, generating $(2n-3)$ rooted trees from each unrooted tree. Thus the number of rooted trees for n species is simply $T_n \times (2n-3) = T_{n+1}$. As we can see from Table 3.1, the number of trees increases explosively with the number of species.

Table 3.1 The number of unrooted (T_n) and rooted (T_{n+1}) trees for n species

n	T_n	T_{n+1}
3	1	3
4	3	15
5	15	105
6	105	945
7	945	10 395
8	10 395	135 135
9	135 135	2 027 025
10	2 027 025	34 459 425
20	$\sim 2.22 \times 10^{20}$	$\sim 8.20 \times 10^{21}$
50	$\sim 2.84 \times 10^{74}$	$\sim 2.75 \times 10^{76}$

3.1.2 Topological distance between trees

Sometimes we would like to measure how different two trees are. For example, we may be interested in the differences among trees estimated from different genes, or the differences between the true tree and the estimated tree in a computer simulation conducted to evaluate a tree-reconstruction method.

A commonly used measure of topological distance between two trees is the *partition distance* defined by Robinson and Foulds (1981) (see also Penny and Hendy 1985). We define this distance for unrooted trees here, but the same definition applies to rooted trees as well, by imagining an outgroup species attached to the root. Note that each branch on the tree defines a *bipartition* or *split* of the species; if we chop the branch, the species will fall into two mutually exclusive sets. For example, branch b in tree T_1 of Fig. 3.5 partitions the eight species into two sets: (1, 2, 3) and (4, 5, 6, 7, 8). This partition is also present on tree T_2. Partitions defined by terminal branches are in all possible trees and are thus not informative for comparison between trees. Thus we focus on internal branches only. Partitions defined by branches b, c, d, and e of tree T_1 are the same as partitions defined by branches b', c', d', and e' of tree T_2, respectively. The partition defined by branch a of tree T_1 is not in tree T_2, nor is the partition defined by branch a' of tree T_2 in tree T_1. The partition distance is defined as the total number of bipartitions in one tree that are not in the other. Thus T_1 and T_2 have a partition distance of 2. As a binary tree of n species has $(n-3)$ internal branches, the partition distance between any two binary trees of n species ranges from 0 (if the two trees are identical) to $2(n-3)$ (if the two trees do not share any bipartition).

The partition distance can be equivalently defined as the number of contractions and expansions needed to transform one tree into the other. Removing an internal branch by reducing its length to zero is a contraction while creating an internal branch is an expansion. Trees T_1 and T_2 of Fig. 3.5 are separated by a contraction (from T_1 to T_0) and an expansion (from T_0 to T_2), so that their partition distance is 2.

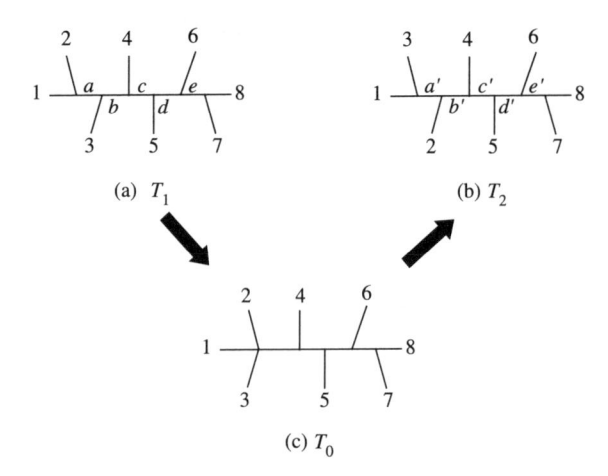

Fig. 3.5 The partition distance between two trees T_1 and T_2 is the total number of bipartitions that are in one tree but not in the other. It is also the number of contractions and expansions needed to change one tree into another. A contraction converts T_1 into T_0 and an expansion converts T_0 into T_2, so the distance between T_0 and T_1 is 1 while the distance between T_1 and T_2 is 2.

The partition distance has limitations. First, the distance does not recognize certain similarities between trees. The three trees in Fig. 3.6 are identical concerning the relationships among species 2–7 but do not share any bipartitions, so that the partition distance between any two of them is the maximum possible. Indeed, the probability that a random pair of unrooted trees achieve the maximum distance is 70–80% for $n = 5$–10, and is even greater for larger n. Figure 3.7 shows the distribution of partition distances for the case of $n = 10$. Second, the partition distance ignores branch lengths in the tree. Intuitively, two trees that are in conflict around short internal branches are less different than two trees that are in conflict around long internal branches. There are no good rules to follow concerning incorporation of branch lengths in defining a distance between two trees; one such measure is suggested by Kuhner and Felsenstein (1994). Third, the partition distance may be misleading if either of the two trees has multifurcations. Suppose we conduct a computer simulation to compare two tree-reconstruction methods, using a binary tree to simulate data sets. We use the partition distance to measure performance: $P = 1 - D/D_{max}$, where $D_{max} = 2(n - 3)$ is the maximum distance and D is the distance between the true tree and the estimated tree. When both the true tree and the estimated tree are binary, P is the proportion of bipartitions in the true tree that are recovered in the estimated tree. If there is no information in the data, the first method returns the star tree as the estimate while the second method returns an arbitrarily resolved binary tree. Now for the first method, $D = (n - 3) = D_{max}/2$, so that $P = 50\%$. The second method has a performance of $P = 1/3$ when $n = 3$ or nearly 0 for large n, since a random tree is very likely not to share any bipartitions with the true tree. However, the two methods clearly have the

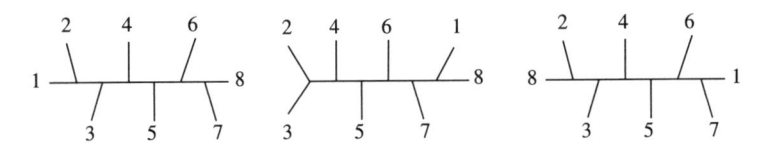

Fig. 3.6 Three trees that do not share any bipartitions and thus achieve the maximum partition distance.

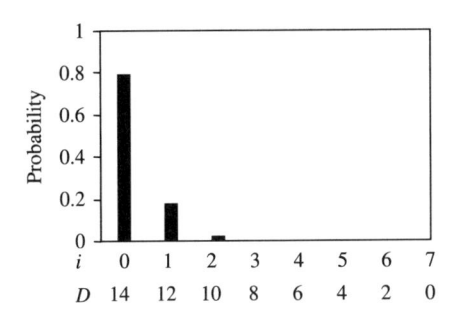

Fig. 3.7 The probability that two random trees from all possible unrooted trees of 10 species share i bipartitions or have partition distance D. Note that $D = 2 \times (10 - 3 - i)$.

same performance, and the measure based on the partition distance is unreasonable for the first method.

3.1.3 Consensus trees

While the partition distance measures how different two trees are, a consensus tree summarizes common features among a collection of trees. Many different consensus trees have been defined; see Bryant (2003) for a comprehensive review. Here we introduce two of them.

The *strict consensus tree* shows only those groups (nodes or clades) that are shared among all trees in the set, with polytomies representing nodes not supported by all trees in the set. Consider the three trees in Fig. 3.8(a). The strict consensus tree is shown in Fig. 3.8(b). The clade (A, B) is in the first and third trees but not in the second, while the clade (A, B, C) is in all three trees. Thus the strict consensus tree shows the clade (A, B, C) as a trichotomy, as is the clade (F, G, H). The strict consensus tree is a conservative way of summarizing the trees and may not be very useful as it often produces the star tree.

The *majority-rule consensus tree* shows nodes or clades that are supported by at least half of the trees in the set. It is also common practice to show the percentage of trees that support every node on the consensus tree (Fig. 3.8c). For example, the clade (A, B) is in two out of the three trees and is thus shown in the majority-rule consensus tree as resolved, with the percentage of support (2/3) shown next to

80 • 3 Phylogeny reconstruction: overview

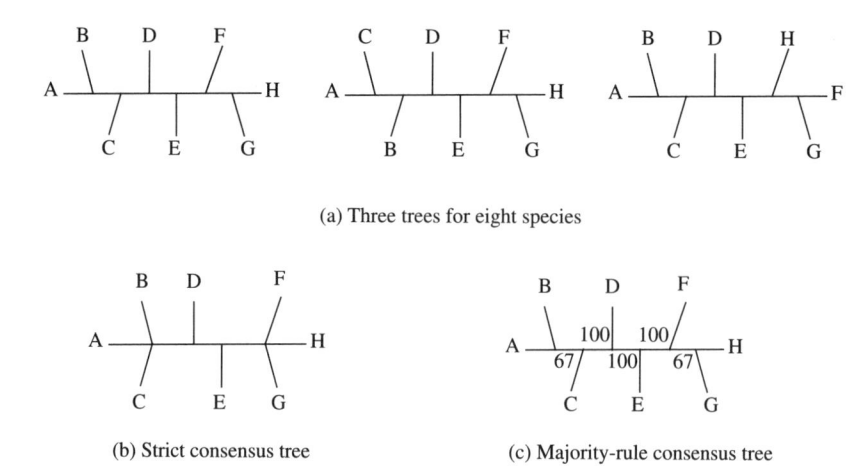

Fig. 3.8 Three trees for eight species (a) and their strict consensus tree (b) and majority-rule consensus tree (c).

the node. It is known that all clades that occur in more than half of the trees in the set can be shown on the same consensus tree without generating any conflict.

Like the partition distance, the majority-rule consensus tree, as a summary of trees in the set, has limitations. Suppose that there are only three distinct trees in the set, which are the trees of Fig. 3.6, each occurring in proportions around 33%. Then the majority-rule consensus tree will be the star tree. In such cases, it appears more informative to report the first few whole trees with the highest support values.

3.1.4 Gene trees and species trees

The phylogeny representing the relationships among a group of species is called the *species tree*. The phylogeny for a set of gene sequences from the species is called the *gene tree*. A number of factors may cause the gene tree to differ from the species tree.

First, estimation errors may cause the estimated gene tree to be different from the species tree even if the (unknown) true gene tree agrees with the species tree. The estimation errors may be either random, due to the limited amount of sequence data, or systematic, due to deficiencies of the tree-reconstruction method. Second, during the early stages of evolution near the root of the universal tree of life, there appears to have been substantial lateral (horizontal) gene transfer (LGT). As a result, different genes or proteins may have different gene trees, in conflict with the species tree. The LGT appears to be so extensive that some researchers question the concept of a universal tree of life (see, e.g., Doolittle 1998). Third, gene duplications, especially if followed by gene losses, can cause the gene tree to be different from the species tree if paralogous copies of the gene are used for phylogeny reconstruction (Fig. 3.9a). Fourth, when the species are closely related, *ancestral polymorphism* or *lineage sorting* can cause

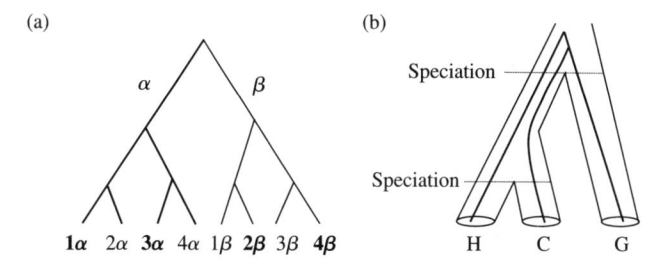

Fig. 3.9 Conflict between species tree and gene tree can be due to gene duplication (a) or ancestral polymorphism (b). In (a), a gene duplicated in the past, creating paralogous copies α and β, followed by divergences of species 1, 2, 3, and 4. If we use gene sequences $1\alpha, 3\alpha, 2\beta, 4\beta$ for phylogeny reconstruction, the true gene tree is $((1\alpha, 3\alpha), (2\beta, 4\beta))$, different from the species tree $((1, 2), (3, 4))$. In (b), the species tree is ((human, chimpanzee), gorilla). However, due to ancestral polymorphism or lineage sorting, the true gene tree is ((human, (chimpanzee, gorilla)).

gene trees to be different from the species tree. An example is shown in Fig. 3.9(b). Here the species tree for human, chimpanzee, and gorilla is ((H, C), G). However, because of sequence variations (polymorphisms) in the extinct ancestral species, the true gene tree is (H, (C, G)). The probability that the gene tree differs from the species tree is greater if the speciation events are closer in time (that is, if the species tree is almost a star tree) and if the long-term population size of the H–C common ancestor is greater. Such information concerning the conflicts between the species tree and the gene trees can be used to estimate the effective population sizes of extinct common ancestors by using sequences from extant species at multiple neutral loci (Takahata 1986; Takahata *et al.* 1995; Yang 2002; Rannala and Yang 2003).

3.1.5 Classification of tree-reconstruction methods

Here we consider some overall features of phylogeny reconstruction methods. First, some methods are *distance based*. In those methods, distances are calculated from pairwise comparison of sequences, forming a distance matrix, which is used in subsequent analysis. A clustering algorithm is often used to convert the distance matrix into a phylogenetic tree (Everitt *et al.* 2001). The most popular methods in this category include UPGMA (unweighted pair-group method using arithmetic averages) (Sokal and Sneath 1963) and neighbour joining (Saitou and Nei 1987). Other methods are *character based*, and attempt to fit the characters (nucleotides or amino acids) observed in all species at every site to a tree. Maximum parsimony (Fitch 1971b; Hartigan 1973), maximum likelihood (ML) (Felsenstein 1981), and Bayesian methods (Rannala and Yang 1996; Mau and Newton 1997; Li *et al.* 2000) are all character based. Distance methods are often computationally faster than character-based methods, and can be easily applied to analyse different kinds of data as long as pairwise distances can be calculated.

Table 3.2 Optimality criteria used for phylogeny reconstruction

Method	Criterion (tree score)
Maximum parsimony	Minimum number of changes, minimized over ancestral states
Maximum likelihood	Log likelihood score, optimized over branch lengths and model parameters
Minimum evolution	Tree length (sum of branch lengths, often estimated by least squares)
Bayesian	Posterior probability, calculated by integrating over branch lengths and substitution parameters

Tree-reconstruction methods can also be classified as being either *algorithmic* (cluster methods) or *optimality based* (search methods). The former include UPGMA and neighbour joining, which use cluster algorithms to arrive at a single tree from the data as the best estimate of the true tree. Optimality-based methods use an optimality criterion (objective function) to measure a tree's fit to data, and the tree with the optimal score is the estimate of the true tree (Table 3.2). In the maximum parsimony method, the tree score is the minimum number of character changes required for the tree, and the *maximum parsimony tree* or *most parsimonious tree* is the tree with the smallest tree score. The ML method uses the log likelihood value of the tree to measure the fit of the tree to the data, and the *maximum likelihood tree* is the tree with the highest log likelihood value. In the Bayesian method, the posterior probability of a tree is the probability that the tree is true given the data. The tree with the maximum posterior probability is the estimate of the true tree, known as the *MAP tree*. In theory, methods based on optimality criteria have to solve two problems: calculation of the criterion for a given tree and search in the space of all trees to identify the tree with the best score. The first problem is often straightforward but the second is virtually impossible when the number of sequences is greater than 20 or 50 because of the huge number of possible trees. As a result, heuristic algorithms are used for tree search. Optimality-based search methods are usually much slower than algorithmic cluster methods.

Some tree-reconstruction methods are model based. Distance methods use nucleotide or amino acid substitution models to calculate pairwise distances. Likelihood and Bayesian methods use substitution models to calculate the likelihood function. These methods are clearly model based. Parsimony does not make explicit assumptions about the evolutionary process. Opinions differ as to whether the method makes any implicit assumptions, and, if so, what they are. We will return to this issue in Chapter 6.

3.2 Exhaustive and heuristic tree search

3.2.1 Exhaustive tree search

For parsimony and likelihood methods of tree reconstruction, which evaluate trees according to an optimality criterion, one should in theory calculate the score for every possible tree and then identify the tree having the best score. Such a strategy is known as *exhaustive search* and is guaranteed to find the best tree. As mentioned above, the stepwise addition algorithm provides a way of enumerating all possible trees for a fixed number of species (Fig. 3.4).

Exhaustive search is, however, computationally unfeasible except for small data sets with, say, fewer than 10 taxa. For the parsimony method, a branch-and-bound algorithm has been developed to speed up the exhaustive search (Hendy and Penny 1982). Even so, the computation is feasible for small data sets only. For the likelihood method, such an algorithm is not available. Thus most computer programs use heuristic algorithms to search in the tree space, and do not guarantee to find the optimal tree.

3.2.2 Heuristic tree search

Heuristic search algorithms may be grouped into two categories. The first includes hierarchical clustering algorithms. These may be subdivided into *agglomerative* methods, which proceed by successive fusions of the n species into groups, and *divisive* methods, which separate the n species successively into finer groups (Everitt et al. 2001). Whether every step involves a fusion or fission, the algorithm involves choosing one out of many alternatives, and the optimality criterion is used to make that choice. The second category of heuristic tree-search algorithms includes *tree-rearrangement* or *branch-swapping* algorithms. They propose new trees through local perturbations to the current tree, and the optimality criterion is used to decide whether or not to move to a new tree. The procedure is repeated until no improvement can be made in the tree score. We describe two cluster algorithms in this subsection and a few branch-swapping algorithms in the next.

Stepwise addition or *sequential addition* is an agglomerative algorithm. It adds sequences one by one, until all sequences are in the tree. When each new sequence is added, all the possible locations are evaluated and the best is chosen using the optimality criterion. Figure 3.10 illustrates the algorithm for the case of five sequences, using parsimony score as the optimality criterion. Note that this algorithm of heuristic tree search is different from the stepwise addition algorithm for enumerating all possible trees explained in Fig. 3.4. In the heuristic search, the locally best subtree is selected at each step, and trees that can be generated from the suboptimal subtrees are ignored. In our example, the 10 five-species trees on the second and third rows of Fig. 3.4 are never visited in the heuristic search. Thus the algorithm is not guaranteed to find the globally optimal tree. It is less clear whether one should add the most similar sequences or the most divergent sequences first. A common practice is to run the algorithm multiple times, adding sequences in a random order.

Star decomposition is a divisive cluster algorithm. It starts from the star tree of all species, and proceeds to resolve the polytomies by joining two taxa at each step. From the initial star tree of n species, there are $n(n-1)/2$ possible pairs, and the pair that results in the greatest improvement in the tree score is grouped together. The root of the tree then becomes a polytomy with $(n-1)$ taxa. Every step of the algorithm

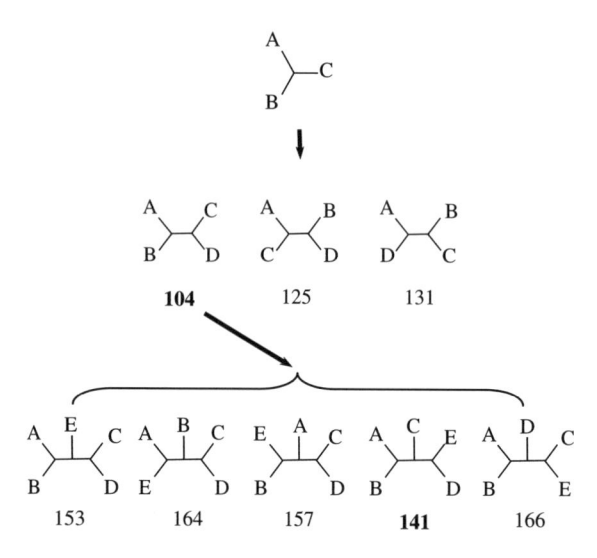

Fig. 3.10 Stepwise addition algorithm under the maximum parsimony criterion. The tree score is the minimum number of changes required by the tree.

reduces the number of taxa connected to the root by one. The procedure is repeated until the tree is fully resolved. Figure 3.11 shows an example of five sequences, using the log likelihood score for tree selection.

For n species, the stepwise-addition algorithm evaluates three trees of four species, five trees of five species, seven trees of six species, with a total of $3 + 5 + 7 + \cdots + (2n - 5) = (n - 1)(n - 3)$ trees in total. In contrast, the star-decomposition algorithm evaluates $n(n - 1)/2 + (n - 1)(n - 2)/2 + \cdots + 3 = n(n^2 - 1)/6 - 7$ trees in total, all of which are for n species. Thus for $n > 4$, the star-decomposition algorithm evaluates many more and bigger trees than the stepwise-addition algorithm and is expected to be much slower. The scores for trees constructed during different stages of the stepwise-addition algorithm are not directly comparable as the trees are of different sizes. Trees evaluated in the star-decomposition algorithm are all of the same size and their tree scores are comparable.

Both the stepwise-addition and star-decomposition algorithms produce resolved trees of all n species. If we stop at the end of either algorithm, we have an algorithmic cluster method for tree reconstruction based on the optimality criterion. However, in most programs, trees generated from these algorithms are treated as starting trees and subjected to local rearrangements. Below are a few such algorithms.

3.2.3 Branch swapping

Branch swapping or tree rearrangements are heuristic algorithms of hill climbing in the tree space. An initial tree is used to start the process. This can be a random tree, or a tree produced by stepwise-addition or star-decomposition algorithms, or by

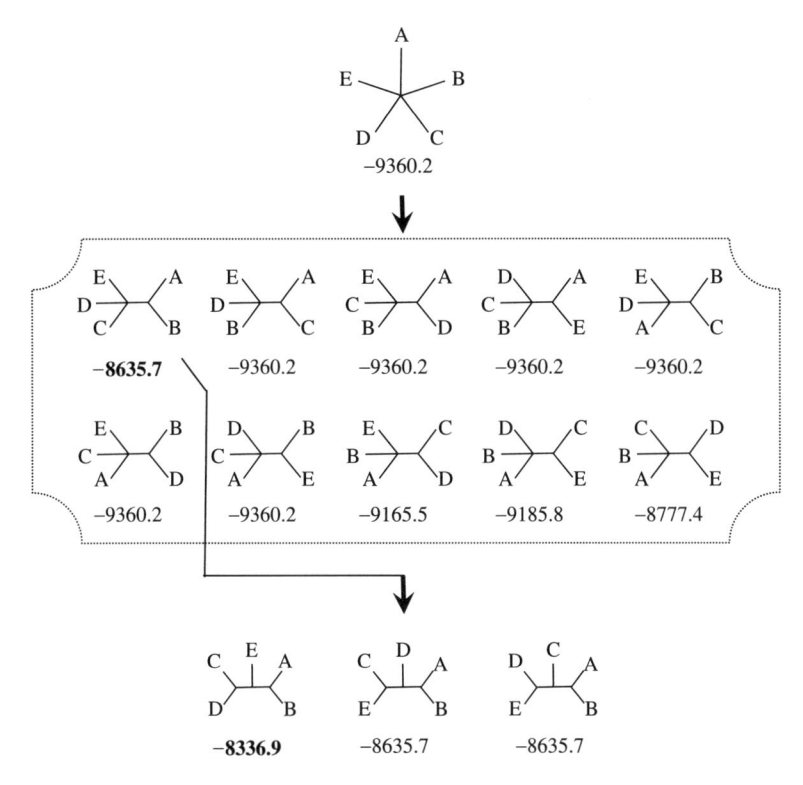

Fig. 3.11 Star-decomposition algorithm under the likelihood criterion. The tree score is the log likelihood value calculated by optimizing branch lengths on the tree.

other faster algorithms such as neighbour joining (Saitou and Nei 1987). The branch-swapping algorithm generates a collection of neighbour trees around the current tree. The optimality criterion is then used to decide which neighbour to move to. The branch-swapping algorithm affects our chance of finding the best tree and the amount of computation it takes to do so. If the algorithm generates too many neighbours, each step will require the evaluation of too many candidate trees. If the algorithm generates too few neighbours, we do not have to evaluate many trees at each step, but there may be many local peaks in the tree space (see below) and the search can easily get stuck at a local peak.

In *nearest-neighbour interchange (NNI)* each internal branch defines a relationship among four subtrees, say, a, b, c, and d (Fig. 3.12). Suppose the current tree is $((a, b), c, d)$ and the two alternative trees are $((a, c), b, d)$ and $((a, d), b, c)$. The NNI algorithm allows us to move from the current tree to the two alternative trees by swapping a subtree on one side of the branch with a subtree on the other side. An unrooted tree for n species has $n - 3$ internal branches. The NNI algorithm thus generates $2(n - 3)$ immediate neighbours. The neighbourhood relationships among the 15 trees for five species are illustrated below in Fig. 3.14.

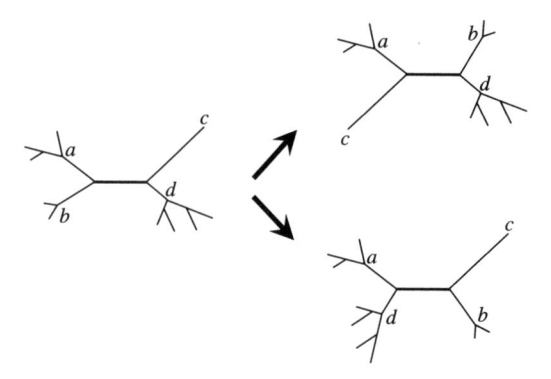

Fig. 3.12 The nearest-neighbour interchange (NNI) algorithm. Each internal branch in the tree connects four subtrees or nearest neighbours (a, b, c, d). Interchanging a subtree on one side of the branch with another on the other side constitutes an NNI. Two such rearrangements are possible for each internal branch.

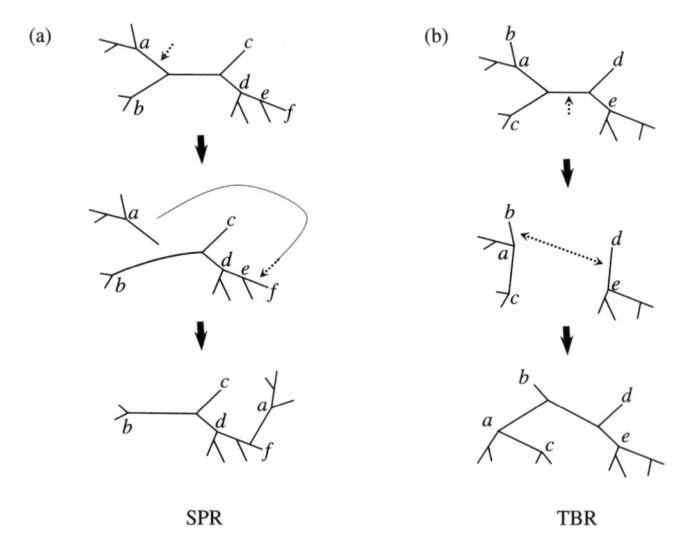

Fig. 3.13 (a) Branch swapping by subtree pruning and regrafting (SPR). A subtree (for example, the one represented by node a) is pruned, and then reattached to a different location on the tree. (b) Branch swapping by tree bisection and reconnection (TBR). The tree is broken into two subtrees by cutting an internal branch. Two branches, one from each subtree, are then chosen and rejoined to form a new tree.

Two other commonly used algorithms are *subtree pruning and regrafting* (SPR) and *tree bisection and reconnection* (TBR). In the former, a subtree is pruned and then reattached to a different location on the tree (Fig. 3.13a). In the latter, the tree is cut into two parts by chopping an internal branch and then two branches, one from

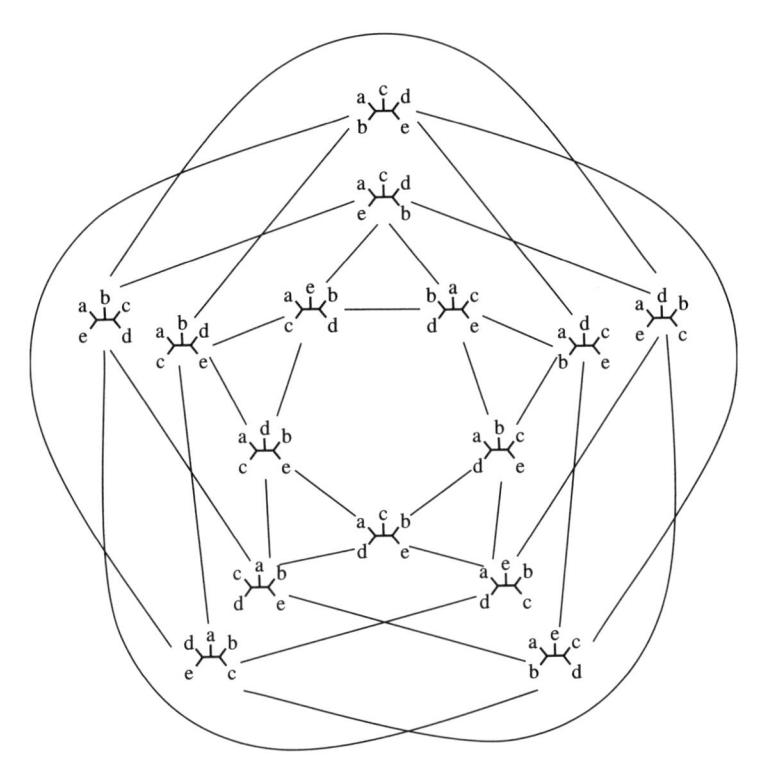

Fig. 3.14 The 15 trees for five species, with neighbourhood relationships defined by the NNI algorithm. Trees that are neighbours under NNI are connected. Note that this visually appealing representation has the drawback that trees close by may not be neighbours. Drawn following Felsenstein (2004).

each subtree, are chosen and rejoined to form a new tree (Fig. 3.13b). TBR generates more neighbours than SPR, which in turn generates more neighbours than NNI.

3.2.4 Local peaks in the tree space

Maddison (1991) and Charleston (1995) discussed local peaks or tree islands in the tree space. Figure 3.15 shows an example for five species and 15 trees. The neighbourhood relationship is defined using the NNI algorithm, with each tree having four neighbours (see Fig. 3.14). The parsimony tree lengths for the two trees on the top of the graph, T_1 and T_2, are 656 and 651, with T_2 being the most parsimonious tree. The eight trees that are neighbours of T_1 and T_2 have tree lengths ranging from 727 to 749, while the five trees that are two steps away from T_1 and T_2 have tree lengths ranging from 824 to 829. Trees T_1 and T_2 are separated from each other by trees of much poorer scores and are thus local peaks. They are local peaks for the SPR and TBR algorithms as well. Also T_1 and T_2 are local peaks when the data are analysed using the likelihood criterion. Indeed for this data set, the rank order of the 15 trees is

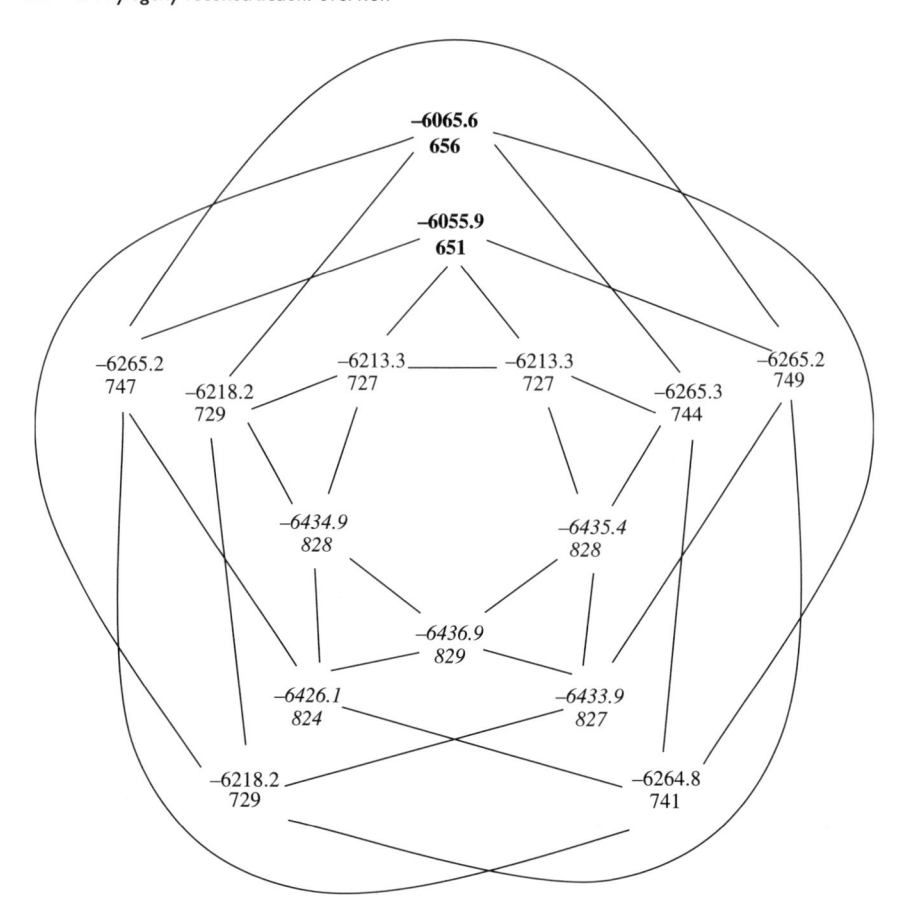

Fig. 3.15 *Local peaks in the tree space.* The log likelihood values (above) and parsimony scores (below) for the 15 trees of Fig. 3.14, shown in the same locations as the trees. The data set was simulated following the construction of Mossel and Vigoda (2005). It consists of 2000 nucleotide sites simulated under the JC69 model (Jukes and Cantor 1969) on the top two trees in Fig. 3.14: T_1: ((a, b), c, (d, e)) and T_2: ((a, e), c, (d, b)), with 1000 sites from each tree. Each external branch had length 0.01 and each internal branch had length 0.1. Trees T_1 and T_2 are two local optima under both parsimony and likelihood criteria.

almost identical under the likelihood and parsimony criteria. Similarly the data set poses serious computational problems for Bayesian algorithms (Huelsenbeck *et al.* 2001; Ronquist and Huelsenbeck 2003), as discussed by Mossel and Vigoda (2005).

One can design a branch-swapping algorithm under which trees T_1 and T_2 are neighbours. However, such an algorithm will define a different neighbourhood relationship among trees, and may have different local peaks or may have local peaks for different data sets. The problem should be more serious for larger trees with more species, as the tree space is much larger. Similarly, in larger data sets with more sites, the peaks tend to be higher and the valleys deeper, making it very difficult to traverse between peaks (Salter 2001).

3.2.5 Stochastic tree search

A hill-climbing algorithm that always goes uphill is called a *greedy algorithm*. Greedy algorithms may easily get stuck at a local peak. Some search algorithms attempt to overcome the problem of local peaks by allowing downhill moves. They can work under either parsimony or likelihood criteria. The first such algorithm is *simulated annealing* (Metropolis *et al.* 1953; Kirkpatrick *et al.* 1983), inspired by annealing in metallurgy, a technique involving heating and controlled cooling of a metal to reduce defects. The heat causes the atoms to move at random, exploring various configurations, while the slow cooling allows them to find configurations with low internal energy. In a simulated-annealing algorithm of optimization, the objective function is modified to have a flattened surface during the early (heated) stage of the search, making it easy for the algorithm to move between peaks. At this stage downhill moves are accepted almost as often as uphill moves. The 'temperature' is gradually reduced as the simulation proceeds, according to some 'annealing schedule'. At the final stage of the algorithm, only uphill moves are accepted, as in a greedy algorithm. Simulated-annealing algorithms are highly specific to the problem, and their implementation is more art than science. The efficiency of the algorithm is affected by the neighbourhood function (branch-swapping algorithms) and the annealing schedule. Implementations in phylogenetics include Goloboff (1999) and Barker (2004) for parsimony, and Salter and Pearl (2001) for likelihood. Fleissner *et al.* (2005) used simulated annealing for simultaneous sequence alignment and phylogeny reconstruction.

A second stochastic tree-search algorithm is the *genetic algorithm*. A 'population' of trees is kept in every generation; these trees 'breed' to produce trees of the next generation. The algorithm uses operations that are similar to mutation and recombination to generate new trees from the current ones. 'Survival' of each tree into the next generation depends on its 'fitness', which is the optimality criterion. Lewis (1998), Katoh *et al.* (2001), and Lemmon and Milinkovitch (2002), among others, implemented genetic algorithms to search for the ML tree.

A third stochastic tree-search algorithm is the Bayesian Markov chain Monte Carlo algorithm. If all trees have the same prior probability, the tree with the highest posterior probability will be the maximum (integrated) likelihood tree. The Bayesian algorithm has a huge advantage over simulated-annealing or genetic algorithms: it is a statistical approach and produces not only a point estimate (the tree with the highest likelihood) but also a measure of uncertainties in the point estimate through posterior probabilities estimated during the search. Chapter 5 discusses Bayesian phylogenetics in detail.

3.3 Distance methods

Distance methods involve two steps: calculation of genetic distances between pairs of species and reconstruction of a phylogenetic tree from the distance matrix. The simplest distance method is perhaps UPGMA (Sokal and Sneath 1963). This method is based on the molecular clock assumption and generates rooted trees. It is applicable to population data and seldom used to analyze species data, as the clock is often violated when the sequences are divergent. Below we describe two other methods that do not require the clock assumption: the least-squares (LS) and neighbour-joining (NJ) methods.

3.3.1 Least-squares method

The least-squares (LS) method takes the pairwise distance matrix as given data and estimates branch lengths on a tree by matching those distances as closely as possible, that is, by minimizing the sum of squared differences between the given and predicted distances. The predicted distance is calculated as the sum of branch lengths along the path connecting the two species. The minimum sum of squared differences then measures the fit of the tree to data (the distances) and is used as the tree score. This method was developed by Cavalli-Sforza and Edwards (1967), who called it the *additive-tree method*.

Let the distance between species i and j be d_{ij}. Let the sum of branch lengths along the path from species i to j on the tree be $\hat{d}_{ij}$. The LS method minimizes the sum, over all distinct pairs i and j, of the squared differences, $(d_{ij} - \hat{d}_{ij})^2$, so that the tree fits the distances as closely as possible. For example, the pairwise distances calculated under the K80 model (Kimura 1980) for the mitochondrial data of Brown et al. (1982) are shown in Table 3.3. These are taken as observed data. Now consider the tree ((human, chimpanzee), gorilla, orangutan) and its five branch lengths t_0, t_1, t_2, t_3, t_4 (Fig. 3.16). The predicted distance in the tree between human and chimpanzee is $t_1 + t_2$, and the predicted distance between human and gorilla is $t_1 + t_0 + t_3$, and so on. The sum of squared differences is then

$$S = \sum_{i<j}(d_{ij} - \hat{d}_{ij})^2$$

$$= (d_{12} - \hat{d}_{12})^2 + (d_{13} - \hat{d}_{13})^2 + (d_{14} - \hat{d}_{14})^2 + (d_{23} - \hat{d}_{23})^2$$
$$+ (d_{24} - \hat{d}_{24})^2 + (d_{34} - \hat{d}_{34})^2. \qquad (3.2)$$

Table 3.3 Pairwise distances for the mitochondrial DNA sequences

1. Human			
2. Chimpanzee	0.0965		
3. Gorilla	0.1140	0.1180	
4. Orangutan	0.1849	0.2009	0.1947
	1. Human	2. Chimpanzee	3. Gorilla	4. Orangutan

As the distances (d_{ij}) are calculated already, S is a function of the five unknown branch lengths t_0, t_1, t_2, t_3, and t_4. The values of branch lengths that minimize S are the LS estimates: $\hat{t}_0 = 0.008840$, $\hat{t}_1 = 0.043266$, $\hat{t}_2 = 0.053280$, $\hat{t}_3 = 0.058908$, $\hat{t}_4 = 0.135795$, with the corresponding tree score $S = 0.00003547$. Similar calculations can be done for the other two trees. Indeed the other two binary trees both converge to the star tree, with the internal branch length estimated to be 0; see Table 3.4. The tree ((human, chimpanzee), gorilla, orangutan) has the smallest S and is called the LS tree. It is the LS estimate of the true phylogeny.

Estimation of branch lengths on a fixed tree by the least-squares criterion uses the same principle as calculating the line of best fit $y = a + bx$ in a scatter plot. If there are no constraints on the branch lengths, the solution is analytical and can be obtained by solving linear equations (Cavalli-Sforza and Edwards 1967). Efficient algorithms that require less computation and less space have also been developed by Rzhetsky and Nei (1993), Bryant and Waddell (1998), and Gascuel (2000). Those algorithms may produce negative branch lengths, which are not meaningful biologically. If the branch lengths are constrained to be nonnegative, the problem becomes one of constrained optimization, which is much more expensive. However, the unconstrained LS criterion may be justified by ignoring the interpretation of branch lengths. The method will select the true tree as the LS tree if and only if the score for the true tree is smaller than the scores for all other trees. If this condition is satisfied in infinite data, the method will be guaranteed to converge to the true tree when more and more data are available. If the condition is satisfied in most finite data sets, the method will recover the true tree with high efficiencies. Thus the unconstrained method can be a well-behaved method of tree reconstruction without a sensible definition of branch lengths. While some simulation studies suggest that constraining branch lengths to be nonnegative

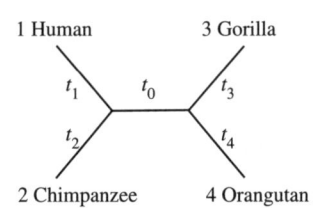

Fig. 3.16 A tree to demonstrate the least-squares criterion for estimating branch lengths.

Table 3.4 Least-squares branch lengths under K80 (Kimura 1980)

Tree	t_0 for internal branch	t_1 for H	t_2 for C	t_3 for G	t_4 for O	S_j
τ_1:((H, C), G, O)	0.008840	0.043266	0.053280	0.058908	0.135795	0.000035
τ_2: ((H, G), C, O)	0.000000	0.046212	0.056227	0.061854	0.138742	0.000140
τ_3: ((H, G), C, O)	as above					
τ_0: (H, G, C, O)	as above					

leads to improved performance in tree reconstruction (e.g. Kuhner and Felsenstein 1994), most computer programs implement the LS method without the constraint. It is noted that when the estimated branch lengths are negative, they are most often close to zero.

The least-squares method described above uses equal weights for the different pairwise distances and is known as the ordinary least squares (OLS). It is a special case of the following generalized or weighted least squares (GLS) with weights $w_{ij} = 1$:

$$S = \sum_{i<j} w_{ij}(d_{ij} - \hat{d}_{ij}). \tag{3.3}$$

Fitch and Margoliash (1967) suggested the use of $w_{ij} = 1/d_{ij}^2$, while Bulmer (1990) used the variance: $w_{ij} = 1/\text{var}(d_{ij})$. However, such weighted LS methods were found not to work well in computer simulations, especially when the distances are large, presumably because the estimated variances are unreliable.

3.3.2 Neighbour-joining method

A criterion used for tree comparison, especially in distance methods, is the amount of evolution measured by the sum of branch lengths in the tree (Kidd and Sgaramella-Zonta 1971; Rzhetsky and Nei 1993). The tree with the smallest sum of branch lengths is known as the *minimum evolution* tree; see Desper and Gascuel (2005) for an excellent review of this class of methods.

Neighbour joining is a cluster algorithm based on the minimum-evolution criterion proposed by Saitou and Nei (1987). Because it is computationally fast and also produces reasonable trees, it is widely used. It is a divisive cluster algorithm (i.e. a star-decomposition algorithm), with the tree length (the sum of branch lengths along the tree) used as the criterion for tree selection at each step. It starts with a star tree and then joins two nodes, choosing the pair to achieve the greatest reduction in tree length. A new node is then created to replace the two nodes joined (Fig. 3.17), reducing the dimension of the distance matrix by one. The procedure is repeated until the tree is fully resolved. The branch lengths on the tree as well as the tree length are updated

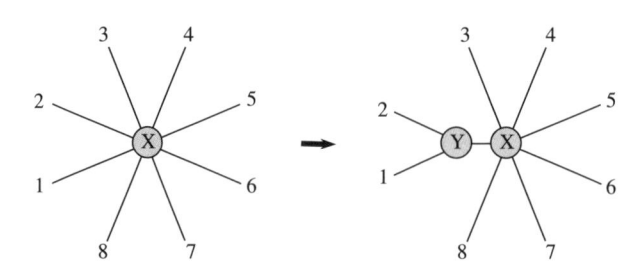

Fig. 3.17 The neighbour-joining method of tree reconstruction is a divisive cluster algorithm, dividing taxa successively into finer groups.

during every step of the algorithm. See Saitou and Nei (1987), Studier and Keppler (1988), and Gascuel (1994) for the update formulae.

A concern with the NJ method, and indeed with any distance-matrix method, is that large distances are poorly estimated, with large sampling errors. For very divergent sequences, the distance formulae may be even inapplicable. Some effort has been made to deal with the problem of large variances in large distance estimates. Gascuel (1997) modified the formula for updating branch lengths in the NJ algorithm to incorporate approximate variances and covariances of distance estimates. This method, called BIONJ, is close to the generalized least-squares method, and was found to outperform NJ, especially when substitution rates are high and variable among lineages. Another modification is the weighted neighbour-joining or WEIGHBOR method of Bruno et al. (2000). This used an approximate likelihood criterion for joining nodes to accommodate the fact that long distances are poorly estimated. Computer simulations suggest that WEIGHBOR produces trees similar to ML, and is more robust to the problem of long-branch attraction (see Subsection 3.4.4 below) than NJ (Bruno et al. 2000). Another idea, due to Ranwez and Gascuel (2002), is to improve distance estimates. When calculating the distance between a pair of sequences, those authors used a third sequence to break the long distance into two even parts and used ML to estimate three branch lengths on the tree of the three sequences; the original pairwise distance is then calculated as the sum of the two branch lengths. Simulations suggest that the improved distance, when combined with the NJ, BIONJ, and WEIGHBOR algorithms, led to improved topological accuracy.

3.4 Maximum parsimony

3.4.1 Brief history

When using allele frequencies (mainly for blood group alleles) to reconstruct the relationships among human populations, Edwards and Cavalli-Sforza (1963) (see also Cavalli-Sforza and Edwards 1967) suggested that a plausible estimate of the evolutionary tree is the one that invokes the minimum total amount of evolution. They called this method the *minimum-evolution method*. In modern terminology, the method discussed by Edwards and Cavalli-Sforza, when applied to discrete data, is identified with parsimony, while minimum evolution nowadays refers to methods minimizing the sum of branch lengths after correcting for multiple hits, as discussed in last section. For discrete morphological characters, Camin and Sokal (1965) suggested use of the minimum number of changes as a criterion for tree selection. For molecular data, minimizing changes on the tree to infer ancestral proteins appears most natural and was practised by many pioneers in the field, for example by Pauling and Zuckerkandl (1963) and Zuckerkandl (1964) as a way of 'restoring' ancestral proteins for 'palaeogenetic' studies of their chemical properties, and by Eck and Dayhoff (1966) to construct empirical matrices of amino acid substitution rates. Fitch (1971b) was the first to present a systematic algorithm to enumerate all and only the most-parsimonious reconstructions. Fitch's algorithm works on binary trees only. Hartigan

94 • *3 Phylogeny reconstruction: overview*

(1973) considered multifurcating trees as well and provided a mathematical proof for the algorithm. Since then, much effort has been made to develop fast algorithms for parsimony analysis of large data sets; see, e.g., Ronquist (1998), Nixon (1999), and Goloboff (1999).

3.4.2 Counting the minimum number of changes given the tree

The minimum number of character changes at a site is often called the *character length* or *site length*. The sum of character lengths over all sites in the sequence is the minimum number of required changes for the entire sequence and is called the *tree length*, *tree score*, or *parsimony score*. The tree with the smallest tree score is the estimate of the true tree, called the *maximum parsimony tree* or the *most parsimonious tree*. It is common, especially when the sequences are very similar, for multiple trees to be equally best; that is, they have the same minimum score and are all shortest trees.

Suppose the data for four species at a particular site are AAGG, and consider the minimum number of changes required by the two trees of Fig. 3.18. We calculate this number by assigning character states to the extinct ancestral nodes. For the first tree, this is achieved by assigning A and G to the two nodes, and one change (A ↔ G on the internal branch) is required. For the second tree, we can assign either AA (shown) or GG (not shown) to the two internal nodes; in either case, a minimum of two changes is required. Note that the set of character states (nucleotides) at a site assigned to ancestral nodes is called an *ancestral reconstruction*. The total number of reconstructions at each site is thus $4^{(n-2)}$ for nucleotides or $20^{(n-2)}$ for amino acids as a binary tree of n species has $n-2$ interior nodes. The reconstruction that achieves the minimum number of changes is called the *most parsimonious reconstruction*. Thus, for the first tree, there is one single most parsimonious reconstruction, while for the second tree, two reconstructions are equally most parsimonious. The algorithm of Fitch (1971b) and Hartigan (1973) calculates the minimum number of changes and enumerates all the most parsimonious reconstructions at a site. We will not describe this algorithm here. Instead we describe in the next subsection a more general algorithm due to Sankoff (1975), which is very similar to the likelihood algorithm to be discussed in Chapter 4.

Some sites do not contribute to the discrimination of trees and are thus non-informative. For example, a constant site, at which the different species have the

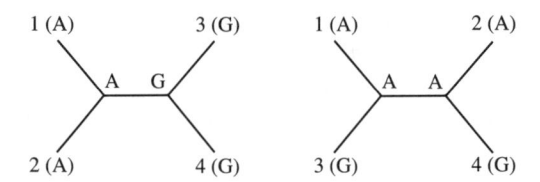

Fig. 3.18 Data AAGG at one site for four species mapped onto two alternative trees ((1, 2), 3, 4) and ((1, 3), 2, 4). The tree on the left requires a minimum of one change while the tree on the right requires two changes to explain the data.

same nucleotide, requires no change for any tree. Similarly a *singleton* site, at which two characters are observed but one is observed only once (e.g. TTTC or AAGA), requires one change for every tree and is thus not informative. Perhaps more strikingly, a site with data AAATAACAAG (for 10 species) is not informative either, as a minimum of three changes are required by any tree, which is also achieved by every tree by assigning A to all ancestral nodes. For a site to be a *parsimony-informative site*, at least two characters have to be observed, each at least twice. Note that the concepts of informative and noninformative sites apply to parsimony only. In distance or likelihood methods, all sites including the constant sites affect the calculation and should be included.

We often refer to the observed character states in all species at a site as a *site configuration* or *site pattern*. The above discussion means that for four species only three *site patterns* are informative: *xxyy*, *xyxy*, and *xyyx*, where *x* and *y* are any two distinct states. It is obvious that those three site patterns 'support' the three trees T_1: ((1, 2), 3, 4); T_2: ((1, 3), 2, 4); and T_3: ((1, 4), 2, 3), respectively. Let the number of sites with those site patterns be n_1, n_2, and n_3, respectively. Then T_1, T_2, or T_3 is the most parsimonious tree if n_1, n_2, or n_3 is the greatest among the three.

3.4.3 Weighted parsimony and transversion parsimony

The algorithm of Fitch (1971b) and Hartigan (1973) assumes that every change has the same cost. In weighted parsimony, different weights are assigned to different types of character changes. Rare changes are penalized more heavily than frequent changes. For example, transitions are known to occur at a higher rate than transversions and can be assigned a lower cost (weight). Weighted parsimony uses a *step matrix* or *cost matrix* to specify the cost of every type of change. An extreme case is *transversion parsimony*, which gives a penalty of 1 for a transversion but no penalty for a transition. Below, we describe Sankoff's (1975) dynamic-programming algorithm, which calculates the minimum cost at a site and enumerates the reconstructions that achieve this minimum given any arbitrary cost matrix.

We first illustrate the basic idea of dynamic-programming algorithms using a fictitious example of a camel caravan travelling on the Silk Route. We start from the source city X, Chang-an in central China, to go to the destination Y, Baghdad in Iraq (Fig. 3.19). The route goes though four countries A, B, C, and D, and has to pass

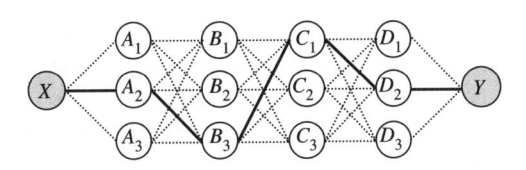

Fig. 3.19 Caravan-travelling example used for illustrating the dynamic-programming algorithm. It is required to determine the shortest route from X to Y, through four countries A, B, C, and D. Stops between neighbouring countries are connected, with their distances known.

one of three caravan stops in every country: A_1, A_2 or A_3 in country A; B_1, B_2 or B_3 in country B; and so on. We know the distance between any two stops in two neighbouring countries, such as XA_2 and A_1B_2. We seek to determine the shortest distance and the shortest route from X to Y. An obvious strategy is to evaluate all possible routes, but this can be expensive as the number of routes (3^4 in the example) grows exponentially with the number of countries. A dynamic-programming algorithm answers many smaller questions, with the new questions building on answers to the old ones. First we ask for the shortest distances (from X) to stops A_1, A_2, and A_3 in country A. These are just the given distances. Next we ask for the shortest distances to stops in country B, and then the shortest distances to stops in country C, and so on. Note that the questions at every stage are easy given the answers to the previous questions. For example, consider the shortest distance to C_1, when the shortest distances to B_1, B_2, and B_3 are already determined. This is just the smallest among the distances of the three routes going through B_1, B_2, or B_3, with the distance through B_j ($j = 1, 2, 3$) being the shortest distance (from X) to B_j plus the distance between B_j and C_1. After the shortest distances to D_1, D_2, and D_3 are determined, it is easy to determine the shortest distance to Y itself. It is important to note that adding another country to the problem will add another stage in the algorithm, so that the amount of computation grows linearly with the increase in number of countries.

We now describe Sankoff's algorithm. We seek to determine the minimum cost for a site on a given tree as well as the ancestral reconstruction that achieves the minimum cost. We use the tree of Fig. 3.20 as an example. The observed nucleotides at the site at the six tips are CCAGAA. Let $c(x, y)$ denote the cost of change from state x to state y, so $c(x, y) = 1$ for a transitional difference and $c(x, y) = 1.5$ for a transversion (Fig. 3.20).

Instead of the minimum cost for the whole tree, we calculate the minimum costs for many subtrees. We refer to a branch on the tree by the node it leads to or by the two nodes it connects. For example branch 10 is also branch 8–10 in Fig. 3.20. We say that each node i on the tree defines a subtree, referred to as subtree i, which consists of branch i, node i, and all its descendant nodes. For example, subtree 3 consists of the single tip branch 10–3 while subtree 10 consists of branch 8–10 and nodes 10, 3, and 4. Define $S_i(x)$ as the minimum cost incurred on subtree i, given that the mother node of node i has state x. Thus $\{S_i(T), S_i(C), S_i(A), S_i(G)\}$ constitute a cost vector for subtree i at node i. They are like the shortest distances to stops in a particular country in the caravan example. We calculate the cost vectors for all nodes on the tree, starting with the tips and visiting a node only after we have visited all its descendant nodes. For a tip node i, the subtree is just the tip branch and the cost is simply read from the cost matrix. For example, tip 3 has the cost vector $\{1.5, 1.5, 0, 1\}$, meaning that the (minimum) cost of subtree 3 is 1.5, 1.5, 0, or 1, if mother node 10 has T, C, A, or G, respectively (Fig. 3.20). If the nucleotide at the tip is undetermined, the convention is to use the minimum cost among all compatible states (Fitch 1971b). For an interior node i, suppose its two daughter nodes are j and k. Then

$$S_i(x) = \min_y [c(x, y) + S_j(y) + S_k(y)]. \tag{3.4}$$

3.4 Maximum parsimony • 97

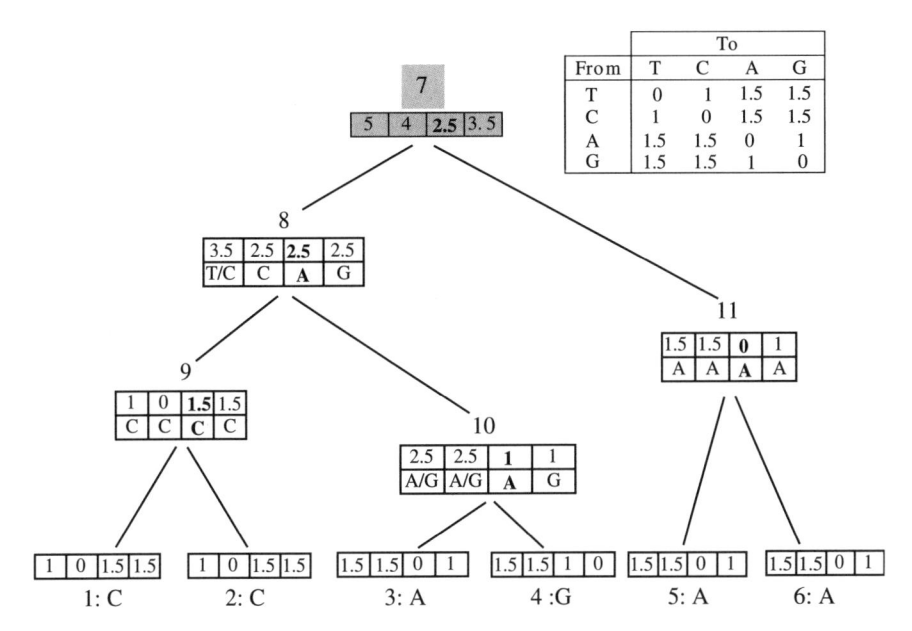

Fig. 3.20 Dynamic-programming algorithm for calculating the minimum cost and enumerating the most parsimonious reconstructions using weighted parsimony. The site has observed data CCAGAA. The cost vector at each node gives the minimum cost of the subtree induced by that node (which includes the node itself, its mother branch, and all its descendants), given that the mother node has nucleotides T, C, A, or G. The nucleotides at the node that achieved the minimum cost are shown below the cost vector. For example, the minimum cost of the subtree induced by node 3 (including the single branch 10–3) is 1.5, 1.5, 0, or 1, if node 10 has T, C, A, or G, respectively. The minimum cost of the subtree induced by node 10 (including branches 8–10 and nodes 10, 3 and 4) is 2.5, 2.5, 1, or 1, if node 8 has T, C, A, or G, respectively; the said minimum is achieved by node 10 having A/G, A/G, A, or G, respectively. The cost vectors are calculated for every node, starting from the tips and proceeding towards the root. At the root (node 7), the cost vector gives the minimum cost of the whole tree as 5, 4, 2.5, or 3.5, if the root has T, C, A, or G, respectively.

Note that subtree i consists of branch i plus subtrees j and k. Thus the minimum cost of subtree i is the cost along branch i, $c(x, y)$, plus the minimum costs of subtrees j and k, minimized over the state y at node i. We use $C_i(x)$ to record the state y that achieved the minimum.

Consider node 10 as an example, for which the cost vector is calculated to be $\{S_{10}(T), S_{10}(C), S_{10}(A), S_{10}(G)\} = \{2.5, 2.5, 1, 1\}$. Here the first entry, $S_{10}(T) = 2.5$, means that the minimum cost of subtree 10, given that mother node 8 has T, is 2.5. To see this, consider the four possible states at node 10: $y = $ T, C, A, or G. The (minimum) cost on subtree 10 is $3 = 0 + 1.5 + 1.5, 4, 2.5$, or 2.5, if node 10 has the state $y = $ T, C, A, or G, respectively (and if node 8 has T). Thus the minimum is 2.5, achieved by node 10 having $y = $ A or G; that is, $S_{10}(T) = 2.5$ and $C_{10}(T) = $ A or G

(Fig. 3.20). This is the minimization over y in equation (3.4). Similarly, the second entry in the cost vector at node 10, $S_{10}(C) = 2.5$, means that the minimum cost for subtree 10, given that node 8 has C, is 2.5. This minimum is achieved by having $C_{10}(C) = A/G$ at node 10.

Similar calculations can be done for nodes 9 and 11. We now consider node 8, which has daughter nodes 9 and 10. The cost vector is calculated to be {3.5, 2.5, 2.5, 2.5}, meaning that the minimum cost of subtree 8 is 3.5, 2.5, 2.5, or 2.5, if mother node 7 has T, C, A, or G, respectively. Here we derive the third entry $S_8(A) = 2.5$, with mother node 7 having A. By using the cost vectors for nodes 9 and 10, we calculate the minimum cost on subtree 8 to be $5 = 1.5 + 1 + 2.5, 4, 2.5$, or 4.5, if node 8 has T, C, A, or G, respectively (and if mother node 7 has A). Thus $S_8(A) = 2.5$ is the minimum, achieved by node 8 having $C_8(A) = A$.

The algorithm is applied successively to all nodes in the tree, starting from the tips and moving towards the root. This upper pass calculates $S_i(x)$ and $C_i(x)$ for all nodes i except the root. Suppose the root has daughter nodes j and k and note that the whole tree consists of subtrees j and k. The minimum cost of the whole tree, given that the root has y, is $S_j(y) + S_k(y)$. This cost vector is {5, 4, 2.5, 3.5}, for $y =$ T, C, A, G at the root (Fig. 3.20). The minimum is 2.5, achieved by having A at the root. In general, if j and k are the daughter nodes of the root, the minimum cost for the whole tree is

$$S = \min_y [S_j(y) + S_k(y)]. \tag{3.5}$$

After calculation of $S_i(x)$ and $C_i(x)$ for all nodes through the upper pass, a down pass reads out the most parsimonious reconstructions. In our example, given A for the root, node 8 achieves the minimum for subtree 8 by having A. Given A at node 8, nodes 9 and 10 should have C and A, respectively. Similarly given A for the root, node 11 should have A. Thus the most parsimonious reconstruction at the site is $y_7 y_8 y_9 y_{10} y_{11} =$ AACAA, with the minimum cost 2.5.

3.4.4 Long-branch attraction

Felsenstein (1978a) demonstrated that the parsimony method can be statistically inconsistent under certain combinations of branch lengths on a four-species tree. In statistics, an estimation method is said to be consistent if the estimate converges to the true value of the parameter when the amount of data (sample size) approaches infinity. Otherwise the method is said to be inconsistent. In phylogenetics, we say that a tree-reconstruction method is inconsistent if the estimated tree topology converges to a wrong tree when the amount of data (the number of sites) increases to infinity.

The tree Felsenstein used has the characteristic shape shown in Fig. 3.21(a), with two long branches separated by a short internal branch. The estimated tree by parsimony, however, tends to group the two long branches together (Fig. 3.21b), a phenomenon known as 'long-branch attraction'. Using a simple model of character evolution, Felsenstein calculated the probabilities of observing sites with the three site patterns $xxyy$, $xyxy$, $xyyx$, where x and y are any two distinct characters, and

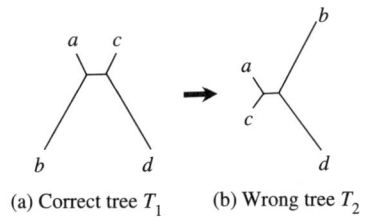

Fig. 3.21 *Long-branch attraction.* When the correct tree (T_1) has two long branches separated by a short internal branch, parsimony tends to recover a wrong tree (T_2) with the two long branches grouped in one clade.

found that $\Pr(xyxy) > \Pr(xxyy)$ when the two long branches are much longer than the three short branches. Thus with more and more sites in the sequence, it will be more and more certain that more sites will have pattern $xyxy$ than pattern $xxyy$, in which case parsimony will recover the wrong tree T_2 instead of the true tree T_1 (Fig. 3.21). The phenomenon has been demonstrated in many real and simulated data sets (see, e.g., Huelsenbeck 1998) and is due to the failure of parsimony to correct for parallel changes on the two long branches. Likelihood and distance methods using simplistic and unrealistic evolutionary models show the same behaviour.

3.4.5 Assumptions of parsimony

Some concerns may be raised about the parsimony reconstruction of ancestral states. First, the method ignores branch lengths. Some branches on the tree are longer than others, meaning that they have accumulated more evolutionary changes than other branches. It is thus illogical to assume that a change is as likely to occur on a long branch as on a short one, as parsimony does, when character states are assigned to ancestral nodes on the tree. Second, the simple parsimony criterion ignores different rates of changes between nucleotides. Such rate differences are taken into account by weighted parsimony through the use of a step matrix, although determining the appropriate weights may be nontrivial. In theory, how likely a change is to occur on a particular branch should depend on the length of the branch as well the relative rate of the change. Attempts to derive appropriate weights for the observed data lead naturally to the likelihood method, which uses a Markov-chain model to describe the nucleotide substitution process, relying on probability theory to accommodate unequal branch lengths, unequal substitution rates between nucleotides, and any other features of the evolutionary process. This is the topic of the next chapter.

4

Maximum likelihood methods

4.1 Introduction

In this chapter, we will discuss likelihood calculation for multiple sequences on a phylogenetic tree. As indicated at the end of the last chapter, this is a natural extension to the parsimony method, when we want to incorporate differences in branch lengths and in substitution rates between nucleotides. Likelihood calculation on a tree is also a natural extension to estimation of the distance between two sequences, discussed in Chapter 1. Indeed Chapter 1 has covered all the general principles of Markov-chain theory and maximum likelihood (ML) estimation needed in this chapter.

A distinction may be made between two uses of likelihood in phylogenetic analysis. The first is to estimate parameters in the evolutionary model and to test hypotheses concerning the evolutionary process when the tree topology is known or fixed. The likelihood method, with its nice statistical properties, provides a powerful and flexible framework for such analysis. The second is to estimate the tree topology. The log likelihood for each tree is maximized by estimating branch lengths and other substitution parameters, and the optimized log likelihood is used as a tree score for comparing different trees. This second use of likelihood involves complexities, which are discussed in Chapter 6.

4.2 Likelihood calculation on tree

4.2.1 Data, model, tree, and likelihood

The likelihood is defined as the probability of observing the data when the parameters are given, although it is considered to be a function of the parameters. The data consist of s aligned homologous sequences, each n nucleotides long, and can be represented as an $s \times n$ matrix $X = \{x_{jh}\}$, where x_{jh} is the hth nucleotide in the jth sequence. Let $\mathbf{x}_h$ denote the hth column in the data matrix. To define the likelihood, we have to specify a model by which the data are generated. Here we use the K80 nucleotide-substitution model (Kimura 1980). We assume that different sites evolve independently of each other and evolution in one lineage is independent of other lineages. We use the tree of five species of Fig. 4.1 as an example to illustrate the likelihood calculation. The observed data at a particular site, TCACC, are shown. The ancestral nodes are numbered 0, 6, 7, and 8, with 0 to be the root. The length of the branch leading to node i is denoted t_i, defined as the expected number of nucleotide substitutions per site. The

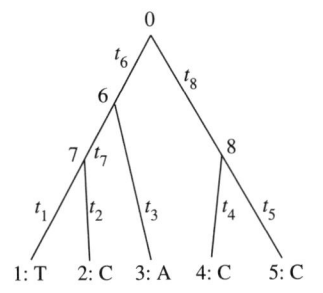

Fig. 4.1 A tree of five species used to demonstrate calculation of the liklihood function. The nucleotides observed at the tips at a site are shown. Branch lengths t_1–t_8 are measured by the expected number of nucleotide substitutions per site.

parameters in the model include the branch lengths and the transition/transversion rate ratio κ, collectively denoted $\theta = \{t_1, t_2, t_3, t_4, t_5, t_6, t_7, t_8, \kappa\}$.

Because of the assumption of independent evolution among sites, the probability of the whole data set is the product of the probabilities of data at individual sites. Equivalently the log likelihood is a sum over sites in the sequence

$$\ell = \log(L) = \sum_{h=1}^{n} \log\{f(\mathbf{x}_h|\theta)\}. \tag{4.1}$$

The ML method estimates θ by maximizing the log likelihood ℓ, often using numerical optimization algorithms (see Section 4.5). Here we consider calculation of ℓ when parameters θ are given.

We focus on one site only. We use x_i to represent the state at ancestral node i, and suppress the subscript h. Since the data at the site, $\mathbf{x}_h = $ TCACC, can result from any combination of ancestral nucleotides $x_0 x_6 x_7 x_8$, calculation of $f(\mathbf{x}_h)$ has to sum over all possible nucleotide combinations for the extinct ancestors.

$$f(\mathbf{x}_h|\theta) = \sum_{x_0}\sum_{x_6}\sum_{x_7}\sum_{x_8} [\pi_{x_0} p_{x_0 x_6}(t_6) p_{x_6 x_7}(t_7) p_{x_7 \mathrm{T}}(t_1) p_{x_7 \mathrm{C}}(t_2)$$
$$\times p_{x_6 \mathrm{A}}(t_3) p_{x_0 x_8}(t_8) p_{x_8 \mathrm{C}}(t_4) p_{x_8 \mathrm{C}}(t_5)]. \tag{4.2}$$

The quantity in the square brackets is the probability of data TCACC for the tips and $x_0 x_6 x_7 x_8$ for the ancestral nodes. This is equal to the probability that the root (node 0) has x_0, which is given by $\pi_{x_0} = 1/4$ under K80, multiplied by eight transition probabilities along the eight branches of the tree.

4.2.2 The pruning algorithm

4.2.2.1 Horner's rule and the pruning algorithm

Summing over all combinations of ancestral states is expensive because there are 4^{s-1} possible combinations for $s-1$ interior nodes. The situation is even worse for amino acid or codon sequences as there will then be 20^{s-1} or 61^{s-1} possible combinations. An important technique useful in calculating such sums is to identify common factors and calculate them only once. This is known as the *nesting rule* or *Horner's rule*, published by the Irish mathematician William Horner in 1830. The rule was also published in 1820 by a London matchmaker, Theophilus Holdred, and had been used in 1303 by the Chinese mathematician Zhu Shijie (朱世杰). By this rule, an nth-order polynomial can be calculated with only n multiplications and n additions. For example, a naïve calculation of $1+2x+3x^2+4x^3$, as $1+2\cdot x+3\cdot x\cdot x+4\cdot x\cdot x\cdot x$, requires six multiplications and three additions. However, by writing it as $1+x\cdot(2+x\cdot(3+4\cdot x))$, only three multiplications and three additions are needed. Application of the nesting rule to equation (4.2) means moving the summation signs to the right as far as possible, to give

$$f(\mathbf{x}_h|\theta) = \sum_{x_0} \pi_{x_0} \left\{ \sum_{x_6} p_{x_0 x_6}(t_6) \left[\left(\sum_{x_7} p_{x_6 x_7}(t_7) p_{x_7 \mathrm{T}}(t_1) p_{x_7 \mathrm{C}}(t_2) \right) p_{x_6 \mathrm{A}}(t_3) \right] \right\}$$
$$\times \left[\sum_{x_8} p_{x_0 x_8}(t_8) p_{x_8 \mathrm{C}}(t_4) p_{x_8 \mathrm{C}}(t_5) \right] \quad (4.3)$$

Thus we sum over x_7 before x_6, and sum over x_6 and x_8 before x_0. In other words, we sum over ancestral states at a node only after we have done so for all its descendant nodes.

The pattern of parentheses and the occurrences of the tip states in equation (4.3), in the form [(T, C), A], [C, C], match the tree of Fig. 4.1. This is no coincidence. Indeed calculation of $f(\mathbf{x}_h|\theta)$ by equation (4.3) constitutes the *pruning algorithm* of Felsenstein (1973b, 1981). This is a variant of the dynamic programming algorithm discussed in Subsection 3.4.3. Its essence is to calculate successively probabilities of data on many subtrees. Define $L_i(x_i)$ to be the probability of observing data at the tips that are descendants of node i, given that the nucleotide at node i is x_i. For example, tips 1, 2, 3 are descendants of node 6, so $L_6(\mathrm{T})$ is the probability of observing $x_1 x_2 x_3 = \mathrm{TCA}$, given that node 6 has the state $x_6 = \mathrm{T}$. With $x_i = \mathrm{T, C, A, G}$, we calculate a vector of conditional probabilities for each node i. In the literature, the conditional probability $L_i(x_i)$ is often referred to as the 'partial likelihood' or 'conditional likelihood'; these are misnomers since likelihood refers to the probability of the whole data set and not probability of data at a single site or part of a single site.

If node i is a tip, its descendant tips include node i only, so that $L_i(x_i) = 1$ if x_i is the observed nucleotide, or 0 otherwise. If node i is an interior node with daughter nodes j and k, we have

$$L_i(x_i) = \left[\sum_{x_j} p_{x_i x_j}(t_j) L_j(x_j)\right] \times \left[\sum_{x_k} p_{x_i x_k}(t_k) L_k(x_k)\right]. \quad (4.4)$$

This is a product of two terms, corresponding to the two daughter nodes j and k. Note that tips that are descendants of node i must be descendants of either j or k. Thus the probability $L_i(x_i)$ of observing all descendant tips of node i (given the state x_i at node i) is equal to the probability of observing data at descendant tips of node j (given x_i) times the probability of observing data at descendant tips of node k (given x_i). These are the two terms in the two pairs of brackets in equation (4.4), respectively. Given the state x_i at node i, the two parts of the tree down node i are independent. If node i has more than two daughter nodes, $L_i(x_i)$ will be a product of as many terms. Now consider the first term, the term in the first pair of brackets, which is the probability of observing data at descendant tips of node j (given the state x_i at node i). This is the probability $p_{x_i x_j}(t_j)$ that x_i will become x_j over branch length t_j times the probability $L_j(x_j)$ of observing the tips of node j given the state x_j at node j, summed over all possible states x_j.

Example. We use the tree of Fig. 4.1 to provide a numerical example of the calculation using the pruning algorithm at one site (Fig. 4.2). For convenience, we fix internal branch lengths at $t_6 = t_7 = t_8 = 0.1$ and external branch lengths at $t_1 = t_2 = t_3 = t_4 = t_5 = 0.2$. We also set $\kappa = 2$. The two transition probability matrices are as follows, in which the ijth element is $p_{ij}(t)$, with the nucleotides ordered T, C, A, and G

$$P(0.1) = \begin{bmatrix} 0.906563 & 0.045855 & 0.023791 & 0.023791 \\ 0.045855 & 0.906563 & 0.023791 & 0.023791 \\ 0.023791 & 0.023791 & 0.906563 & 0.045855 \\ 0.023791 & 0.023791 & 0.045855 & 0.906563 \end{bmatrix},$$

$$P(0.2) = \begin{bmatrix} 0.825092 & 0.084274 & 0.045317 & 0.045317 \\ 0.084274 & 0.825092 & 0.045317 & 0.045317 \\ 0.045317 & 0.045317 & 0.825092 & 0.084274 \\ 0.045317 & 0.045317 & 0.084274 & 0.825092 \end{bmatrix}.$$

Consider node 7, which has daughter nodes 1 and 2. Using equation (4.4), we obtain the first entry in the probability vector as $L_7(T) = p_{TT}(0.2) \times p_{TC}(0.2) = 0.825092 \times 0.084274 = 0.069533$. This is the probability of observing T and C at tips 1 and 2, given that node 7 has T. The other entries, $L_7(C)$, $L_7(A)$, and $L_7(G)$, can be calculated similarly, as can the vector for node 8. Next the vector at node 6 can be calculated, by using the conditional probability vectors at daughter nodes 7 and 3. Finally we calculate the vector for node 0, the root. The first entry, $L_0(T) = 0.000112$, is the probability of observing the descendant tips (1, 2, 3, 4, 5) of node 0, given that node 0 has $x_0 = $ T. Equation (4.4) gives this as the product of two terms. The first

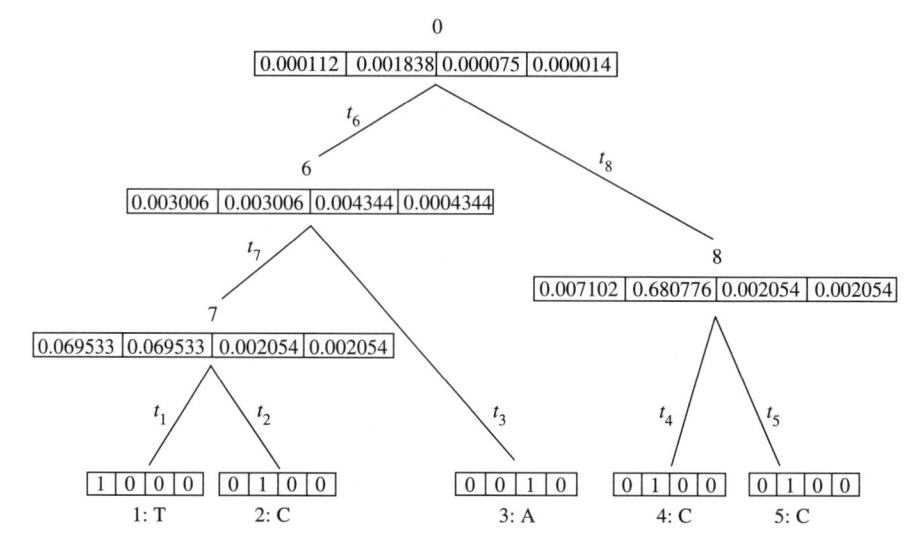

Fig. 4.2 Demonstration of the pruning algorithm for likelihood calculation when the branch lengths and other model parameters are fixed. The tree of Fig. 4.1 is reproduced, showing the vector of conditional probabilities at each node. The four elements in the vector at each node are the probabilities of observing data at the descendant tips, given that the node has T, C, A, or G, respectively. For example, 0.007102 for node 8 is the probability of observing data $x_4 x_5 = \text{CC}$ at tips 4 and 5, given that node 8 has T. The K80 model is assumed, with $\kappa = 2$. The branch lengths are fixed at 0.1 for the internal branches and 0.2 for the external branches. The transition probability matrices are shown in the text.

term, $\sum_{x_6} p_{x_0 x_6}(t_6) L_6(x_6)$, sums over x_6 and is the probability of observing data TCA at the tips 1, 2, 3, given that node 0 has T. This is $0.906563 \times 0.003006 + 0.045855 \times 0.003006 + 0.023791 \times 0.004344 + 0.023791 \times 0.004344 = 0.003070$. The second term, $\sum_{x_8} p_{x_0 x_8}(t_8) L_8(x_8)$, is the probability of observing data CC at tips 4 and 5, given that node 0 has T. This is $0.906563 \times 0.007102 + 0.045855 \times 0.680776 + 0.023791 \times 0.002054 + 0.023791 \times 0.002054 = 0.037753$. The product of the two terms gives $L_0(\text{T}) = 0.00011237$. Other entries in the vector for node 0 can be similarly calculated. □

To summarize, we calculate the probabilities of data $x_1 x_2$ down node 7, then the probabilities of data $x_1 x_2 x_3$ down node 6, then the probabilities of data $x_4 x_5$ down node 8, and finally the probabilities of the whole data $x_1 x_2 x_3 x_4 x_5$ down node 0. The calculation proceeds from the tips of the tree towards the root, visiting each node only after all its descendant nodes have been visited. In computer science, this way of visiting nodes on the tree is known as the *post-order tree traversal* (as opposed to *pre-order tree traversal*, in which ancestors are visited before descendants). After visiting all nodes on the tree and calculating the probability vector for the root, the probability of data at the site is given as

$$f(\mathbf{x}_h|\theta) = \sum_{x_0} \pi_{x_0} L_0(x_0). \tag{4.5}$$

Note that π_{x_0} is the (prior) probability that the nucleotide at the root is x_0, given by the equilibrium frequency of the nucleotide under the model. In our example, this gives $f(\mathbf{x}_h|\theta) = 0.000509843$, with $\log\{f(\mathbf{x}_h|\theta)\} = -7.581408$.

4.2.2.2 Savings on computation

The pruning algorithm is a major time-saver. Like the dynamic programming algorithm discussed in Chapter 3, the amount of computation required by the pruning algorithm for one calculation of the likelihood increases linearly with the number of nodes or the number of species, even though the number of combinations of ancestral states increases exponentially.

Some other obvious savings may be mentioned here as well. First, the same transition-probability matrix is used for all sites or site patterns in the sequence and may be calculated only once for each branch. Second, if two sites have the same data, the probabilities of observing them will be the same and need be calculated only once. Collapsing sites into *site patterns* thus leads to a saving in computation, especially if the sequences are highly similar so that many sites have identical patterns. Under JC69, some sites with different data, such as TCAG and TGCA for four species, also have the same probability of occurrence and can be collapsed further (Saitou and Nei 1986). The same applies to K80, although the saving is not as much as under JC69. It is also possible to collapse partial site patterns corresponding to subtrees (e.g. Kosakovsky Pond and Muse 2004). For example, consider the tree of Fig. 4.1 and two sites with data TCACC and TCACT. The conditional probability vectors for interior nodes 6 and 7 are the same at the two sites (because the data for species 1, 2, and 3 are the same) and can be calculated only once. However, such collapsing of partial site patterns depends on the tree topology and involves an overhead for bookkeeping. Reports vary as to its effectiveness.

4.2.2.3 Hadamard conjugation

It is fitting to mention here an alternative method, called *Hadamard conjugation*, for calculating the site-pattern probabilities and thus the likelihood. The Hadamard matrix is a square matrix consisting of -1 and 1 only. With -1 and 1 representing grey and dark grey, the matrix is useful for designing pavements. Indeed it was invented by the English mathematician James Sylvester (1814–1897) under the name 'anallagmatic pavement' and later studied by the French mathematician Jacques Hadamard (1865–1963). It was introduced to molecular phylogenetics by Hendy and Penny (1989), who used it to transform branch lengths on an unrooted tree to the site pattern probabilities, and vice versa. The transformation or conjugation works for binary characters or under Kimura's (1981) '3ST' model of nucleotide substitution, which assumes three substitution types: one rate for transitions and two rates for transversions. It is

4.2.3 Time reversibility, the root of the tree and the molecular clock

As discussed in Chapter 1, most substitution models used in molecular phylogenetics describe time-reversible Markov chains. For such chains, the transition probabilities satisfy $\pi_i p_{ij}(t) = \pi_j p_{ji}(t)$ for any i, j, and t. Reversibility means that the chain will look identical probabilistically whether we view the chain with time running forward or backward. An important consequence of reversibility is that the root can be moved arbitrarily along the tree without affecting the likelihood. This was called the *pulley principle* by Felsenstein (1981). For example, substituting $\pi_{x_6} p_{x_6 x_0}(t_6)$ for $\pi_{x_0} p_{x_0 x_6}(t_6)$ in equation (4.2), and noting $\sum_{x_0} p_{x_6 x_0}(t_6) p_{x_0 x_8}(t_8) = p_{x_6 x_8}(t_6 + t_8)$, we have

$$f(\mathbf{x}_h | \theta) = \sum_{x_6} \sum_{x_7} \sum_{x_8} \pi_{x_6} p_{x_6 x_7}(t_7) p_{x_6 x_8}(t_6 + t_8) p_{x_7 T}(t_1) p_{x_7 C}(t_2)$$

$$\times p_{x_6 A}(t_3) p_{x_8 C}(t_4) p_{x_8 C}(t_5). \tag{4.6}$$

This is the probability of the data if the root is at node 6, and the two branches 0–6 and 0–8 are merged into one branch 6–8, of length $t_6 + t_8$. The resulting tree is shown in Fig. 4.3.

Equation (4.6) also highlights the fact that the model is over-parametrized in Fig. 4.1, with one branch length too many. The likelihood is the same for any combinations of t_6 and t_8 as long as $t_6 + t_8$ is the same. The data do not contain information to estimate t_6 and t_8 separately and only their sum is estimable. Thus another consequence of time reversibility is that only unrooted trees can be identified if the molecular clock (rate constancy over time) is not assumed and every branch is allowed to have its own rate.

Under the clock, however, the root of the tree can indeed be identified. With a single rate throughout the tree, every tip is equidistant from the root, and the natural

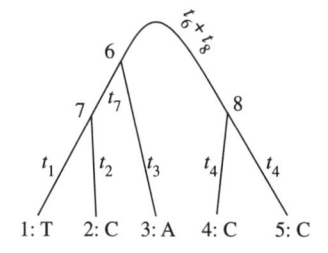

Fig. 4.3 The ensuing unrooted tree when the root is moved from node 0 to node 6 in the tree of Fig. 4.1.

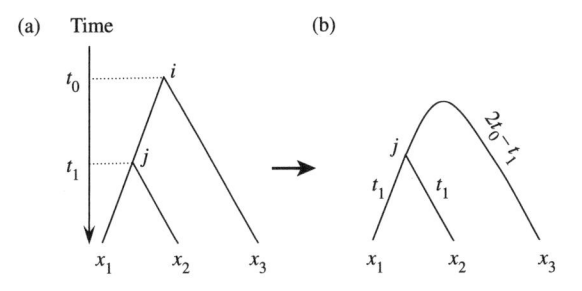

Fig. 4.4 (a) On a rooted tree for three species, the likelihood calculation has to sum over ancestral states i and j at two ancestral nodes. Under the clock, the model involves two parameters t_0 and t_1, measured by the expected number of substitutions per site from the ancestral node to the present time. (b) The same calculation can be achieved by moving the root to the ancestor of species 1 and 2, and summing over the state j at the new root; the tree then becomes a star tree with three branches of lengths t_1, t_1, and $2t_0 - t_1$.

parameters are the ages of the ancestral nodes, measured by the expected number of substitutions per site. A binary tree of s species has $s - 1$ internal nodes and thus $s - 1$ branch length parameters under the clock model. An example for three species is shown in Fig. 4.4(a).

Even under the clock, the pulley principle may be used to simplify the likelihood calculation in theoretical studies of small trees. For example, likelihood calculation on the tree of Fig. 4.4(a) involves summing over the ancestral states i and j at the two ancestral nodes. However, it is simpler to move the root to the common ancestor of species 1 and 2, so that one has to sum over ancestral states at only one node; the probability of data $x_1 x_2 x_3$ at a site then becomes

$$f(x_1 x_2 x_3 | \theta) = \sum_i \sum_j \pi_i p_{ij}(t_0 - t_1) p_{jx_1}(t_1) p_{jx_2}(t_1) p_{ix_3}(t_0)$$

$$= \sum_j \pi_j p_{jx_1}(t_1) p_{jx_2}(t_1) p_{jx_3}(2t_0 - t_1). \quad (4.7)$$

4.2.4 Missing data and alignment gaps

It is straightforward to deal with missing data in a likelihood calculation. For example, suppose the nucleotides observed at a site for three species (Fig. 4.4) are YTR, where Y stands for either T or C and R for A or G. The probability of observing such ambiguous data is a sum over all site patterns that are compatible with the observed data; that is, $\Pr(\text{YTR}) = \Pr(\text{TTA}) + \Pr(\text{TTG}) + \Pr(\text{CTA}) + \Pr(\text{CTG})$. One could calculate and sum over the probabilities of these four distinct site patterns, but this summation can be achieved efficiently in the pruning algorithm by setting up the conditional probability vectors for the tips as follows: $(1, 1, 0, 0)$ for tip 1 and $(0, 0, 1, 1)$ for tip 3. In other words, $L_i(x_i)$ for a tip node i is set to 1 for any nucleotide

x_i that is compatible with the observed state. This idea is due to Felsenstein (2004), who also pointed out that random sequencing errors in the data can be dealt with in a similar way.

Alignment gaps pose a difficulty for all current likelihood programs. One option is to treat an alignment gap as the fifth nucleotide or the 21st amino acid, different from all other character states. This idea is used in some parsimony algorithms, but is not common in likelihood implementations (but see McGuire *et al.* 2001). Two worse options are commonly used: (i) to delete all sites at which there are alignment gaps in any sequence and (ii) to treat them as if they were undetermined nucleotides (ambiguous characters). The information loss caused by deleting sites with alignment gaps can be substantial. For example, noncoding regions often contain many indels (insertions and deletions), which contain phylogenetic information and should ideally be used. The approach of treating alignment gaps as ambiguous characters is problematic as well since gaps mean that the nucleotides do not exist and not that they exist but are unknown. Furthermore, a stretch of five gaps may well represent one evolutionary event (either an insertion or deletion of five nucleotides), but is treated as five independent events. It is less clear what effects those two approaches have on phylogenetic tree reconstruction, but both clearly tend to underestimate the amount of sequence divergence. It seems reasonable to delete a site if it contains alignment gaps in most species and to keep the site if it contains alignment gaps in very few species. In analysis of highly divergent species, it is common practice to remove regions of the protein for which the alignment is unreliable.

In theory, it is advantageous to implement models that incorporate insertions and deletions as well as substitutions. Kishino *et al.* (1990) implemented such a model, useful for aligned sequences with gaps. There is also much interest in constructing sequence alignment and inferring phylogenies at the same time. An advantage of such a statistical approach over heuristic procedures of sequence alignment such as CLUSTAL (Thompson *et al.* 1994) is that the method can estimate the insertion and deletion rates from the data, rather than requiring the user to specify *gap-opening* and *gap-extension* penalties. Early models, developed by Bishop and Thompson (1986) and Thorne *et al.* (1991, 1992), are somewhat simplistic and also involve intensive computations even on only two sequences. However, improvements are being made, both to the biological realism of the model and to the computational efficiency. This is currently an area of active research (see, e.g., Hein *et al.* 2000, 2003; Metzler 2003; Lunter *et al.* 2003, 2005; Holmes 2005; Fleissner *et al.* 2005; Redelings and Suchard 2005).

4.2.5 An numerical example: phylogeny of apes

We use the sequences from the 12 proteins encoded by the heavy strand of the mitochondrial genome from seven ape species. The data are a subset of the mammalian sequences analysed by Cao *et al.* (1998). The 12 proteins are concatenated into one long sequence and analysed as one data set as they appear to have similar substitution patterns. The other protein in the genome, ND6, is not included as it is encoded by

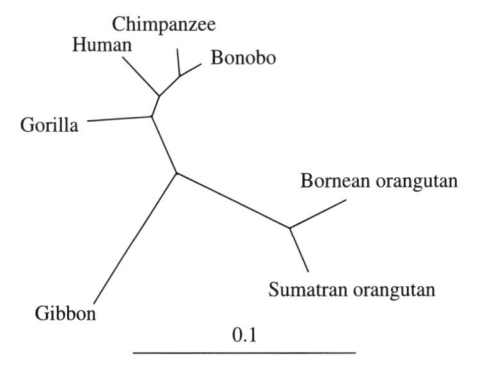

[Phylogenetic tree figure showing: Chimpanzee, Human, Bonobo, Gorilla, Bornean orangutan, Sumatran orangutan, Gibbon, with scale bar 0.1]

Fig. 4.5 The ML tree for seven ape species estimated from the 12 mitochondrial proteins. Branches are drawn in proportion to their lengths, measured by the number of amino acid substitutions per site. The MTMAM model is assumed.

the opposite strand of the DNA with quite different base compositions. The species and the GenBank accession numbers for the sequences are human (*Homo sapiens*, D38112), common chimpanzee (*Pan troglodytes*, D38113), bonobo chimpanzee (*Pan paniscus*, D38116), gorilla (*Gorilla gorilla*, D38114), Bornean orangutan (*Pongo pygmaeus pygmaeus*, D38115), Sumatran orangutan (*Pongo pygmaeus abelii*, X97707), and gibbon (*Hylobates lar*, X99256). Alignment gaps are removed, with 3331 amino acids in the sequence.

There are 945 binary unrooted trees for seven species, so we evaluate them exhaustively. We assume the empirical MTMAM model for mammalian mitochondrial proteins (Yang *et al.* 1998). The ML tree is shown in Fig. 4.5, which has the log likelihood score $\ell = -14\,558.59$. The worst binary tree has the score $-15\,769.00$, while the star tree has the score $-15\,777.60$. Figure 4.6 shows that the same tree (of Fig. 4.5) has the highest log likelihood, the shortest tree length by parsimony and also the shortest likelihood tree length (the sum of MLEs of branch lengths). Thus maximum likelihood, maximum parsimony, and minimum evolution selected the same best tree for this data set. (Note that minimum evolution normally estimates branch lengths by applying least squares to a distance-matrix method while here I used ML to estimate the branch lengths for convenience.) Similarly, use of the Poisson model in the likelihood analysis, assuming the same rate between any two amino acids, gives the same tree as the ML tree, with $\ell = -16\,566.60$.

4.3 Likelihood calculation under more complex models

The description above assumes that all sites in the sequence evolve according to the same substitution-rate matrix. This may be a very unrealistic assumption for real sequences. Much progress has been made in the last two decades in extending models used in likelihood analysis. A few important extensions are discussed in this section.

110 • 4 Maximum likelihood methods

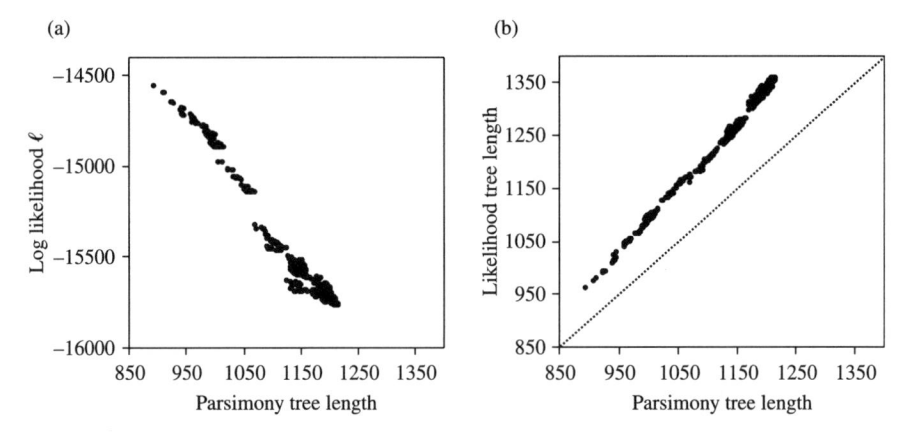

Fig. 4.6 Different criteria for tree selection calculated for all 945 binary unrooted trees for the mitochondrial protein data. (a) The log likelihood score ℓ is plotted against the parsimony tree length. (b) The likelihood tree length is plotted against the parsimony tree length. The likelihood tree length is measured by the estimated number of amino acid substitutions on the tree, calculated as the sum of estimated branch lengths multiplied by the number of sites. The parsimony tree length is the minimum number of changes and is thus an under-count. All three criteria select the same tree (the tree of Fig. 4.5) as the best estimate.

4.3.1 Models of variable rates among sites

In real sequences, the substitution rates are often variable across sites. Ignoring rate variation among sites can have a major impact on phylogenetic analysis (e.g. Tateno *et al.* 1994; Huelsenbeck 1995a; Yang 1996c; Sullivan and Swofford 2001). To accommodate variable rates in a likelihood model, one should not in general use a rate parameter for every site, as otherwise there will be too many parameters to estimate and the likelihood method may misbehave. A sensible approach is to use a statistical distribution to model the rate variation. Both discrete- and continuous-rate distributions have been used.

4.3.1.1 Discrete-rates model

In this model, sites are assumed to fall into several (say, K) classes with different rates (Table 4.1). The rate at any site in the sequence takes a value r_k with probability p_k, with $k = 1, 2, \ldots, K$. The rs and ps are parameters to be estimated by ML from the data. There are two constraints. First the frequencies sum to 1: $\Sigma p_k = 1$. Second, the average rate is fixed at $\Sigma p_k r_k = 1$, so that the branch length is measured as the expected number of nucleotide substitutions per site averaged over the sequence. The latter constraint is necessary to avoid the use of too many parameters. The rates rs are thus relative multiplication factors. The model with K rate classes thus involves $2(K - 1)$ free parameters. Every site will in effect be evolving along the same tree

4.3 Likelihood calculation under more complex models

Table 4.1 The discrete-rates model

Site class	1	2	3	...	K
Probability	p_1	p_2	p_3	...	p_K
Rate	r_1	r_2	r_3	...	r_K

topology, with proportionally elongated or shrunken branches. In other words, the substitution-rate matrix at a site with rate r is rQ, with Q shared across all sites.

As we do not know to which rate class each site belongs, the probability of observing data at any site is a weighted average over the site classes

$$f(\mathbf{x}_h|\theta) = \sum_k p_k \times f(\mathbf{x}_h|r = r_k; \theta). \tag{4.8}$$

The likelihood is again calculated by multiplying the probabilities across sites. The conditional probability, $f(\mathbf{x}_h|r;\theta)$, of observing data $\mathbf{x}_h$ given the rate r, is just the probability under the one-rate model, with all branch lengths multiplied by r. It can be calculated using the pruning algorithm for each site class. A variable-rates model with K rate classes thus takes K times as much computation as the one-rate model.

As an example, consider the tree of Fig. 4.4(b). We have

$$f(x_1 x_2 x_3 | r; \theta) = \sum_j \pi_j p_{jx_1}(t_1 r) p_{jx_2}(t_1 r) p_{jx_3}((2t_0 - t_1)r) \tag{4.9}$$

(compare with equation 4.7).

Discrete-rates models are known as *finite-mixture models*, since the sites are a mixture from K classes. As in any mixture model, one can only fit a few site classes to practical data sets, so K should not exceed 3 or 4. The general model of Table 4.1 is implemented by Yang (1995a). A special case is the *invariable-sites* model, which assumes two site classes: the class of *invariable sites* with rate $r_0 = 0$ and another class with a constant rate r_1 (Hasegawa *et al.* 1985). As the average rate is 1, we have $r_1 = 1/(1 - p_0)$, where the proportion of invariable sites p_0 is the only parameter in the model. Note that a variable site in the alignment cannot have rate $r_0 = 0$. Thus the probability of data at a site is

$$f(\mathbf{x}_h|\theta) = \begin{cases} p_0 + p_1 \times f(\mathbf{x}_h|r = r_1; \theta), & \text{if the site is constant,} \\ p_1 \times f(\mathbf{x}_h|r = r_1; \theta), & \text{if the site is variable.} \end{cases} \tag{4.10}$$

4.3.1.2 Gamma-rates model

A second approach is to use a continuous distribution to approximate variable rates at sites. The most commonly used distribution is the gamma (see Fig. 1.6). See Section 1.3 for a discussion of the use of the gamma model in calculating pairwise distances. The gamma density is

$$g(r; \alpha, \beta) = \frac{\beta^\alpha r^{\alpha-1} e^{-\beta r}}{\Gamma(\alpha)}, \tag{4.11}$$

with mean α/β and variance α/β^2. We let $\beta = \alpha$ so that the mean is 1. The shape parameter α is inversely related to the extent of rate variation among sites. As in the discrete-rates model, we do not know the rate at the site, and have to average over the rate distribution

$$f(\mathbf{x}_h|\theta) = \int_0^\infty g(r) f(\mathbf{x}_h|r; \theta) dr. \tag{4.12}$$

Here the collection of parameters θ includes α as well as branch lengths and other parameters in the substitution model (such as κ). Note that the sum in equation (4.8) for the discrete model is now replaced by an integral for the continuous model.

An algorithm for calculating the likelihood function under a continuous-rates model, such as equation (4.12), was described by Yang (1993), Gu et al. (1995), and Kelly and Rice (1996). Note that the conditional probability $f(\mathbf{x}_h|r)$ is a sum over all combinations of ancestral states (equation 4.2). Each term in the sum is a product of transition probabilities across the branches. Each transition probability has the form $p_{ij}(tr) = \sum_k c_{ijk} e^{\lambda_k tr}$, where t is the branch length, λ_k is the eigenvalue, and c_{ijk} is a function of the eigenvectors (see equation 1.39). Thus after expanding all the products, $f(\mathbf{x}_h|r)$ is a sum of many terms of the form ae^{br}, and then the integral over r can be obtained analytically (Yang 1993). However, this algorithm is very slow because of the huge number of terms in the sum, and is practical for small trees with fewer than 10 sequences only.

4.3.1.3 Discrete-gamma model

One may use the discrete-rates model as an approximation to the continuous gamma, known as the *discrete-gamma model*. Yang (1994a) tested this strategy, using K equal-probability rate classes, with the mean or median for each class used to represent all rates in that class (Fig. 4.7). Thus $p_k = 1/K$ while r_k is calculated as a function of the gamma shape parameter α. The probability of data at a site is then given by equation (4.8). The model involves one single parameter α, just like the continuous gamma. The discrete gamma may be considered a crude way of calculating the integral of equation (4.12). Yang's (1994a) test on small data sets suggested that as few as $K = 4$ site classes provided good approximation. In large data sets with hundreds of sequences, more categories may be beneficial (Mayrose et al. 2005). Because of the discretization, the discrete gamma means less rate variation among sites than the continuous gamma for the same parameter α. Thus, estimates of α under the discrete gamma are almost always smaller than estimates under the continuous gamma.

Felsenstein (2001a) and Mayrose et al. (2005) discussed the use of numerical integration (quadrature) algorithms to calculate the probability of equation (4.12).

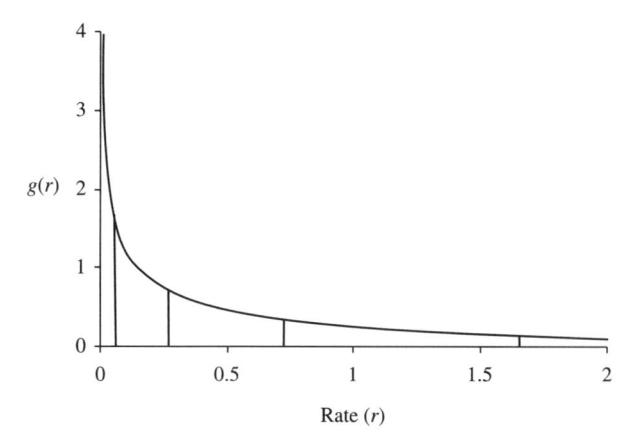

Fig. 4.7 The discrete-gamma model of variable rates across sites uses K equal-probability categories to approximate the continuous gamma distribution, with the mean rate in each category used to represent all rates in that category. Shown here is the probability density function $g(r)$ for the gamma distribution with shape parameter $\alpha = 0.5$ (and mean 1). The four vertical lines are at $r = 0.06418, 0.27500, 0.70833$, and 1.64237. These are the 20th, 40th, 60th, and 80th percentiles of the distribution and cut the density into five categories, each of proportion 1/5. The mean rates in the five categories are $0.02121, 0.15549, 0.49708, 1.10712$, and 3.24910.

These calculate the values of the integrand for a fixed set of values of r and then approximate the integrand by using simpler functions such as the polynomial. A concern is that the integrand in equation (4.12) may peak at 0 for constant sites and at large values for highly variable sites, so that an algorithm using fixed points is unlikely to approximate the integral well for all sites. Adaptive algorithms are more reliable as they sample more densely in regions where the integrand is large. However, they are too expensive as different points are used for different sites, which effectively means that the transition-probability matrix has to be calculated for every site along every branch for every rate category. Numerical integration by quadrature, if accurate and stable, may be very useful in other contexts (for example, in models of codon substitution with variable ω ratios among sites), and more work is needed to evaluate its reliability.

4.3.1.4 Other rate distributions

Besides the gamma model, several other statistical distributions have been implemented and tested. Waddell *et al.* (1997) tested the log-normal model. Kelly and Rice (1996) took a nonparametric approach, assuming a general continuous distribution without specifying any particular distributional form. With such a general model, not much inference is possible. Gu *et al.* (1995) added a proportion of invariable sites to the gamma distribution, so that a proportion p_0 of sites are invariable while the other sites (with proportion $p_1 = 1 - p_0$) have rates drawn from the gamma. The model is known as 'I + Γ' and has been widely used. This model is somewhat pathological as the gamma distribution with $\alpha \leq 1$ already allows for sites with very low rates; as a result, adding a proportion of invariable sites creates a strong correlation between p_0 and α, making it impossible to estimate both parameters reliably (Yang 1993; Sullivan et al. 1999; Mayrose et al. 2005). Another drawback of the model is that the estimate of p_0 is very sensitive to the number and divergences of the sequences included in the data. The proportion p_0 is never larger than the observed proportion of constant sites; with the addition of more and divergent sequences, the proportion of constant sites drops, and the estimate of p_0 tends to go down as well. The I + Γ model may be useful if the true rate distribution has two peaks, one near 0 and another at an intermediate rate. Mayrose et al. (2005) suggested a gamma mixture model, which assumes that the rate for any site is a random variable from a mixture of two gamma distributions with different parameters. This appears more stable than the I + Γ model. For estimating branch lengths and phylogenies, the different distributions tend to produce similar results, so that the simple gamma appears adequate. For estimating rates at sites from large data sets with many sequences, the mixture models may be preferable.

Note that under both continuous- and discrete-rates models, data at different sites are independent and identically distributed (*i.i.d.*). While the model allows different sites to evolve at different rates, it does not specify *a priori* which sites should have which rates. Instead the rate for any site is a random draw from a common distribution. As a result, data at different sites have the same distribution.

4.3.1.5 Bayesian estimation of substitution rates at sites

Large data sets with many sequences may provide opportunities for multiple changes at individual sites, making it possible to estimate the relative substitution rate at each site. In both discrete- and variable-rates models, the rates for sites are random variables and are integrated out in the likelihood function. Thus to estimate the rate we use the conditional (posterior) distribution of the rate given the data at the site:

$$f(r|\mathbf{x}_h; \theta) = \frac{f(r|\theta) f(\mathbf{x}_h|r; \theta)}{f(\mathbf{x}_h|\theta)}. \tag{4.13}$$

Parameters θ may be replaced by their estimates, such as the maximum likelihood estimates (MLEs). This is known as the empirical Bayes approach. Under the continuous-rates model, one can use the posterior mean as the estimate of the rate at the site (Yang and Wang 1995). Under the discrete-rates model, the rate r in equation (4.13) takes one of K possible values, with $f(r|\theta) = p_k$. One may calculate the posterior mean or use the rate with the highest posterior probability as the best estimate (Yang 1994a).

The empirical Bayes approach ignores sampling errors in the parameter estimates, which may be a source of concern in small data sets. One can assign a prior on the parameters and use a full Bayesian approach to deal with uncertainties in the parameters. Mateiu and Rannala (2006) developed an interesting Markov-chain Monte Carlo (MCMC) algorithm for estimating rates at sites, using a uniformized Markov chain

(Kao 1997, pp. 273–277) to implement the continuous gamma model. This strategy allows large data sets with hundreds of sequences to be analysed under the continuous model. The authors' simulations show that the discrete gamma model provides good estimates of branch lengths, but tends to underestimate high rates unless a very large number of rate classes are used (with $K \geq 40$, say).

Another likelihood approach to estimating rates at sites was implemented by Nielsen (1997), who treated the rate at every site as a parameter, estimated by ML. This method suffers from estimating too many parameters, so that the estimates are often zero or infinity.

4.3.1.6 Correlated rates at adjacent sites

Yang (1995a) and Felsenstein and Churchill (1996) implemented models that allow rates to be correlated across adjacent sites. In the *auto-discrete-gamma* model (Yang 1995a), the rates at two adjacent sites have a bivariate gamma distribution (that is, the marginal distributions of both rates are gamma), discretized to make the computation feasible. This is an extension to the discrete-gamma model and includes a parameter ρ, which measures the strength of autocorrelation. The model is implemented through a hidden Markov chain, which describes the transition from one rate class to another along the sequence. Felsenstein and Churchill (1996) described a similar hidden Markov-chain model, in which a segment of nucleotide sites are assumed to have the same rate, with the length of the segment reflecting the strength of the correlation. Tests on real data suggest that substitution rates are indeed highly correlated. Nevertheless, estimation of branch lengths or other parameters in the model did not seem to be seriously affected by ignoring the correlation of rates at adjacent sites.

4.3.1.7 Covarion models

In both the discrete- and continuous-rates models, the rate for a site is applied to all branches in the tree, so that a fast-evolving site is fast-evolving throughout the phylogeny. This assumption is relaxed in the *covarion* (for COncomitantly VARIable codON) models, which are based on Fitch's (1971a) idea of coevolving codons in a protein-coding gene, with substitutions in one codon affecting substitutions in other codons. Such models allow a site to switch from one rate class to another. As a result, a site may be fast-evolving along some lineages while slowly evolving along others. Tuffley and Steel (1998) discussed an extension to the invariable-sites model, in which a site switches between two states: the 'on' state ($+$), in which the nucleotide has a constant rate of evolution and the 'off' state ($-$), in which the nucleotide is invariable. A likelihood implementation of this model is provided by Huelsenbeck (2002). Similarly Galtier (2001) implemented an extension to the discrete-gamma model, in which a site switches from one rate class to another over evolutionary time, with K rate classes from the discrete gamma.

Here we use the simpler model of Tuffley and Steel (1998) and Huelsenbeck (2002) to describe the implementation of such models. A simple approach is to construct a Markov chain with an expanded state space. Instead of the four nucleotides, we consider eight states: $T_+, C_+, A_+, G_+, T_-, C_-, A_-$, and G_-, with '+' and '−' representing the 'on' and 'off' states, respectively. Nucleotides of the '+' state can change between themselves, according to a substitution model such as JC69 or HKY85. A nucleotide of the '−' state can change to the same nucleotide of the '+' state only. In the likelihood calculation, one has to sum over all the eight states for each ancestral node to calculate the probability of data at each site. Furthermore, a nucleotide observed at a tip node can be in either the '+' or the '−' state, so that the probability of data at a site will be a sum over all compatible patterns of the expanded states. Suppose the observed data at a site for three species are TCA. Then Pr(TCA) will be a sum over eight site patterns: $T_+C_+A_+, T_+C_+A_-, T_+C_-A_+, \ldots$, and $T_-C_-A_-$. This summation can be achieved efficiently using likelihood treatment of missing data in the pruning algorithm, as discussed in Subsection 4.2.4; we let the conditional probability $L_i(x_i)$ for the tip node i be 1 for both states $x_i = T_+$ and $x_i = T_-$ if the observed nucleotide for tip i is T, say. The model of Galtier (2001) can be implemented in the same way, although the Markov chain has $4K$ states instead of eight states.

An interesting use of the same idea is made by Guindon *et al.* (2004), who implemented a codon-based switching model, which allows a codon to switch between site classes with different ω ratios. The model thus allows the selective pressure on the protein indicated by the ω ratio to vary both among sites and among lineages, and is perhaps close to Fitch's (1971a) original idea of coevolving codons. Chapter 8 provides further discussions of codon-based models. It may be noted that all the covarion models discussed above assume that data at different sites are *i.i.d.*

4.3.1.8 Numerical example: phylogeny of apes

We use the MTMAM + Γ_5 model to analyse the data set of mitochondrial protein sequences, with a discrete-gamma model used to accommodate variable rates among sites, with $K = 5$ rate classes used. The shape parameter is fixed at $\alpha = 0.4$, which is the estimate from mammalian species (Yang *et al.* 1998). The ML tree is the same as under the MTMAM model (Fig. 4.5). The log likelihood value for the tree is $\ell = -14413.38$, much higher than that under the one-rate model. If α is estimated from the data, the same tree is found to be the ML tree, with $\hat{\alpha} = 0.333$ and $\ell = -14,411.90$. □

4.3.2 Models for combined analysis of multiple data sets

How we model rate variation among sites should depend on whether we know *a priori* which sites are likely to be evolving fast and which sites evolving slowly. If we lack such information, it is natural to assume a random statistical distribution of rates, as in the last subsection. However, if such information is available, it should be advantageous to use it. For example, different genes may be evolving at different rates, and one would like to accommodate such rate heterogeneity in a combined analysis of multiple data sets from different gene loci. A similar situation is analysis of the three codon positions in a protein-coding gene, which evolve at very different rates. Thus we use 'multiple data sets' to refer to any such *a priori* partitions of sites in the alignment, which are expected to have different evolutionary dynamics. For example, one may assign different rate parameters r_1, r_2, and r_3 for the three codon positions. The first rate or the average rate may be fixed at 1, so that the branch length is measured by the number of substitutions per site in the first codon position or averaged over all three positions. Even though such heterogeneous nucleotide-substitution models cannot capture the complexity of codon substitution, they appear effective in practical data analysis, without the cost of codon-substitution models (Ren et al. 2005; Shapiro et al. 2006). A few such models were implemented by Yang (1995a). Besides the rates, parameters reflecting other features of the evolutionary process, such as the transition/transversion rate ratio or base compositions, can be allowed to differ among genes or site partitions as well (Yang 1995a, 1996b).

Such models may be called *fixed-rates models*, as opposed to the discrete- and continuous-rates models discussed above, which may be called *random-rates models*. They correspond to the *fixed-effects* and *random-effects models* in statistics. Suppose there are K site partitions (e.g. genes or codon positions) with rates $r_1, r_2, \ldots, r_K$. Let $I(h)$ label the site partition that site h belongs to. The log likelihood is then

$$\ell(\theta, r_1, r_2, \ldots, r_K; X) = \sum_h \log\{f(\mathbf{x}_h | r_{I(h)}; \theta)\}. \quad (4.14)$$

Here the rates for site partitions are parameters in the model while θ include branch lengths and other substitution parameters. The probability of data at a site is calculated by using the correct rate parameter $r_{I(h)}$ for the site, in contrast to the random-rates model, in which the probability for a site is an average over the site classes (equations 4.8 and 4.12). Likelihood calculation under the fixed-rates model therefore takes about the same amount of computation as under the one-rate model.

The fixed-rates and random-rates models have several differences. In the former, one knows which partition or site class each site is from, and the rates for site partitions are parameters, estimated by ML. In the latter, one does not know which rate each site has; instead one treats the rate as a random draw from a statistical distribution and estimates parameters of that distribution (such as α for the gamma model) as a measure of the variability in the rates. In a random-rates model, data at different sites are *i.i.d.* In a fixed-rates model, data at different sites within the same partition are *i.i.d.*, but sites from different partitions have different distributions. This distinction should be taken into account in statistical tests based on resampling sites.

Models that include both fixed and random effects are called *mixed-effects models*, and these are implemented for rate variation among sites as well. For example, one can use fixed rate parameters to accommodate large-scale variations among site partitions (codon positions or genes), and use the discrete gamma to account for the remaining rate variation within each partition (Yang 1995a, 1996b). This general framework is useful for analysing sequence data sets from multiple heterogeneous loci, to assemble information concerning common features of the evolutionary process among genes while accommodating their heterogeneity. It allows estimation of gene-specific rates and parameters, and hypothesis testing concerning similarities and differences among genes. Both likelihood (Yang 1996b; Pupko et al. 2002b) and Bayesian

(Suchard et al. 2003; Nylander et al. 2004; Pagel and Meade 2004) approaches can be taken. A similar use of this strategy is in estimation of species divergence times under local-clock models, in which the divergence times are shared across loci while the multiple loci may have different evolutionary characteristics (Kishino et al. 2001; Yang and Yoder 2003).

In the literature there has been a debate between the *combined analysis* and *separate analysis*. The former, also called the *total evidence* approach, concatenates sequences from multiple loci and then treats the resulting sequence as one 'super-gene', ignoring possible differences among the genes. The latter analyses different genes separately. The separate analysis can reveal differences among genes but does not provide a natural way of assembling information from multiple heterogeneous data sets. It often over-fits the data as each locus is allowed to have its own set of parameters. This debate has more recently transformed into a debate between the *supermatrix* and *supertree* approaches, especially when sequences are missing for some species at some loci (see, e.g., Bininda-Emonds 2004). The supermatrix approach concatenates sequences, patching up missing sequences with question marks, and is equivalent to the combined analysis. It fails to accommodate possible differences among loci. The supertree approach reconstructs phylogenies using data from different genes, and then uses heuristic algorithms to assemble the subtrees into a supertree for all species. The approach does not properly accommodate the uncertainties in the individual subtrees and makes *ad hoc* decisions for dealing with conflicts among them. Under some situations both supermatrix and supertree approaches may be useful. However, from a statistical viewpoint, both are inferior to a likelihood-based approach to combined analysis, which accommodates the heterogeneity among multiple data sets.

4.3.3 Nonhomogeneous and nonstationary models

In analysis of distant sequences, one often observes considerable variation in nucleotide or amino acid compositions among sequences. The assumption of a homogeneous and stationary Markov-chain model is clearly violated. One can test whether base compositions are homogeneous by using a contingency table of nucleotide or amino acid counts in the sequences to construct a X^2 statistic (e.g. Tavaré 1986). However, such a formal test is hardly necessary because typical molecular data sets are large and such a test can reject the null hypothesis with ease. Empirical studies (e.g. Lockhart et al. 1994) suggest that unequal base compositions can mislead tree-reconstruction methods, causing them to group sequences according to the base compositions rather than genetic relatedness.

Dealing with the drift of base compositions over time in a likelihood model is difficult. Yang and Roberts (1995) implemented a few nonhomogeneous models, in which every branch in the tree is assigned a separate set of base frequency parameters ($\pi_T, \pi_C, \pi_A, \pi_G$) in the HKY85 model. Thus the sequences drift to different base compositions during the evolutionary process, after they diverged from the root sequence. This model involves many parameters and is useable for small trees with a few species only. Previously, Barry and Hartigan (1987b) described a model with even more parameters, in which a general transition-probability matrix P with 12 parameters is estimated for every branch. The model does not appear to have ever been used in data analysis. Galtier and Gouy (1998) implemented a simpler version of the Yang and Roberts model. Instead of the HKY85 model, they used the model of Tamura (1992), which assumes that G and C have the same frequency and A and T have the same frequency, so that only one frequency parameter for the GC content is needed in the substitution-rate matrix. Galtier and Gouy estimated different GC content parameters for branches on the tree. Because of the reduced number of parameters, this model was successfully used to analyse relatively large data sets (Galtier *et al.* 1999).

The problem of too many parameters may be avoided by constructing a prior on them, using a stochastic process to describe the drift of base compositions over branches or over time. The likelihood calculation then has to integrate over the trajectories of base frequencies. This integral is daunting, and it is unclear whether it can be calculated using Bayesian MCMC algorithms. More research is needed to test the feasibility of such an approach.

4.3.4 Amino acid and codon models

The discussions up to now assume models of nucleotide substitution applied to noncoding DNA sequences. The same theory, including the pruning algorithm, can be applied in a straightforward manner to analyse protein sequences under models of amino acid substitution (Bishop and Friday 1985, 1987; Kishino *et al.* 1990) or protein-coding DNA sequences under models of codon substitution (Goldman and Yang 1994; Muse and Gaut 1994). Discussions of such models are given in Chapter 2. A difference is that the substitution-rate and transition-probability matrices are of sizes 20×20 for amino acids or 61×61 for codons (as there are 61 sense codons in the universal genetic code), instead of 4×4 for nucleotides. Furthermore, the summation over ancestral states is now over all ancestral amino acids or ancestral codons. As a result, likelihood computation under amino acid and codon models is much more expensive than under nucleotide models. However, no new principles are involved in such extensions.

4.4 Reconstruction of ancestral states

4.4.1 Overview

Evolutionary biologists have had a long tradition of reconstructing traits in extinct ancestral species and using them to test interesting hypotheses. The MacClade program (Maddison and Maddison 2000) provides a convenient tool for ancestral reconstruction using different variants of the parsimony method. Maddison and Maddison (2000) also provided an excellent review of the many uses (and misuses) of ancestral reconstruction. The *comparative method* (e.g. Felsenstein 1985b; Harvey and Pagel 1991; Schluter 2000) uses reconstructed ancestral states to uncover associated changes between two characters. Although association does not necessarily mean a cause–effect relationship, establishing an evolutionary association is the first step in inferring the adaptive significance of a trait. For example, butterfly larvae may be palatable (P_+) or unpalatable (P_-) and they may be solitary (S_+) or gregarious (S_-). If we can establish a significant association between character states P_+ and S_+, and if in particular P_+ always appears before S_+ on the phylogeny based on ancestral state reconstructions, a plausible explanation is that palatability drives the evolution of solitary behaviour (Harvey and Pagel 1991). In such analysis, ancestral reconstruction is often the first step.

For molecular data, ancestral reconstruction has been used to estimate the relative rates of substitution between nucleotides or amino acids (e.g. Dayhoff *et al.* 1978; Gojobori *et al.* 1982), to count synonymous and nonsynonymous substitutions on the tree to infer adaptive protein evolution (e.g. Messier and Stewart 1997; Suzuki and Gojobori 1999), to infer changes in nucleotide or amino acid compositions (Duret *et al.* 2002), to detect coevolving nucleotides or amino acids (e.g. Shindyalov *et al.* 1994; Tuff and Darlu 2000), and to conduct many other analyses. Most of these procedures have been superseded by more rigorous likelihood analyses. A major application of ancestral sequence reconstruction is in the so-called chemical palaeogenetic restoration studies envisaged by Pauling and Zuckerkandl (1963; see also Zuckerkandl 1964). Ancestral proteins inferred by parsimony or likelihood were synthesized using site-directed mutagenesis and their chemical and physiological properties were examined (e.g. Malcolm *et al.* 1990; Stackhouse *et al.* 1990; Libertini and Di Donato 1994; Jermann *et al.* 1995; Thornton *et al.* 2003; Ugalde *et al.* 2004). Golding and Dean (1998), Chang and Donoghue (2000), Benner (2002) and Thornton (2004) should be consulted for reviews.

The parsimony method for reconstructing ancestral states was developed by Fitch (1971b) and Hartigan (1973). A more general algorithm for weighted parsimony was described in Subsection 3.4.3. While discussing the parsimony method for ancestral reconstruction, Fitch (1971b) and Maddison and Maddison (1982) emphasized the advantage of a probabilistic approach to ancestral reconstruction and the importance of quantifying the uncertainty in the reconstruction. Nevertheless, early studies (and some recent ones) failed to calculate the right probabilities. When a Markov-chain model is used to describe the evolution of characters, the ancestral states are random variables and cannot be estimated from the likelihood function, which averages over all possible ancestral states (see Section 4.2). To infer ancestral states, one should calculate the conditional (posterior) probabilities of ancestral states given the data. This is the empirical Bayes (EB) approach, implemented by Yang *et al.* (1995a) and Koshi and Goldstein (1996a). It is empirical as parameter estimates (such as MLEs) are used in the calculation of posterior probabilities of ancestors. Many statisticians consider EB to be a likelihood approach. To avoid confusion, I will refer to the approach as empirical Bayes (instead of likelihood).

Compared with parsimony reconstruction, the EB approach takes into account different branch lengths and different substitution rates between nucleotides or amino acids. It also provides posterior probabilities as a measure of the accuracy of the reconstruction. The empirical Bayes approach has the drawback of not accommodating sampling errors in parameter estimates, which may be problematic in small data sets which lack information to estimate parameters reliably. Huelsenbeck and Bollback (2001) and Pagel *et al.* (2004) implemented full (hierarchical) Bayesian approaches to ancestral state reconstruction, which assign priors on parameters and averages over their uncertainties through MCMC algorithms (see Chapter 5). Another approach, due to Nielsen (2001a), samples substitutions on branches of the tree. This is equivalent to reconstructing states for ancestral nodes, and its suitability depends on how the ancestral states are used. The approach of reconstructing substitutions on the tree is computationally expensive in species data due to the many substitutions that may have occurred at a site, but may be useful for analysis of population data in which the sequences are highly similar.

In this subsection, I will describe the empirical Bayes approach to ancestral sequence reconstruction, and discuss the modifications in the hierarchical Bayesian approach. I will also briefly discuss the reconstruction of ancestral states of a discrete morphological character, which can be achieved using the same theory, but the analysis is much more difficult due to lack of information for estimating model parameters.

4.4.2 Empirical and hierarchical Bayes reconstruction

A distinction can be made between the *marginal* and *joint reconstructions*. The former assigns a character state to a single node, while the latter assigns a set of character states to all ancestral nodes. Marginal reconstruction is more suitable when one wants the sequence at a particular node, as in the molecular restoration studies. Joint reconstruction is more suitable when one counts changes at each site.

Here we use the example of Fig. 4.2 to demonstrate the empirical Bayes approach to ancestral reconstruction. We pretend that the branch lengths and the transition/transversion rate ratio κ, used in calculating the conditional probabilities (i.e. the $L_i(x_i)$'s shown on the tree) are the true values. In real data analysis, they are replaced by the MLEs from the sequence data set.

4.4.2.1 Marginal reconstruction

We calculate the posterior probabilities of character states at one ancestral node. Consider node 0, the root. The posterior probability that node 0 has the nucleotide x_0, given the data at the site, is

$$f(x_0|\mathbf{x}_h;\theta) = \frac{f(x_0|\theta)f(\mathbf{x}_h|x_0;\theta)}{f(\mathbf{x}_h|\theta)} = \frac{\pi_{x_0}L_0(x_0)}{\sum_{x_0}\pi_{x_0}L_0(x_0)}. \tag{4.15}$$

Note that $\pi_{x_0}L_0(x_0)$ is the joint probability of the states at the tips $\mathbf{x}_h$ and the state at the root x_0. This is calculated by summing over all other ancestral states except x_0 (see Subsection 4.2.2 for the definition of $L_0(x_0)$). Figure 4.2 shows $L_0(x_0)$, while the prior probability $f(x_0|\theta) = \pi_{x_0} = 1/4$ for any nucleotide x_0 under the K80 model. The probability of data at the site is $f(\mathbf{x}_h|\theta) = 0.000509843$. Thus the posterior probabilities at node 0 are 0.055 ($= 0.25 \times 0.00011237/0.000509843$), 0.901, 0.037, and 0.007, for T, C, A, and G, respectively. C is the most probable nucleotide at the root, with posterior probability 0.901.

Posterior probabilities at any other interior node can be calculated by moving the root to that node and then applying the same algorithm. These are 0.093 (T), 0.829 (C), 0.070 (A), 0.007 (G) for node 6; 0.153 (T), 0.817 (C), 0.026 (A), 0.004 (G) for node 7; and 0.010 (T), 0.985 (C), 0.004 (A), 0.001 (G) for node 8. From these marginal reconstructions, one might guess that the best joint reconstruction is $x_0 x_6 x_7 x_8 =$ CCCC, with posterior probability $0.901 \times 0.829 \times 0.817 \times 0.985 = 0.601$. However, this calculation is incorrect, since the states at different nodes are not independent. For example, given that node 0 has $x_0 = $ C, the probability that nodes 6, 7, and 8 will have C as well will be much higher than when the state at node 0 is unknown.

The above description assumes the same rate for all sites, but applies to the fixed-rates models where the sites are from fixed site partitions. In a random-rates model, both $f(\mathbf{x}_h|\theta)$ and $f(\mathbf{x}_h|x_0;\theta)$ are sums over the rate categories and can similarly be calculated using the pruning algorithm.

4.4.2.2 Joint reconstruction

With this approach, we calculate the posterior probability for a set of character states assigned to all interior nodes at a site. Let $\mathbf{y}_h = (x_0, x_6, x_7, x_8)$ be such an assignment or reconstruction

$$f(\mathbf{y}_h|\mathbf{x}_h;\theta) = \frac{f(\mathbf{x}_h, \mathbf{y}_h|\theta)}{f(\mathbf{x}_h|\theta)} \tag{4.16}$$

$$= \frac{\pi_{x_0} p_{x_0 x_6}(t_6) p_{x_6 x_7}(t_7) p_{x_7 \text{T}}(t_1) p_{x_7 \text{C}}(t_2) p_{x_6 \text{A}}(t_3) p_{x_0 x_8}(t_8) p_{x_8 \text{C}}(t_4) p_{x_8 \text{C}}(t_5)}{f(\mathbf{x}_h|\theta)}.$$

The numerator $f(\mathbf{x}_h, \mathbf{y}_h|\theta)$ is the joint probability of the tip states $\mathbf{x}_h$ and the ancestral states $\mathbf{y}_h$, given parameters θ. This is the term in the square brackets in equation (4.2). The probability of data at a site, $f(\mathbf{x}_h|\theta)$, is a sum over all possible reconstructions $\mathbf{y}_h$, while the percentage of contribution from a reconstruction $\mathbf{y}_h$ is the posterior probability for that reconstruction.

The difficulty with a naïve use of this formula, as in Yang et al. (1995a) and Koshi and Goldstein (1996a), is the great number of ancestral reconstructions (all combinations of x_0, x_6, x_7, x_8). Note that only the numerator is used to compare the different reconstructions, as $f(\mathbf{x}_h|\theta)$ is fixed. Instead of maximizing the product of the transition probabilities, we can maximize the sum of the logarithms of the transition probabilities. The dynamical programming algorithm for determining the best reconstructions for weighted parsimony, described in Subsection 3.4.3, can thus be used after the following minor modifications. First, each branch now has its own cost matrix while a single cost matrix was used for all branches for parsimony. Second, the score for each reconstruction involves an additional term $\log(\pi_{x_0})$. The resulting algorithm is equivalent to that described by Pupko et al. (2000). It works under the one-rate and fixed-rates models but not under random-rates models. For the latter, Pupko *et al.* (2002a) implemented an approximate branch-and-bound algorithm.

Application of the dynamical-programming algorithm to the problem of Fig. 4.2 leads to $x_0 x_6 x_7 x_8 = $ CCCC as the best reconstruction, with posterior probability 0.784. This agrees with the marginal reconstruction, according to which C is the most probable nucleotide for every node, but the probability is higher than the incorrect value 0.601, mentioned above. The next few best reconstructions can be obtained by a slight extension of the dynamic programming algorithm, as TTTC (0.040), CCTC (0.040), CTTC (0.040), AAAC (0.011), and CAAC (0.011). The above calculations are for illustration only. In real data analysis without the clock, one should use an unrooted tree and estimate branch lengths and other parameters by ML.

The marginal and joint reconstructions use slightly different criteria. They normally produce consistent results, with the most probable joint reconstruction for a site consisting of character states that are also the best in the marginal reconstructions. Conflicting results may arise when the competing reconstructions have similar probabilities, in which case neither reconstruction is very reliable.

4.4.2.3 Similarities and differences from parsimony

If we assume the JC69 model with symmetrical substitution rates and also equal branch lengths, the EB and parsimony approaches will give exactly the same rankings of the joint reconstructions. Under JC69, the off-diagonal elements of the transition-probability matrix are all equal and smaller than the diagonals: $P_{ij}(t) < P_{ii}(t)$, so that a reconstruction requiring fewer changes will have a higher posterior probability than one requiring more changes (see equation 4.16). When branch lengths are allowed to differ, as they are in a likelihood analysis, and the substitution rates are unequal, as under more complex substitution models, parsimony and EB may produce different results. In both approaches, the main factor influencing the accuracy of ancestral reconstruction is the sequence divergence level. The reconstruction is less reliable at more variable sites or for more divergent sequences (Yang *et al.* 1995a; Zhang and Nei 1997).

4.4.2.4 Hierarchical Bayesian approach

In real data analysis, parameters θ in equations (4.15) and (4.16) are replaced by their estimates, say the MLEs. In large data sets, this may not be a problem as the parameters are reliably estimated. In small data sets, the parameter estimates may involve large sampling errors, so that the EB approach may suffer from inaccurate parameter estimates. For example, if a branch length is estimated to be zero no change will be possible along that branch during ancestral reconstruction, even though the zero estimate may not be reliable. In such a case, it is advantageous to use a hierarchical (full) Bayesian approach, which assigns a prior on parameters θ and integrates them out. Such an approach is implemented by Huelsenbeck and Bollback (2001).

Uncertainties in the phylogeny are a more complex issue. If the purpose of ancestral reconstruction is for use in further analysis, as in comparative methods (Felsenstein

1985b; Harvey and Purvis 1991), one can average over uncertainties in the phylogenies, in substitution parameters, as well as in the ancestral states by sampling from the posterior of these quantities in an MCMC algorithm (Huelsenbeck et al. 2000c; Huelsenbeck and Bollback 2001). If the purpose is to reconstruct the sequence at a particular ancestral node, the best thing to do appears to be to get a phylogeny as reliable as possible and use it as fixed. Obviously reconstructing sequences for nodes that never existed is not meaningful. Even a shared node between two trees may have different meanings depending on the tree topology. Suppose that species 1, 2, and 3 form a monophyletic clade, but the relationship inside the clade is unresolved. Two approaches are then possible for reconstructing the ancestral sequence at the root of the clade. The first is to infer a binary tree, say, ((12)3), and use it as fixed to reconstruct the sequence at the root. The second is to average over the uncertainties in the phylogeny, assuming that the same root is shared by all three trees. The first approach appears preferable as the roots in the three binary trees do not have the same biological meaning.

*4.4.3 Discrete morphological characters

It is convenient to discuss here the similar problem of reconstructing morphological characters. Parsimony was the predominant method for such analysis until recently. In their pioneering work, Schluter (1995), Mooers and Schluter (1999), and Pagel (1999) emphasized the importance of quantifying uncertainties in ancestral reconstruction and championed the likelihood reconstruction. This effort has met with two difficulties. First, their formulation of the likelihood method was not correct. Second, a single morphological character has little information for estimating model parameters, such as branch lengths on the tree and relative substitution rates between characters, and this lack of information makes the analysis highly unreliable. Below I will fix the first problem but can only lament on the second.

The problem considered by Schluter and Pagel is reconstruction of a binary morphological character evolving under a Markov model with rates q_{01} and q_{10} (see Exercise 1.3 in Chapter 1). To reduce the number of parameters to be estimated, all branches on the tree are assumed to have the same length. The correct solution to this problem is the EB approach (equation 4.16) (Yang et al. 1995a; Koshi and Goldstein 1996a). Schluter (1995), Mooers and Schluter (1999), and Pagel (1999) used $f(\mathbf{x}_h, \mathbf{y}_h|\theta)$ of equation (4.16) as the 'likelihood function' for comparing ancestral reconstructions $\mathbf{y}_h$. This is equivalent to the EB approach except that Mooers and Schluter neglected the π_{x_0} term, and Pagel used 1/2 for π_{x_0}; the prior probabilities at the root should be given by the substitution model as $\pi_0 = q_{10}/(q_{01} + q_{10})$ and $\pi_1 = q_{01}/(q_{01} + q_{10})$. Note that $f(\mathbf{x}_h, \mathbf{y}_h|\theta)$ is the joint probability of $\mathbf{x}_h$ and $\mathbf{y}_h$ and is not the likelihood of $\mathbf{y}_h$. The 'likelihood ratio' discussed by those authors cannot be interpreted in the sense of Edwards (1992) but is the ratio of the posterior probabilities for the two states; a 'log likelihood ratio' of 2 means posterior probabilities $0.88 (= e^2/(e^2 + 1))$ and 0.12 for the two states. Pagel (1999) described the EB calculation as the 'global' approach and preferred an alternative 'local approach', in which both substitution parameters θ and ancestral states $\mathbf{y}_h$ are estimated from the joint probability $f(\mathbf{x}_h, \mathbf{y}_h|\theta)$. This local approach is invalid, as θ should be estimated from the likelihood $f(\mathbf{x}_h|\theta)$, which sums over all possible ancestral states $\mathbf{y}_h$.

If one insisted on estimating ancestral states from the likelihood function, it would be possible to do so only for the marginal reconstruction at the root. The likelihood function would be $L_0(x_0)$, the probability of data at the site given the state x_0 at the root. The placement of the root would then affect the likelihood and the rooted tree should be used. For molecular data, this approach suffers from the problem that the number of parameters grows without bound when the sample size increases. It is not possible to use the likelihood function for the joint reconstruction. For the example of Fig. 4.2 for nucleotides,

$$f(\mathbf{x}_h|\mathbf{y}_h;\theta) = p_{x_7 \text{T}}(t_1) p_{x_7 \text{C}}(t_2) p_{x_6 \text{A}}(t_3) p_{x_8 \text{C}}(t_4) p_{x_8 \text{C}}(t_5). \tag{4.17}$$

This is a function of x_6, x_7, x_8 (for nodes that are connected to the tips), and not of x_0 (for nodes that are not directly connected to the tips); under the Markov model, given the states at the immediate ancestors of the tips, the states at the tips are independent of states at older ancestors. Thus it is not possible to reconstruct ancestral states from this 'likelihood' function.

Even with the EB approach, the substitution rates q_{01} and q_{10} cannot be estimated reliably from only one character. One might hope to use a likelihood ratio test to compare the one-rate ($q_{01} = q_{10}$) and two-rates ($q_{01} \neq q_{10}$) models. However, this test is problematic as the asymptotic χ^2 distribution cannot be used due to the small sample size (which is one) and more importantly, the failure to reject the one-rate model may reflect a lack of power of the test rather than a genuine symmetry of the rates. Analyses by a number of authors (e.g. Mooers and Schluter 1999) suggest that ancestral reconstruction is sensitive to assumptions about relative rates. Another worrying assumption is that of equal branch lengths. This means that the expected amount of evolution is the same for every branch on the tree, and is more unreasonable than the clock assumption (rate constancy over time). An alternative may be to use estimates of branch lengths obtained from molecular data, but this is open to the criticism that the molecular branch lengths may not reflect the amount of evolution in the morphological character.

An approach to dealing with uncertainties in the parameters is to use a hierarchical Bayesian approach, averaging over uncertainties in substitution rates and branch lengths. Schultz and Churchill (1999) implemented such an algorithm, and found that the posterior probabilities for ancestral states are very sensitive to priors on relative substitution rates, even in seemingly ideal situations where the parsimony reconstruction is unambiguous. Studies such as these highlight the misleading overconfidence of parsimony as well as the extreme difficulty of reconstructing ancestral states for a single character. The Bayesian approach does offer the consolation that if different inferences are drawn from the same data under the same model they must be due to differences in the prior.

In general, classical statistical approaches such as ML are not expected to work well in data sets consisting of one sample point, or in problems involving as many

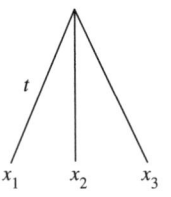

Fig. 4.8 A tree of three species for demonstrating the bias in ancestral reconstruction. The three branch lengths are equal, at $t = 0.2$ substitutions per site.

parameters as the observed data. The ML method applied to analyse one or two morphological characters (Pagel 1994; Lewis 2001) may not have good statistical properties, as the asymptotic theory for MLEs and likelihood ratio tests is not expected to apply.

4.4.4 Systematic biases in ancestral reconstruction

The ancestral character states reconstructed by parsimony or likelihood (EB) are our best guesses under the respective criteria. However, if they are used for further statistical analysis or in elaborate tests, one should bear in mind that they are inferred pseudo-data instead of real observed data. They involve random errors due to uncertainties in the reconstruction. Worse still, use of only the best reconstructions while ignoring the suboptimal ones can cause systematic biases. Collins *et al.* (1994), Perna and Kocher (1995), and Eyre-Walker (1998) discussed such biases with parsimony reconstruction. They exist also with likelihood (EB) reconstruction. The problem lies not in the use of parsimony instead of likelihood for ancestral reconstruction, but in the use of the optimal reconstructions while ignoring the suboptimal ones.

As an example, consider the star tree of three sequences of Fig. 4.8. Suppose that the substitution process has been stationary and followed the model of Felsenstein (1981), with parameters $\pi_T = 0.2263$, $\pi_C = 0.3282$, $\pi_A = 0.3393$, and $\pi_G = 0.1062$ (these frequencies are from the human mitochondrial D loop hypervariable region I). Suppose that each branch length is 0.2 nucleotide substitutions per site, so that the transition-probability matrix is

$$P(0.2) = \begin{bmatrix} 0.811138 & 0.080114 & 0.082824 & 0.025924 \\ 0.055240 & 0.836012 & 0.082824 & 0.025924 \\ 0.055240 & 0.080114 & 0.838722 & 0.025924 \\ 0.055240 & 0.080114 & 0.082824 & 0.781821 \end{bmatrix}.$$

We use the correct model and branch lengths to calculate posterior probabilities for ancestral states at the root, and examine the frequencies of A and G in the reconstructed ancestral sequence. This mimics studies which use ancestral reconstruction to detect possible drift in base compositions. At sites with data AAG, AGA, and GAA, the posterior probabilities for the states at the root are 0.006 (T), 0.009 (C), 0.903 (A), and 0.083 (G), so that A is much more likely than G. However, if we use A and ignore

G at every such site, we will over-count A and under-count G. Similarly, at sites with data GGA, GAG, and AGG, the posterior probabilities are 0.002 (T), 0.003 (C), 0.034 (A), and 0.960 (G). The best reconstruction is G, so that we over-count G and under-count A at such sites. The greater frequency of A than of G in the data means that there are more sites with data AAG, AGA, and GAA (with probability 0.02057) than sites with data GGA, GAG, and AGG (with probability 0.01680). The net bias is then an over-count of A and under-count of G in the ancestor. Indeed, the base compositions in the reconstructed sequence for the root are 0.212 (T), 0.324 (C), 0.369 (A), and 0.095 (G), more extreme than the frequencies in the observed sequences. Thus ancestral reconstruction indicates an apparent gain of the rare nucleotide G over the time of evolution from the root to the present. As the process is in fact stationary, this apparent drift in base compositions is an artefact of the EB (and parsimony) reconstruction, caused by ignoring the suboptimal reconstructions at every site. In a recent study, Jordan *et al.* (2005) used parsimony to reconstruct ancestral protein sequences and observed a systematic gain of rare amino acids (and loss of common ones) over evolutionary time. The trend is exactly the same as discussed above, although it is unclear to what extent the pattern is due to artefact in ancestral reconstruction.

Perna and Kocher (1995) studied the use of parsimony to infer ancestral states and then to count changes along branches to estimate the substitution-rate matrix. The bias involved in such an analysis appears even greater than in counts of nucleotides discussed above. Clearly, the problem is more serious for more divergent sequences since the reliability of ancestral reconstruction degrades with the increase in sequence divergence. However, the bias is considerable even in data sets of closely related sequences such as the human mitochondrial D-loop sequences (Perna and Kocher 1995).

Despite these caveats, the temptation to infer ancestors and use them to perform all sorts of statistical tests appears too great to resist. Ancestral reconstruction is thus used frequently, with many interesting and spurious discoveries being made all the time.

Instead of ancestral reconstruction, one should try to rely on a likelihood-based approach, which sums over all possible ancestral states, weighting them appropriately according to their probabilities of occurrence (see Section 4.2). For example, the relative rates of substitutions between nucleotides or amino acids can be estimated by using a likelihood model (Yang 1994b; Adachi and Hasegawa 1996a; Yang *et al.* 1998). Possible drifts in nucleotide compositions may be tested by implementing models that allow different base frequency parameters for different branches (Yang and Roberts 1995; Galtier and Gouy 1998). Chapter 8 provides a few more examples in which both ancestral reconstruction and full likelihood-based approaches are used to analyse protein-coding genes to detect positive selection.

If a likelihood analysis under the model is too complex and one has to resort to ancestral reconstruction, a heuristic approach to reducing the bias may be to use the suboptimal as well as the optimal reconstructions in the analysis. One may use a simpler existing likelihood model to calculate posterior probabilities for ancestral states and use them as weights to accommodate both optimal and suboptimal reconstructions. In the example above, the posterior probabilities for the root state at a site with data AAG are 0.006 (T), 0.009 (C), 0.903 (A), and 0.083 (G). Instead of using A for the site and ignoring all other states, one can use both A and G, with weights 0.903 and 0.083 (rescaled so that they sum to one). If we use all four states at every site in this way, we will recover the correct base compositions for the root sequence with no bias at all, since the posterior probabilities are calculated under the correct model. If the assumed likelihood model is too simplistic and incorrect, the posterior probabilities (weights) will be incorrect as well. Even so this approach may be less biased than ignoring suboptimal reconstructions entirely. Akashi *et al.* (2006) applied this approach to count changes between preferred (frequently used) and unpreferred (rarely used) codons in protein-coding genes in the *Drosophila melanogaster* species subgroup and found that it helped to reduce the bias in ancestral reconstruction. Similarly Dutheil *et al.* (2005) used reconstructed ancestral states to detect coevolving nucleotide positions, indicated by an excess of substitutions at two sites that occur along the same branches. They calculated posterior probabilities for ancestral states under an independent-site model and used them as weights to count substitutions along branches at the two sites under test.

*4.5 Numerical algorithms for maximum likelihood estimation

The likelihood method estimates parameters θ by maximizing the log likelihood ℓ. In theory, one may derive the estimates by setting to zero the first derivatives of ℓ with respect to θ and solving the resulting system of equations, called the *likelihood equations*:

$$\frac{\partial \ell}{\partial \theta} = 0. \tag{4.18}$$

This approach leads to analytical solutions to pairwise distance estimation under the JC69 and K80 models, as discussed in Section 1.4. For three species, analytical solution is possible only in the simplest case of binary characters evolving under the molecular clock (Yang 2000a). The problem becomes intractable as soon as it is extended to nucleotides with four states (Exercise 4.2), or to four species. The latter case, of four species under the clock, was studied by Chor and Snir (2004), who derived analytical estimates of branch lengths for the 'fork' tree (two subtrees each with two species) but not for the 'comb' tree (one subtree with one species and another with three species).

In general, numerical iterative algorithms have to be used to maximize the log likelihood. Developing a reliable and efficient optimization algorithm for practical problems is a complicated task. This section gives only a flavour of such algorithms. Interested readers should consult a textbook on nonlinear programming or numerical optimization, such as Gill *et al.* (1981) or Fletcher (1987).

Maximization of a function (called the *objective function*) is equivalent to minimization of its negative. Below, we will follow the convention and describe our problem of likelihood maximization as a problem of minimization. The objective

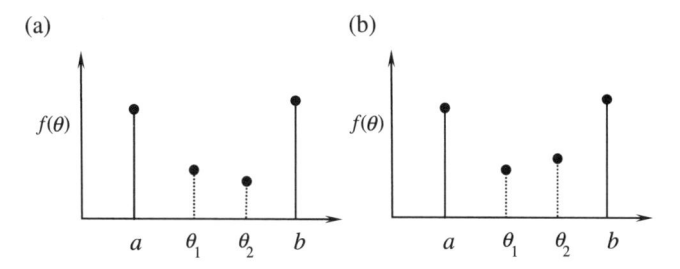

Fig. 4.9 Reduction of the interval of uncertainty $[a, b]$, which contains the minimum. The objective function f is evaluated at two interior points θ_1 and θ_2. (a) If $f(\theta_1) \geq f(\theta_2)$, the minimum must lie in the interval $[\theta_1, b]$. (b) Otherwise if $f(\theta_1) < f(\theta_2)$, the minimum must lie in the interval $[a, \theta_2]$.

function is thus the negative log likelihood: $f(\theta) = -\ell(\theta)$. Note that algorithms that reach the minimum with fewer function evaluations are more efficient, since $f(\theta)$ is expensive to calculate.

4.5.1 Univariate optimization

If the problem is one-dimensional, the algorithm is called *line search* since the search is along a line. Suppose we determine that the minimum is in the interval $[a, b]$. This is called the *interval of uncertainty*. Most line search algorithms reduce this interval successively, until its width is smaller than a pre-specified small value. Assuming that the function is unimodal in the interval (i.e. there is only one valley between a and b), we can reduce the interval of uncertainty by comparing the function values at two interior points θ_1 and θ_2 (Fig. 4.9). Different schemes exist concerning choices of the points. Here we describe the golden section search.

4.5.1.1 Golden section search

Suppose the interval of uncertainty is $[0, 1]$; this can be rescaled to become $[a, b]$. We place two interior points at γ and $(1 - \gamma)$, where the golden ratio $\gamma \approx 0.6180$ satisfies $\gamma/(1 - \gamma) = 1/\gamma$. The new interval becomes $[1 - \gamma, 1]$ if $f(\gamma) < f(1 - \gamma)$ or $[0, \gamma]$ otherwise (Fig. 4.10). No matter how the interval is reduced, one of the two points will be in the correct position inside the new interval. With the golden section search, the interval of uncertainty is reduced by γ at each step. The algorithm is said to have a linear convergence rate.

4.5.1.2 Newton's method and polynomial interpolation

For smooth functions, more efficient algorithms can be implemented by approximating f using simple functions whose minimum can be obtained analytically. For

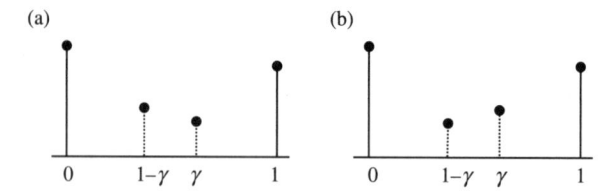

Fig. 4.10 The golden section search. Suppose the minimum is inside the interval $[0, 1]$. Two points are placed at γ and $(1 - \gamma)$ of the interval, where $\gamma = 0.6180$. (a) If $f(\gamma) < f(1 - \gamma)$, the new interval will be $[1 - \gamma, 1]$. (b) If $f(\gamma) \geq f(1 - \gamma)$, the new interval will be $[0, \gamma]$. No matter how the interval is reduced, one of the two points will be in the correct position inside the new interval.

example, if we approximate f by a parabola (quadratic), of the form

$$\tilde{f} = a\theta^2 + b\theta + c, \tag{4.19}$$

with $a > 0$, then $\tilde{f}$ has a minimum at $\theta^* = -b/(2a)$. If the function value and its first and second derivatives at the current point θ_k are known, we can use the first three terms of the Taylor expansion to approximate $f(\theta)$:

$$\tilde{f}(\theta) = f(\theta_k) + f'(\theta_k)(\theta - \theta_k) + \tfrac{1}{2}f''(\theta_k)(\theta - \theta_k)^2. \tag{4.20}$$

This is a quadratic function in θ, in the form of equation (4.19), with $a = f''(\theta_k)/2$ and $b = f'(\theta_k) - f''(\theta_k)\theta_k$. If $f''(\theta_k) > 0$, the quadratic (4.20) achieves its minimum at

$$\theta_{k+1} = \theta_k - \frac{f'(\theta_k)}{f''(\theta_k)}. \tag{4.21}$$

As f may not be a quadratic, θ_{k+1} may not be the minimum of f. We thus use θ_{k+1} as the new current point to repeat the algorithm. This is Newton's method, also known as the Newton–Raphson method.

Newton's method is highly efficient. Its rate of convergence is quadratic, meaning that, roughly speaking, the number of correct figures in θ_k doubles at each step (e.g. Gill et al. 1981, p. 57). A problem, however, is that it requires the first and second derivatives, which may be expensive, troublesome, or even impossible to compute. Without the derivatives, a quadratic approximation can be constructed by using the function values at three points. Similarly a cubic polynomial can be constructed by using the function values and first derivatives (but not the second derivatives) at two points. It is in general not worthwhile to fit high-order polynomials. Another serious problem with Newton's method is that its fast convergence rate is only local, and if the iterate is not close to the minimum the algorithm may diverge hopelessly. The iteration may also encounter numerical difficulties if $f''(\theta_k)$ is zero or too small. Thus it is important to obtain good starting values for Newton's method, and certain safeguards are necessary to implement the algorithm.

4.5 Numerical algorithms for maximum likelihood estimation • 131

A good strategy is to combine a guaranteed reliable method (such as golden section) with a rapidly converging method (such as Newton's quadratic interpolation), to yield an algorithm that will converge rapidly if f is well-behaved, but is not much less efficient than the guaranteed method in the worst case. Suppose a point $\tilde{\theta}$ is obtained by quadratic interpolation. We can check to make sure that $\tilde{\theta}$ lies in the interval of uncertainty $[a, b]$ before evaluating $f(\tilde{\theta})$. If $\tilde{\theta}$ is too close to the current point or too close to either end of the interval, one may revert to the golden section search. Another idea of safeguarding Newton's method is to redefine

$$\theta_{k+1} = \theta_k - \alpha f'(\theta_k)/f''(\theta_k), \tag{4.22}$$

where the step length α, which equals 1 initially, is repeatedly halved until the algorithm is non-increasing, that is, until $f(\theta_{k+1}) < f(\theta_k)$.

4.5.2 Multivariate optimization

Most models used in likelihood analysis in molecular phylogenetics include multiple parameters, and the optimization problem is multidimensional. A naïve approach is to optimize one parameter at a time with all other parameters fixed. However, this is inefficient when the parameters are correlated. Figure 4.11 shows an example in which the two parameters are (positively) correlated. A search algorithm that updates one parameter at a time makes impressive improvements to the objective function

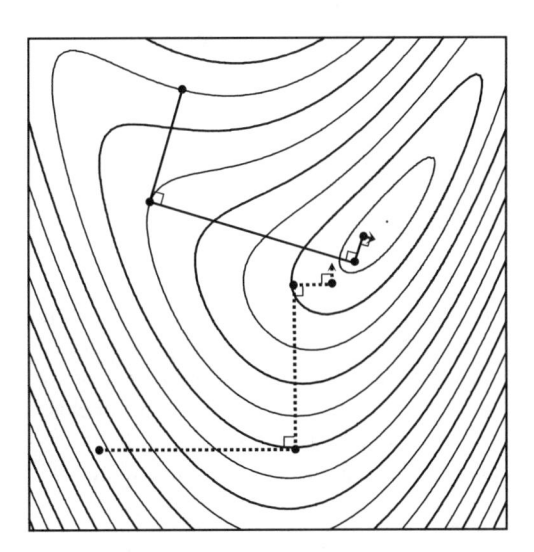

Fig. 4.11 The log likelihood contour when two parameters are positively correlated. A search algorithm changing one parameter at a time (the dotted lines and arrow) is very inefficient. Similarly the steepest-ascent search (the solid lines and arrow) is inefficient because its search direction is always perpendicular to the previous search direction.

132 • 4 Maximum likelihood methods

initially, but becomes slower and slower when it is close to the optimum. As every search direction is at a 90° angle to the previous search direction, the algorithm zig-zags in tiny baby steps. Standard optimization algorithms update all variables simultaneously.

4.5.2.1 Steepest-descent search

Many optimization algorithms use the first derivatives, $g = df(\theta)$, called the *gradient*. The simplest among them is the *steepest-descent* algorithm (or *steepest-ascent* for maximization). It finds the steepest-descent direction, locates the minimum along that direction, and repeats the procedure until convergence. The gradient g is the direction that the function increases the quickest locally and is thus *the steepest-ascent direction*, while $-g$ is the *steepest-descent direction*. Note that the minimum along a search direction occurs when the search direction becomes a tangent line to a contour curve. At that point, the new gradient is perpendicular to the tangent line or the previous search direction (Fig. 4.11). Thus the steepest-descent algorithm suffers from the same problem as the naïve algorithm of changing one variable at a time: every search direction forms a 90° angle to the previous search direction. The algorithm is known to be inefficient and unstable.

4.5.2.2 Newton's method

The multivariate version of the Newton algorithm relies on quadratic approximation to the objective function. Let $G = d^2 f(\theta)$ be the *Hessian matrix* of second partial derivatives. With p variables, both θ and g are $p \times 1$ vectors while G is a $p \times p$ matrix. A second-order Taylor expansion of f around the current point θ_k gives

$$f(\theta) \approx f(\theta_k) + g_k^T (\theta - \theta_k) + \tfrac{1}{2}(\theta - \theta_k)^T G_k (\theta - \theta_k), \qquad (4.23)$$

where the superscript T means transpose. By setting the gradient of the right-hand side of equation (4.23) to 0, one obtains the next iterate as

$$\theta_{k+1} = \theta_k - G_k^{-1} g_k. \qquad (4.24)$$

Note that this is very similar to equation (4.21) for the univariate case. Similarly, the multivariate version shares the fast convergence rate as well as the major drawbacks of the univariate algorithm; that is, the method requires calculation of the first and second derivatives and may diverge when the iterate is not close enough to the minimum. A common strategy is to take $s_k = -G_k^{-1} g_k$ as a search direction, called the *Newton direction*, and to perform a line search to determine how far to go along that direction.

$$\theta_{k+1} = \theta_k + \alpha s_k = \theta_k - \alpha G_k^{-1} g_k. \qquad (4.25)$$

Here α is called the *step length*. It is often too expensive to optimize α in this way. A simpler version is to try $\alpha = 1, 1/2, 1/4, \ldots$, until $f(\theta_{k+1}) \leq f(\theta_k)$. This is sometimes known as the *safe-guided Newton algorithm*. Special treatments are also needed when G_k is not *positive definite*.

When the objective function is the negative log likelihood, $f = -\ell$, the Hessian matrix $G = -d^2\ell(\theta)$ is also called the observed information matrix. In some simple statistical problems, the expected information, $E[d^2\ell(\theta)]$, may be easier to calculate, and can be used in Newton's algorithm. The method is then known as *scoring*. Both Newton's method and scoring have the benefit that the approximate variance–covariance matrix of the MLEs are readily available at the end of the iteration. See Subsection 1.4.1 for a discussion of these concepts.

4.5.2.3 Quasi-Newton methods

Quasi-Newton methods include a class of methods that require first derivatives but not second derivatives. While Newton's method calculates the second derivatives G at every current point, quasi-Newton methods build up information about G or its inverse from the calculated values of the objective function f and the first derivatives g during the iteration. If the first derivatives are not available, they may be calculated using the difference approximation. Without the need for second or even first derivatives, quasi-Newton algorithms greatly increased the range of problems that can be solved. The basic algorithm can be sketched as follows:

a. Supply an initial guess θ_0.
b. For $k = 0, 1, 2, \ldots$, until convergence:
   1. Test θ_k for convergence;
   2. Calculate a search direction $s_k = -B_k g_k$;
   3. Perform a line search along s_k to determine the step length α_k: $\theta_{k+1} = \theta_k + \alpha_k s_k$;
   4. Update B_k to give B_{k+1}.

Here B_k is a symmetric positive definite matrix, which can be interpreted as an approximation to G_k^{-1}, the inverse of the Hessian matrix. Note the similarity of this algorithm to Newton's method. In step b3, the scalar $\alpha_k > 0$ is chosen to minimize $f(\theta_{k+1})$, by using a line search algorithm as discussed above. A number of strategies have been developed to update the matrix B. Well-known ones include the Broyden–Fletcher–Goldfard–Shanno (BFGS) and Davidon–Fletcher–Powell (DFP) formulae. See Gill et al. (1981) and Fletcher (1987) for details.

When the first derivatives are impossible or expensive to calculate, an alternative approach is to use a *derivative-free method*. See Brent (1973) for discussions of such methods. According to Gill et al. (1981), quasi-Newton methods, with the first derivatives calculated using the difference approximation, are more efficient than derivative-free methods.

4.5.2.4 Optimization under linear inequality constraints

The discussion up to now has assumed that the parameters are unconstrained and can take values over the whole real line. Practical problems are rarely of this kind. For example, branch lengths should be nonnegative, and the nucleotide frequency parameters π_T, π_C, π_A should satisfy the following constraints: $\pi_T, \pi_C, \pi_A \geq 0$ and $\pi_T + \pi_C + \pi_A \leq 1$. Algorithms for constrained optimization are much more complex, with simple lower and upper bounds being easier than general linear inequality constraints (Gill et al. 1981). Variable transformation is sometimes an effective approach for dealing with linear inequality constraints.

4.5.3 Optimization on a fixed tree

In a phylogenetic problem, the parameters to be estimated include branch lengths in the tree and all parameters in the substitution model. Given the values of parameters, one can use the pruning algorithm to calculate the log likelihood. In theory, one can then apply any of the general-purpose optimization algorithms discussed above to find the MLEs iteratively. However, this will almost certainly produce an inefficient algorithm. Consider the recursive calculation of the conditional probabilities $L_i(x_i)$ in the pruning algorithm. When a branch length changes, $L_i(x_i)$ for only those nodes ancestral to that branch are changed, while those for all other nodes are not affected. Direct application of a multivariate optimization algorithm thus leads to many duplicated calculations of the same quantities.

To take advantage of such features of likelihood calculation on a tree, one can optimize one branch length at a time, keeping all other branch lengths and substitution parameters fixed. Suppose one branch connects nodes a and b. By moving the root to coincide with node a, we can rewrite equation (4.5) as

$$f(\mathbf{x}_h|\theta) = \sum_{x_a} \sum_{x_b} \pi_{x_a} p_{x_a x_b}(t_b) L_a(x_a) L_b(x_b). \tag{4.26}$$

The first and second derivatives of ℓ with respect to t_b can then be calculated analytically (Adachi and Hasegawa 1996b; Yang 2000b), so that t_b can be optimized efficiently using Newton's algorithm. One can then estimate the next branch length by moving the root to one of its ends. A change to any substitution parameter, however, typically changes the conditional probabilities for all nodes, so saving is not possible. To estimate substitution parameters, two strategies appear possible. Yang (2000b) tested an algorithm that cycles through two phases. In the first phase, branch lengths are optimized one by one while all substitution parameters are held fixed. Several cycles through the branch lengths are necessary to achieve convergence. In the second phase, the substitution parameters are optimized using a multivariate optimization algorithm such as BFGS, with branch lengths fixed. This algorithm works well when the branch lengths and substitution parameters are not correlated, for example, under the HKY85 or GTR (REV) models, in which the transition/transversion rate ratio κ for HKY85 or the rate ratio parameters in GTR are not strongly correlated with the branch lengths. However, when there is strong correlation, the algorithm can be very inefficient. This is the case with the gamma model of variable rates at sites, in which the branch lengths and the gamma shape parameter α often have strong negative correlations. A second strategy (Swofford 2000) is to embed the first phase of the algorithm into the second phase. One uses a multivariate optimization algorithm (such as BFGS) to estimate the substitution parameters, with the log likelihood for any given values of the substitution parameters calculated by optimizing the branch lengths. It may be necessary to optimize the branch lengths to a high precision, as inaccurate calculations of the log likelihood (for given substitution parameters) may cause problems for the BFGS algorithm, especially if the first derivatives are calculated numerically using difference approximation.

4.5.4 Multiple local peaks on the likelihood surface for a fixed tree

Numerical optimization algorithms discussed above are all local hill-climbing algorithms. They converge to a local peak, but may not reach the globally highest peak if multiple local peaks exist. Fukami and Tateno (1989) presented a proof that under the F81 model (Felsenstein 1981) and on a tree of any size, the log likelihood curve for one branch length has a single peak when other branch lengths are fixed. Tillier (1994) suggested that the result applies to more general substitution models as well. However, Steel (1994a) pointed out that this result does not guarantee one single peak in the whole parameter space. Consider a two-parameter problem and imagine a peak in the northwest region and another peak in the southeast region of the parameter space. Then there is always one peak if one looks only in the north–south direction or west-east direction, although in fact two local peaks exist. Steel (1994a) and Chor et al. (2000) further constructed counter examples to demonstrate the existence of multiple local peaks for branch lengths even on small trees with four species.

How common local peaks are in real data analysis is less clear. A symptom of the problem is that different runs of the same analysis starting from different initial values may lead to different results. Rogers and Swofford (1999) used computer simulation to examine the problem, and reported that local peaks were less common for the ML tree than for other poorer trees. It is hard to imagine that the likelihood surfaces are qualitatively different for the different trees, so one possible reason for this finding may be that more local peaks exist at the boundary of the parameter space (say, with zero branch lengths) for the poor trees than for the ML tree. Existence of multiple local peaks is sometimes misinterpreted as an indicator of the unrealistic nature of the assumed substitution model. If such a connection exists, the opposite is generally true: multiple local peaks are much less common under simplistic and unrealistic models like JC69 than under more realistic parameter-rich models.

There are no foolproof remedy to the problem of local peaks. A simple strategy is to run the iteration algorithm multiple times, starting from different initial values. If multiple local peaks exist, one should use the estimates corresponding to the highest peak. Stochastic search algorithms that allow downhill moves, such as simulated annealing and genetic algorithms, are useable as well, although they are often very expensive computationally; see Subsection 3.2.5.

4.5.5 Search for the maximum likelihood tree

The above discussion concerns estimation of branch lengths and substitution parameters for a given tree topology. If our purpose is to estimate parameters in the substitution model or to test hypotheses concerning them, we should normally consider the tree topology as known or fixed, and then one or two optimizations will suffice. This is a straightforward application of ML estimation, which concerns estimation of parameters θ when the likelihood (probability density) function $L(\theta; X) = f(X|\theta)$ is fully specified.

If our interest is in reconstruction of the phylogeny, we can take the optimized log likelihood for the tree as the score for tree selection under the likelihood criterion. One can then in theory repeat the process for other trees and solve as many optimization problems as the number of tree topologies. This is distinctly more complex than the conventional estimation problem. We have two levels of optimization: one of optimizing branch lengths (and substitution parameters) on each fixed tree to calculate the tree score and the other of searching for the best tree in the tree space. Because the neighbouring trees generated during branch swapping share subtrees, tree-search algorithms should avoid repeated computations of the same quantities. Much effort has been taken to develop fast algorithms for likelihood tree search, such as the parallel algorithm of Olsen *et al.* (1994) and the genetic algorithms of Lewis (1998) and Lemmon and Milinkovitch (2002).

A few recent algorithms have explored the idea that optimizing branch lengths to a high precision may be a waste of time if the tree is very poor and that instead it is sensible to change the tree topology before the branch lengths are fully optimized. Guindon and Gascuel (2003) developed an efficient algorithm that adjusts the tree topology and branch lengths simultaneously. Candidate trees are generated by local rearrangements of the current tree, for example by using the NNI algorithm, and in calculating their likelihood scores only the branch lengthes affected by the local rearrangements are optimized, while branch lengths within clades unaffected by the rearrangements are not always optimized. Vinh and von Haeseler (2004) developed the IQPNNI algorithm for likelihood tree search, combining so-called importance quartet puzzling (IQP; see below) with the NNI branch-swapping algorithm. To avoid being tracked at a local optimum, a random fraction of the species are deleted from the tree and then reattached to see whether the tree score can be improved, with the quartets used to help decide on the location for reattachment.

Likelihood tree search is currently a very active area of research. Efficient algorithms for analysing large data sets with hundreds or thousands of sequences are likely to result from this effort. One difficulty is that we lack sensible ways of generating good search directions when the algorithm traverses the tree space. In contrast, numerical optimization algorithms such as BFGS are far better than stabbing in the dark. They accumulate local curvature information during the iteration and are able to propose efficient search directions, resulting in superlinear convergence. It is unclear whether similar ideas may be developed for ML tree search.

4.6 Approximations to likelihood

The computational burden of the likelihood method has prompted the development of approximate methods. One idea is to use other methods to estimate branch lengths on a given tree rather than optimizing branch lengths by ML. For example, Adachi and Hasegawa (1996b) used least-squares estimates of branch lengths calculated from a pairwise distance matrix to calculate approximate likelihood scores. Similarly Rogers and Swofford (1998) used parsimony reconstruction of ancestral states to estimate approximate branch lengths. The approximate branch lengths provide good starting values for a proper likelihood optimization, but the authors suggested that they could also be used to calculate the approximate likelihood values for the tree without further optimization, leading to approximate likelihood methods of tree reconstruction.

Strimmer and von Haeseler (1996) and Schmidt et al. (2002) implemented an approximate likelihood algorithm for tree search called *quartet puzzling*. This uses ML to evaluate the three trees for every possible quartet of species. The full tree for all s species is then constructed by a majority-rule consensus of those quartet trees. This method may not produce the ML tree but is very fast. Ranwez and Gascuel (2002) implemented an algorithm that combines features of NJ and ML. It is based on triplets of taxa, and shares the divide-and-conquer strategy of the quartet approach. In this regard, NJ may be viewed as an approximate ML method based on doublets of species, since the pairwise distances are ML estimates, while the cluster algorithm is used to assemble the 'trees' of two species into a tree of all s species. Indeed Ota and Li (2000) took this view ever further and used ML to improve the NJ tree. The authors developed a divide-and-conquer heuristic algorithm in which an initial NJ tree is subjected to local branch swapping under the likelihood criterion. The branch swapping is applied only to clades with low bootstrap values while other branches are fixed to save computation. Simulations suggested that this NJ–ML hybrid method is more efficient than NJ, and comparable with heuristic ML tree search.

4.7 Model selection and robustness

4.7.1 LRT, AIC, and BIC

4.7.1.1 The likelihood ratio test (LRT)

The likelihood ratio test is a powerful and general method for testing hypotheses concerning model parameters in the likelihood framework. It compares two nested hypotheses formulated *a priori*. In other words, one hypothesis is a special case of the other. Suppose the null hypothesis H_0 involves p_0 parameters and the alternative hypothesis H_1 is an extension of H_0, with p_1 parameters. Let the maximum log likelihood values under the two models be $\ell_0 = \log\{L(\hat{\theta}_0)\} = \ell(\hat{\theta}_0)$ and $\ell_1 = \log\{L(\hat{\theta}_1)\} = \ell(\hat{\theta}_1)$, where $\hat{\theta}_0$ and $\hat{\theta}_1$ are the MLEs under the two models, respectively. Then under certain regularity conditions, the likelihood ratio test statistic

$$2\Delta\ell = 2\log(L_1/L_0) = 2(\ell_1 - \ell_0) \qquad (4.27)$$

is asymptotically distributed as $\chi^2_{p_1-p_0}$ if H_0 is true. In other words, if the null model is true, twice the log likelihood difference between the null and alternative models is approximately χ^2 distributed with the degree of freedom equal to the difference in the number of parameters between the two models. The approximation applies to large samples. The regularity conditions required for the χ^2 approximation are similar to those for the asymptotic properties of the MLEs, and will not be discussed here (see Stuart *et al.* 1999, pp. 245–246). When the χ^2 approximation is invalid or unreliable, Monte Carlo simulation can be used to derive the correct null distribution (Goldman 1993).

Example. Application to rbcL genes. We apply the LRT to the data of the plastid *rbcL* genes from 12 plant species. The sequence alignment was kindly provided by Dr Vincent Savolainen. There are 1428 nucleotide sites in the sequence. The tree topology is shown in Fig. 4.12, which will be used to compare models. Table 4.2 shows the log likelihood values and parameter estimates under three sets of nucleotide-substitution models. The first set includes JC69, K80, and HKY85, in which the same Markov model is applied to all sites in the sequence. The second set includes the '+Γ_5' models, which use the discrete gamma to accommodate the variable rates among sites, using five rate classes (Yang 1994a). The third set includes the '+C' models (Yang 1996b). They assume that different codon positions have different substitution rates (r_1, r_2, r_3), and, for K80 and HKY85, different substitution parameters as well. Parameters such as branch lengths in the tree, κ in K80 and HKY85, and the relative rates for codon positions in the '+C' models, are estimated by ML. The base frequency parameters in the HKY85 models are estimated by using the observed frequencies, averaged over the sequences (Table 4.3).

Here we consider three tests in detail. First, the JC69 and K80 models can be compared using an LRT to test the null model H_0: $\kappa = 1$. Note that K80 will be reduced to JC69 when parameter $\kappa = 1$ is fixed. Thus JC69 is the null model, and includes $p_0 = 21$ branch lengths as parameters. The (optimized) log likelihood is $\ell_0 = -6262.01$. K80 is the alternative model, with one extra parameter κ. The log likelihood is $\ell_1 = -6113.86$. The test statistic is $2\Delta\ell = 2(\ell_1 - \ell_0) = 296.3$. This is much greater than the critical value $\chi^2_{1,1\%} = 6.63$, indicating that JC69 is rejected by a big margin. The transition and transversion rates are very different, as is clear from the MLE $\hat{\kappa} = 3.56$ under K80. Similarly comparison between K80 and HKY85 using an LRT with three degrees of freedom leads to rejection of the simpler K80 model.

A second test compares the null model JC69 against JC69+Γ_5, to test the hypothesis that different sites in the sequence evolve at the same rate. The one-rate model is a special case of the gamma model when the shape parameter $\alpha = \infty$ is fixed. The test statistic is $2\Delta\ell = 648.42$. In this test, the regularity conditions mentioned above are not all satisfied, as the value ∞ is at the boundary of the parameter space in the alternative model. As a result, the null distribution is not χ^2_1, but is a 50:50 mixture of

4.7 Model selection and robustness • 139

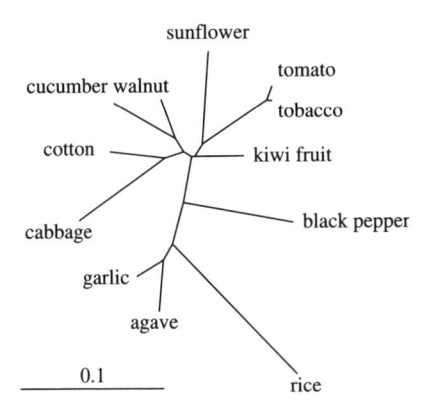

Fig. 4.12 The ML tree for the plastic *rbcL* genes from 12 plant species, estimated under the HKY85 + Γ_5 model, using a discrete-gamma model with five rate classes. The branches are drawn in proportion to their estimated lengths.

Table 4.2 Likelihood ratio test to examine the model's fit to the data of 12 *rbcL* genes

Model	p	ℓ	MLEs
JC69	21	−6 262.01	
K80	22	−6 113.86	$\hat{\kappa} = 3.561$
HKY85	25	−6 101.76	$\hat{\kappa} = 3.620$
JC69 + Γ_5	22	−5 937.80	$\hat{\alpha} = 0.182$
K80 + Γ_5	23	−5 775.40	$\hat{\kappa} = 4.191, \hat{\alpha} = 0.175$
HKY85 + Γ_5	26	−5 764.26	$\hat{\kappa} = 4.296, \hat{\alpha} = 0.175$
JC69 + C	23	−5 922.76	$\hat{r}_1 : \hat{r}_2 : \hat{r}_3 = 1 : 0.556 : 5.405$
K80 + C	26	−5 728.76	$\hat{\kappa}_1 = 1.584, \hat{\kappa}_2 = 0.706, \hat{\kappa}_3 = 5.651,$ $\hat{r}_1 : \hat{r}_2 : \hat{r}_3 = 1 : 0.556 : 5.611$
HKY85 + C	35	−5 624.70	$\hat{\kappa}_1 = 1.454, \hat{\kappa}_2 = 0.721, \hat{\kappa}_3 = 6.845$ $\hat{r}_1 : \hat{r}_2 : \hat{r}_3 = 1 : 0.555 : 5.774$

p is the number of parameters in the model, including 21 branch lengths in the tree of Fig. 4.12. The base frequency parameters under the HKY85 models are estimated using the observed frequencies (see Table 4.3).

Table 4.3 Observed base compositions at the three codon positions in the data set of 12 *rbcL* genes (Fig. 4.12)

Position	π_T	π_C	π_A	π_G
1	0.1829	0.1933	0.2359	0.3878
2	0.2659	0.2280	0.2998	0.2063
3	0.4116	0.1567	0.2906	0.1412
All	0.2867	0.1927	0.2754	0.2452

the point mass at 0 and χ_1^2 (Chernoff 1954; Self and Liang 1987). The critical values are 2.71 at 5% and 5.41 at 1%, rather than 3.84 at 5% and 6.63 at 1% according to χ_1^2. This null mixture distribution may be intuitively understood by considering the MLE of the parameter in the alternative model. If the true value is inside the parameter space, its estimate will have a normal distribution around the true value, and will be less or greater than the true value, each half of the times. If the true value is at the boundary, half of the times the estimate would be outside the space if there were no constraint; in such cases, the estimate will be forced to the true value and the log likelihood difference will be 0. Note that use of χ_1^2 makes the test too conservative; if the test is significant under χ_1^2, it will be significant when the mixture distribution is used. For the *rbcL* data set, the observed test statistic is huge, so that the null model is rejected whichever null distribution is used. The rates are highly variable among sites, as indicated also by the estimate of α. Similar tests using the K80 and HKY85 models also suggest significant variation in rates among sites.

A third test compares JC69 and JC69 + C. In the alternative model JC69 + C, the three codon positions are assigned different relative rates $r_1 (= 1)$, r_2, and r_3, while the null model JC69 is equivalent to constraining $r_1 = r_2 = r_3$, reducing the number of parameters by two. The test statistic is $2\Delta\ell = 678.50$, to be compared with χ_2^2. The null model is clearly rejected, suggesting that the rates are very different at the three codon positions. The same conclusion is reached if the K80 or HKY95 models are used in the test.

The two most complex models in Table 4.2, HKY85 + Γ_5 and HKY85 + C, are the only models not rejected in such likelihood ratio tests. (The two models themselves are not nested and a χ^2 approximation to the LRT is not applicable.) This pattern is typical in the analysis of molecular data sets, especially large data sets; we seem to have no difficulty in rejecting an old simpler model whenever we develop a new model by adding a few extra parameters, sometimes even if the biological justification for the new parameters is dubious. The pattern appears to reflect the fact that most molecular data sets are very large and the LRT tends to favour parameter-rich models in large data sets. □

4.7.1.2 Akaike information criterion (AIC)

The LRT is applicable for comparing two hypotheses that are nested. Although Cox (1961; 1962; see also Atkinson 1970; Lindsey 1974a; 1974b; Sawyer 1984) discussed the use of LRT to compare non-nested models, the idea does not appear to have become popular or practical. The Akaike Information Criterion (AIC; Akaike 1974) can be used to compare multiple models that are not necessarily nested. The AIC score is calculated for each model, defined as

$$\text{AIC} = -2\ell + 2p, \qquad (4.28)$$

where $\ell = \ell(\hat{\theta})$ is the optimum log likelihood under the model, and p is the number of parameters. Models with small AICs are preferred. According to this criterion, an extra parameter is worthwhile if it improves the log likelihood by more than one unit.

4.7.1.3 Bayesian information criterion (BIC)

In large data sets, both LRT and AIC are known to favour complex parameter-rich models and to reject simpler models too often (Schwarz 1978). The Bayesian Information Criterion (BIC) is based on a Bayesian argument and penalizes parameter-rich models more severely. It is defined as

$$\text{BIC} = -2\ell + p\log(n), \tag{4.29}$$

where n is the sample size (sequence length) (Schwarz 1978). Again models with small BIC scores are preferred.

Qualitatively, LRT, AIC, and BIC are all mathematical formulations of the *parsimony principle* of model building. Extra parameters are deemed necessary only if they bring about significant or considerable improvements to the fit of the model to data, and otherwise simpler models with fewer parameters are preferred. However, in large data sets, these criteria can differ markedly. For example, if the sample size $n > 8$, BIC penalizes parameter-rich models far more severely than does AIC.

Model selection is a controversial and active area of research in statistics. Posada and Buckley (2004) provided a nice overview of methods and criteria for model selection in molecular phylogenetics. For automatic model selection, Posada and Crandall (1998) developed MODELTEST, a collection of command scripts for use with the PAUP program (Swofford 2000). Well-known substitution models are compared hierarchically using the LRT, although other criteria such as AIC and BIC are included as well. The program enables the investigator to avoid making thoughtful decisions concerning the model to be used in phylogeny reconstruction.

4.7.1.4 Application to the ape mitochondrial protein data

We use the three model selection criteria to compare a few models applied to the data set analysed in Subsection 4.2.5. The ML tree of Fig. 4.5 is assumed. The three empirical amino acid substitution models DAYHOFF (Dayhoff *et al.* 1978), JTT (Jones *et al.* 1992), and MTMAM (Yang *et al.* 1998) are fitted to the data, with either one rate for all sites or gamma rates for sites. In the discrete-gamma model, five rate categories are used; the estimates of the shape parameter α range from 0.30 to 0.33 among the three models. The results are shown in Table 4.4. The LRT can be used to compare nested models only, so each empirical model (e.g. DAYHOFF) is compared with its gamma counterpart (e.g. DAYHOFF $+ \Gamma_5$). The LRT statistics are very large, so there is no doubt that the substitution rates are highly variable among sites, whether χ_1^2 or the 50: 50 mixture of 0 and χ_1^2 is used as the null distribution. The AIC and BIC scores can be used to compare nonnested models, such as the three empirical models. As they involve the same number of parameters, the ranking using AIC or BIC is the same as using the log likelihood. MTMAM fits the data better than the

Table 4.4 Comparison of models for the mitochondrial protein sequences from the apes

Model	p	ℓ	LRT	AIC	BIC
DAYHOFF	11	−15 766.72		31 555.44	31 622.66
JTT	11	−15 332.90		30 687.80	30 755.02
MTMAM	11	−14 558.59		29 139.18	29 206.40
DAYHOFF+Γ_5	12	−15 618.32	296.80	31 260.64	31 333.97
JTT+Γ_5	12	−15 192.69	280.42	30 409.38	30 482.71
MTMAM+Γ_5	12	−14 411.90	293.38	28 847.80	28 921.13

p is the number of parameters in the model. The sample size is $n = 3331$ amino acid sites for the BIC calculation. The LRT column shows the test statistic $2\Delta\ell$ for comparing each empirical model with the corresponding gamma model.

other two models, which is expected since the data are mitochondrial proteins while DAYHOFF and JTT were derived from nuclear proteins. The best model for the data according to all three criteria is MTMAM+Γ_5.

4.7.2 Model adequacy and robustness

All models are wrong but some are useful.

(George Box, 1979)

Models are used for different purposes. Sometimes they are used to test an interesting biological hypothesis, so that the models themselves are our focus. For example, the molecular clock (rate constancy over time) is an interesting hypothesis predicted by certain theories of molecular evolution, and it can be examined by using an LRT to compare a clock model and a nonclock model (see Chapter 7 for details). Often the model, or at least some aspects of the model assumptions, is not our main interest, but has to be dealt with in the analysis. For example, in testing the molecular clock, we need a Markov model of nucleotide substitution. Similarly, in phylogeny reconstruction, our interest is in the tree, but we have to assume an evolutionary model to describe the mechanism by which the data are generated. The model is then a nuisance, but its impact on our analysis cannot be ignored.

It should be stressed that a model's fit to data and its impact on inference are two different things. Often model robustness is even more important than model adequacy. It is neither possible nor necessary for a model to match up with the biological reality in every detail. Instead, a model should capture important features of the biological process to enable us to make reference. What features are important will depend on the question being asked, and one has to use one's knowledge of the subject matter to make the judgement. Structural biologists tend to emphasize the uniqueness of every residue in the protein. Similarly one has every reason to believe that every species is unique, as distinct as humans are from other species. However, by no means should one use one separate parameter for every site and every branch in formulating a statistical model to describe the evolution of the protein sequence. Such a model, saturated with parameters, is not workable. Biologists often fail to appreciate the power of so-called *i.i.d.* models, which assume that the sites in the sequence are independent and identically distributed. A common misconception is that *i.i.d.* models assume that every site evolves at the same rate, following the same pattern. It should be noted that almost all models implemented in molecular phylogenetics, such as models of variable rates among sites (Yang 1993, 1994a), models of variable selective pressures among sites (Nielsen and Yang 1998), and the covarion models that allow the rate to vary both among sites and among lineages (Tuffley and Steel 1998; Galtier 2001; Guindon *et al.* 2004) are *i.i.d.* models. The *i.i.d.* assumption is a statistical device useful for reducing the number of parameters.

Some features of the process of sequence evolution are both important to the fit of the model and critical to our inference. They should be incorporated in the model. Variable rates among sites appear to be such a factor for phylogeny reconstruction or estimation of branch lengths (Tateno *et al.* 1994; Huelsenbeck 1995a; Gaut and Lewis 1995; Sullivan *et al.* 1995). Some factors may be important to the model's fit, as judged by the likelihood, but may have little impact on the analysis. For example, adding the transition/transversion rate ratio κ to the JC69 model almost always leads to a huge improvement to the log likelihood, but often has minimal effect on estimates of branch lengths. The difference between HKY85 and GTR (REV) is even less important, even though HKY85 is rejected in most data sets when compared against GTR. The most troublesome factors are those that have little impact on the fit of the model but a huge impact on our inference. For example, in estimation of species divergence times under local molecular-clock models, different models for lineage rates appear to fit the data almost equally well, judged by their log likelihood values, but they can produce very different time estimates (see Chapter 7). Such factors have to be carefully assessed even if the statistical test does not indicate their importance.

For phylogeny reconstruction, a number of computer simulations have been conducted to examine the robustness of different methods to violations of model assumptions. Such studies have in general found that model-based approaches were quite robust to the underlying substitution model (e.g. Hasegawa *et al.* 1991; Gaut and Lewis 1995). However, the importance of model assumptions appears to be dominated by the shape of the tree reflected in the relative branch lengths, which determines the overall level of difficulty of tree reconstruction. 'Easy' trees, with long internal branches or with long external branches clustered together, are successfully reconstructed by all methods and models; indeed wrong simplistic models tend to show even better performance than the more complex true model (see Section 6.2 for a discussion of such complexities). 'Hard' trees, with short internal branches and with long external branches spread over different parts of the tree, are difficult to reconstruct by all methods. For such trees, simplistic models may not even be statistically consistent, and use of complex and realistic models is critical.

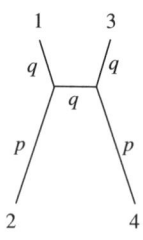

Fig. 4.13 A tree of four species with two branch lengths p and q, defined as the probability that any site is different at the two ends of the branch. For a binary character, this probability is $p = (1-e^{-2t})/2$, where t is the expected number of character changes per site (see Exercise 1.3 in Chapter 1).

4.8 Exercises

*4.1 Collapsing site patterns for likelihood calculation under the JC69 model. Under JC69, the probability of data at a site depends on whether the nucleotides are different in different species, but not on what the nucleotides are. For example, sites with data TTTC, TTTA, AAAG all have the same probability of occurrence. Show that if such sites are collapsed into patterns, there is a maximum of $(4^{s-1} + 3 \times 2^{s-1} + 2)/6$ site patterns for s sequences (Saitou and Nei 1986).

*4.2 Try to estimate the single branch length under the JC69 model for the star tree of three sequences under the molecular clock (see Saitou (1988) and Yang (1994c, 2000a), for discussions of likelihood tree reconstruction under this model). The tree is shown in Fig. 4.8, where t is the only parameter to be estimated. Note that there are only three site patterns, with one, two, or three distinct nucleotides, respectively. The data are the observed numbers of sites with such patterns: n_0, n_1, and n_2, with the sum to be n. Let the proportions be $f_i = n_i/n$. The log likelihood is $\ell = n \sum_{i=0}^{2} f_i \log(p_i)$, with p_i to be the probability of observing site pattern i. Derive p_i by using the transition probabilities under the JC69 model, given in equation (1.3). You can calculate $p_0 = \Pr(\text{TTT})$, $p_1 = \Pr(\text{TTC})$, and $p_2 = \Pr(\text{TCA})$. Then set $d\ell/dt = 0$. Show that the transformed parameter $z = e^{-4/3t}$ is a solution to the following quintic equation:

$$36z^5 + 12(6 - 3f_0 - f_1)z^4 + (45 - 54f_0 - 42f_1)z^3 + (33 - 60f_0 - 36f_1)z^2$$
$$+ (3 - 30f_0 - 2f_1)z + (3 - 12f_0 - 4f_1) \equiv 0. \tag{4.30}$$

4.3 Calculate the probabilities of sites with data $xxyy$, $xyyx$, and $xyxy$ in four species for the unrooted tree of Fig. 4.13, using two branch lengths p and q under a symmetrical substitution model for binary characters (Exercise 1.3). Here it is more convenient to define the branch length as the proportion of different sites at the two ends of the branch. Show that $\Pr(xxyy) < \Pr(xyxy)$ if and only if $q(1 - q) < p^2$. With such branch lengths, parsimony for tree reconstruction is inconsistent (Felsenstein 1978a).

5

Bayesian methods

5.1 The Bayesian paradigm

5.1.1 Overview

There are two principal philosophies in statistical data analysis: the *classical* or *frequentist* and the *Bayesian*. The *frequentist* defines the probability of an event as the expected frequency of occurrence of that event in repeated random draws from a real or imaginary population. The performance of an inference procedure is judged by its properties in repeated sampling from the data-generating model (i.e. the likelihood model), with the parameters fixed. Important concepts include bias and variance of an estimator, confidence intervals, and *p* values; these are all covered in a typical biostatistics course. Maximum likelihood (ML) estimation and the likelihood ratio test figure prominently in classical statistics; for example, *t* tests of population means, the χ^2 test of association in a contingency table, analysis of variance, linear correlation, and regression are either likelihood methods or approximations to likelihood.

Bayesian statistics is not mentioned in most biostatistics courses. The key feature of *Bayesian* methods is the notion of a probability distribution for the parameter. Here probability cannot be interpreted as the frequency in random draws from a population but instead is used to represent uncertainty about the parameter. In classical statistics, parameters are unknown constants and cannot have distributions. Bayesian proponents argue that since the value of the parameter is unknown, it is sensible to specify a probability distribution to describe its possible values. The distribution of the parameter before the data are analysed is called the *prior distribution*. This can be specified by using either an objective assessment of prior evidence concerning the parameter or the researcher's subjective opinion. The Bayes theorem is then used to calculate the *posterior distribution* of the parameter, that is, the conditional distribution of the parameter given the data. All inferences about the parameter are based on the posterior.

While probability theory has been a subject of study for several hundred years, most conspicuously as related to gambling, statistics is a relatively young field. Concepts of regression and correlation were invented by Francis Galton and Karl Pearson around 1900 in studies of human inheritance. The field blossomed in the 1920s and 1930s when Ronald A. Fisher developed many of the techniques of classical statistics, such as analysis of variance, experimental design, and likelihood. The theories of hypothesis testing and confidence intervals, including such concepts as simple and composite hypotheses, type-I and type-II errors, etc. were developed by Jerzy Neyman and Egon

Pearson (Karl Pearson's son) at about the same time. These contributions completed the foundation of classical statistics.

In contrast, Bayesian ideas are much older, dating to a paper published posthumously by Thomas Bayes in 1763, in which he calculated the posterior distribution of a binomial probability using an implicit $U(0, 1)$ prior. The idea was developed further by Laplace and others in the 19th century but was not popular among statisticians of the early 20th century. There are two major reasons for this lack of acceptance. The first is philosophical: the method's reliance and indeed insistence on a prior for unknown parameters was fiercely criticized by prominent statisticians, such as Fisher. The second is computational: except for simple toy examples that are analytically tractable, computation of posterior probabilities in practical applications almost always requires numerical calculation of integrals, often high-dimensional integrals. Nowadays, controversies concerning the prior persist, but Bayesian computation has been revolutionized by Markov-chain Monte Carlo algorithms (MCMC Metropolis *et al.* 1953; Hastings 1970). Incidentally, the power of MCMC was not generally appreciated until the publication of Gelfand and Smith (1990), which demonstrates the feasibility of such algorithms to solve nontrivial practical problems. MCMC has made it possible to implement sophisticated parameter-rich models for which the likelihood analysis would not be feasible. The 1990s were exciting years for Bayesian statistics; it appeared that every statistician was a Bayesian and everybody was developing MCMC algorithms in various subject areas where statistics is applied. Recent years have seen the overzealous excitement about the method subsiding to a more reasonable level; partly the method is now too popular to need any more justification, and partly frustrated programmers have started to appreciate the complexities and difficulties of implementing and validating MCMC algorithms.

This chapter provides a brief introduction to the theory and computation of Bayesian statistics and its applications to molecular evolution. The reader is encouraged to consult any of the excellent textbooks on Bayesian theory (e.g. Leonard and Hsu 1999; O'Hagan and Forster 2004) and computation (e.g. Gilks *et al.* 1996). Here I will use simple examples, such as distance estimation under the JC69 model, to introduce the general principles. I will discuss the application of Bayesian inference to reconstruction of phylogenetic trees and to population genetics analysis under the coalescent. Another application, to estimation of species divergence times under relaxed molecular clocks, is discussed in Chapter 7. Several excellent reviews of Bayesian phylogenetics have been published, including Huelsenbeck *et al.* (2001) and Holder and Lewis (2003).

5.1.2 Bayes's theorem

Suppose the occurrence of a certain event B may depend on whether another event A has occurred. Then the probability that B occurs is given by the law of total probabilities

$$P(B) = P(A) \times P(B|A) + P(\bar{A}) \times P(B|\bar{A}). \tag{5.1}$$

Here $\bar{A}$ stands for 'non A' or 'A does not occur'. Bayes's theorem, also known as the *inverse-probability theorem*, gives the conditional probability that B occurs given that A occurs.

$$P(A|B) = \frac{P(A) \times P(B|A)}{P(B)} = \frac{P(A) \times P(B|A)}{P(A) \times P(B|A) + P(\bar{A}) \times P(B|\bar{A})}. \qquad (5.2)$$

We illustrate the theorem using an example.

Example. (*False positives of a clinical test*). Suppose a new test has been developed to screen for an infection in the population. If a person has the infection, the test accurately reports a positive 99% of the time, and if a person does not have the infection, the test falsely reports a positive only 2% of the time. Suppose that 0.1% of the population have the infection. What is the probability that a person who has tested positive actually has the infection?

Let A be the event that a person has the infection, and $\bar{A}$ no infection. Let B stand for test-positive. Then $P(A) = 0.001, P(\bar{A}) = 0.999, P(B|A) = 0.99, P(B|\bar{A}) = 0.02$. The probability that a random person from the population tests positive is, according to equation (5.1),

$$P(B) = 0.001 \times 0.99 + 0.999 \times 0.02 = 0.2097. \qquad (5.3)$$

This is close to the proportion among the noninfected individuals of the population. Equation (5.2) then gives the probability that a person who has tested positive has the infection as

$$P(A|B) = \frac{0.001 \times 0.99}{0.2097} = 0.0472. \qquad (5.4)$$

Thus among the positives, only 4.7% are true positives while 95.3% ($= 1 - 0.0472$) are false positives. Despite the apparent high accuracy of the test, the incidence of the infection is so low (0.1%) that most people who test positive (95.3%) do not have the infection. □

When Bayes's theorem is used in Bayesian statistics, A and $\bar{A}$ correspond to different hypotheses H_1 and H_2, while B corresponds to the observed data (X). Bayes's theorem then specifies the conditional probability of hypothesis H_1 given the data as

$$P(H_1|D) = \frac{P(H_1) \times P(X|H_1)}{P(X)} = \frac{P(H_1) \times P(X|H_1)}{P(H_1) \times P(X|H_1) + P(H_2) \times P(X|H_2)}. \qquad (5.5)$$

$P(H_2|X) = 1 - P(H_1|X)$ is given similarly. Here $P(H_1)$ and $P(H_2)$ are called *prior probabilities*. They are probabilities assigned to the hypotheses before the data are observed or analysed. The conditional probabilities $P(H_1|X)$ and $P(H_2|X)$

are called *posterior probabilities*. $P(X|H_1)$ and $P(X|H_2)$ are the likelihood under each hypothesis. The hypotheses may correspond to different values of a parameter; that is, H_1: $\theta = \theta_1$ and H_2: $\theta = \theta_2$. The extension to more than two hypotheses is obvious.

Note that in the disease testing example discussed above, $P(A)$ and $P(\bar{A})$ are frequencies of infected and noninfected individuals in the population. There is no controversy concerning the use of Bayes's theorem in such problems. However, in Bayesian statistics, the prior probabilities $P(H_1)$ and $P(H_2)$ often do not have such a frequentist interpretation. The use of Bayes's theorem in such a context is controversial. We will return to this controversy in the next subsection.

When the hypothesis concerns unknown continuous parameters, probability densities are used instead of probabilities. Bayes's theorem then takes the following form

$$f(\theta|X) = \frac{f(\theta)f(X|\theta)}{f(X)} = \frac{f(\theta)f(X|\theta)}{\int f(\theta)f(X|\theta)\,d\theta}. \tag{5.6}$$

Here $f(\theta)$ is the *prior distribution*, $f(X|\theta)$ is the likelihood (the probability of data X given parameter θ), and $f(\theta|X)$ is the *posterior distribution*. The *marginal probability* of the data, $f(X)$, is a normalizing constant, to make $f(\theta|X)$ integrate to 1. Equation (5.6) thus says that the posterior is proportional to the prior times the likelihood, or equivalently, the posterior information is the sum of the prior information and the sample information. In the following, we focus on the continuous version, but take it for granted that the theory applies to the discrete version as well. Also, when the model involves more than one parameter, θ will be a vector of parameters.

The posterior distribution is the basis for all inference concerning θ. For example, the mean, median, or mode of the distribution can be used as the point estimate of θ. For interval estimation, one can use the 2.5% and 97.5% quantiles of the posterior density to construct the 95% *equal-tail credibility interval* (CI) (Fig. 5.1a). When the posterior

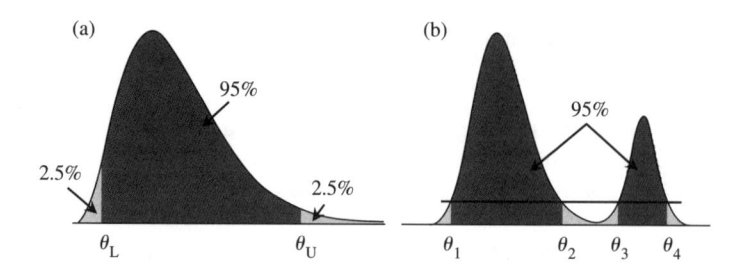

Fig. 5.1 (a) The 95% equal-tail credibility interval (θ_L, θ_U) is constructed by locating the 2.5% and 97.5% quantiles of the posterior probability distribution. (b) The 95% highest posterior density (HPD) interval includes values of θ that have the highest density and that cover 95% of the probability mass. With multiple peaks in the density, the HPD region may consist of disconnected intervals: (θ_1, θ_2) and (θ_3, θ_4).

density is skewed or has multiple peaks, this interval has the drawback of including values of θ that are less supported than values outside the interval. One can then use the 95% *highest posterior density* (HPD) interval, which includes values of θ of the highest posterior density that encompass 95% of the density mass. In the example of Fig. 5.1b, the HPD region consists of two disconnected intervals. In general, the posterior expectation of any function $h(\theta)$ of parameters θ is constructed as

$$E(h(\theta)|X) = \int h(\theta) f(\theta|X) \, d\theta. \tag{5.7}$$

An important strength of the Bayesian approach is that it provides a natural way of dealing with nuisance parameters (see Subsection 1.4.3) through integration or marginalization. Let $\theta = (\lambda, \eta)$, with λ to be the parameters of interest and η the nuisance parameters. The joint posterior density of λ and η is

$$f(\lambda, \eta|X) = \frac{f(\lambda, \eta) f(X|\lambda, \eta)}{f(X)} = \frac{f(\lambda, \eta) f(X|\lambda, \eta)}{\int f(\lambda, \eta) f(X|\lambda, \eta) \, d\lambda \, d\eta}, \tag{5.8}$$

from which the (marginal) posterior density of λ can be obtained as

$$f(\lambda|X) = \int f(\lambda, \eta|X) \, d\eta. \tag{5.9}$$

Example. Consider the use of the JC69 model to estimate the distance θ between the human and orangutan 12s rRNA genes from the mitochondrial genome (see Subsection 1.4.1). The data are summarized as $x = 90$ differences out of $n = 948$ sites. The MLE was found to be $\hat{\theta} = 0.1015$, with the 95% confidence (likelihood) interval to be $(0.0817, 0.1245)$. To apply the Bayesian approach, one has to specify a prior. Uniform priors are commonly used in Bayesian analysis. In this case, one could specify, say, $U(0, 100)$, with a large upper bound. However, the uniform prior is not very reasonable since sequence distances estimated from real data are often small (say, < 1). Instead we use an exponential prior

$$f(\theta) = \frac{1}{\mu} e^{-\theta/\mu}, \tag{5.10}$$

with mean $\mu = 0.2$. The posterior distribution of θ is

$$f(\theta|x) = \frac{f(\theta) f(x|\theta)}{f(x)} = \frac{f(\theta) f(x|\theta)}{\int f(\theta) f(x|\theta) \, d\theta}. \tag{5.11}$$

We consider the data to have a binomial distribution, with probability

$$p = \tfrac{3}{4} - \tfrac{3}{4} e^{-4\theta/3} \tag{5.12}$$

for a difference and $1-p$ for an identity. The likelihood is thus

$$f(x|\theta) = p^x(1-p)^{n-x} = \left(\tfrac{3}{4} - \tfrac{3}{4}e^{-4\theta/3}\right)^x \left(\tfrac{1}{4} + \tfrac{3}{4}e^{-4\theta/3}\right)^{n-x}. \qquad (5.13)$$

This is the same as the likelihood of equation (1.43) except for a scale constant $4^n/3^x$. Analytical calculation of the integral in the denominator of equation (5.11) is awkward, so I use Mathematica to calculate it numerically, which gives $f(x) = 5.16776 \times 10^{-131}$. Figure 5.2 shows the posterior density, plotted together with the prior and scaled likelihood. In this case the posterior is dominated by the likelihood, and the prior is nearly flat at the neighbourhood of the peak of the likelihood. The mean of the posterior distribution is found by numerical integration to be $E(\theta|x) = \int \theta f(\theta|x) \, d\theta = 0.10213$, with standard deviation 0.01091. The mode is at $\theta = 0.10092$. The 95% equal-tail credibility interval can be constructed by calculating numerically the 2.5% and 97.5% percentiles of the posterior density, to be $(0.08191, 0.12463)$. This is very similar to the confidence interval in the likelihood analysis, despite their different interpretations (see below). The 95% HPD interval, $(0.08116, 0.12377)$, is found by lowering the posterior density function from the maximum (at 36.7712) until the resulting interval encompasses 95% of the density mass (Fig. 5.2). As the posterior density has a single peak and is nearly symmetrical, the HPD and the equal-tail intervals are nearly identical. □

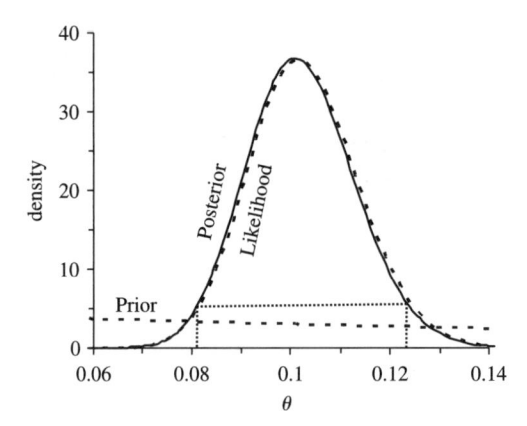

Fig. 5.2 Prior and posterior densities for sequence distance θ under the JC69 model. The likelihood is shown as well, rescaled to match up the posterior density. Note that the posterior density is the prior density times the likelihood, followed by a change of scale to make the area under the posterior density curve equal to 1. The data analysed here are the human and orangutan mitochondrial 12s rRNA genes, with $x = 90$ differences at $n = 948$ sites. The 95% HPD interval, $(0.08116, 0.12377)$, is indicated on the graph.

5.1.3 Classical versus Bayesian statistics

There is a large body of statistics literature on the controversy between the frequentist/likelihood methods and Bayesian methods. See, for example, Lindley and Phillips (1976) for a Bayesian introduction and Efron (1986) from the classical viewpoint. Here I will mention some of the major criticisms from both schools.

5.1.3.1 Criticisms of frequentist statistics

A major Bayesian criticism of classical statistics is that it does not answer the right question. Classical methods provide probability statements about the data or the method for analysing the data, but not about the parameter, even though the data have already been observed and our interest is in the parameter. We illustrate these criticisms using the concepts of confidence intervals and p values. While discussed in every biostatistics course, these concepts are known to be confusing to biologists (and, allegedly, to beginning statisticians as well); in molecular phylogenetics at least, published misinterpretations are embarrassingly common.

Consider first the confidence interval. Suppose the data are a sample $(x_1, x_2, \ldots, x_n)$ from the normal distribution $N(\mu, \sigma^2)$, with unknown parameters μ and σ^2. Provided n is large (say, > 50), a 95% confidence interval for μ is

$$(\bar{x} - 1.96s/\sqrt{n}, \ \bar{x} + 1.96s/\sqrt{n}), \tag{5.14}$$

where $s = \left[\sum(x_i - \bar{x})^2/(n-1)\right]^{1/2}$ is the sample standard deviation. What does this mean? Many of us would want to say that a 95% confidence interval means that there is a 95% chance that the interval includes the true parameter value. But this interpretation is wrong.

To appreciate the correct interpretation, consider the following example. Suppose we take two random draws x_1 and x_2 from a discrete distribution $f(x|\theta)$, in which the random variable x takes two values $\theta - 1$ and $\theta + 1$, each with probability $1/2$. The unknown parameter θ is in the range $-\infty < \theta < \infty$. Then

$$\hat{\theta} = \begin{cases} (x_1 + x_2)/2, & \text{if } x_1 \neq x_2, \\ x_1 - 1, & \text{if } x_1 = x_2 \end{cases} \tag{5.15}$$

defines a 75% confidence set for θ. Here we consider $\hat{\theta}$ as an interval or set even though it contains a single value. Let $a = \theta - 1$ and $b = \theta + 1$ be the two values that x_1 and x_2 can take. There are four possible data outcomes, each occurring with probability $1/4$: aa, bb, ab, and ba. Only when the observed data are aa does $\hat{\theta}$ differ from θ. Thus in 75% of the data sets, all sampled from the distribution $f(x|\theta)$, the confidence set $\hat{\theta}$ includes (equals) the true parameter value. In other words, $\hat{\theta}$ is a 75% confidence set for θ. The probability 75% is called the *confidence level* or *coverage probability*. It means that if we sample repetitively from the data-generating model and construct a confidence interval for each data set, then 75% of those confidence intervals will include the true parameter value.

152 • 5 Bayesian methods

Now suppose in the observed data set, the two values x_1 and x_2 are distinct. We then know for certain that $\hat{\theta} = \theta$. It appears counterintuitive, if not absurd, for the confidence set to have only 75% coverage probability when we know for certain that it includes the true parameter value. At any rate, the confidence level (which is 0.75) is not the probability that the interval includes the true parameter value (which is 1 for the observed data). One can also construct examples in which a 95% confidence interval clearly does not include the true parameter value.

To return to the normal distribution example of equation (5.14), the correct interpretation of the confidence interval is based on repeated sampling (Fig. 5.3). With μ and σ^2 fixed, imagine taking many samples (of the same size as the observed data) from the same population that the observed data are from, and constructing a confidence interval for each sample according to equation (5.14). Then 95% of these confidence intervals will contain the true parameter value μ. Note that the confidence intervals vary among data sets while parameters μ and σ^2 are fixed. An equivalent interpretation is that confidence intervals are *pre-trial* evaluations; before the data are collected, the confidence interval we will construct using the procedure of equation (5.14) will include the true parameter value with probability 0.95. This statement is true whatever the true value of μ is. However, after the data were observed, the 95% confidence interval we did construct using the said procedure may or may not include the true value, and in general no probability (except 0 or 1) can be attached to the event that our interval includes the true parameter value. The criticism is that the confidence-interval theory dodges the problem of what information is available about the parameter and makes a round-about probability statement about the data or the procedure instead; the biologist is interested not in the procedure of constructing confidence intervals but in the parameter and the particular confidence interval constructed from her data. In contrast, the Bayesian credibility interval gives a straightforward answer to this question. Given the data, a 95% Bayesian credibility interval includes the true parameter value with probability 95%. Bayesian methods are *post-trial* evaluations, making probabilistic statements about the parameters conditioned on the observed data.

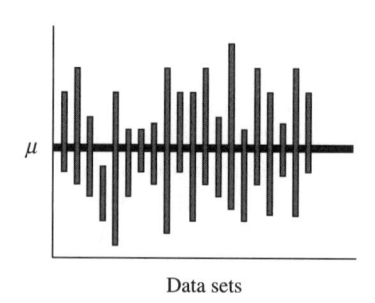

Fig. 5.3 Interpretation of the confidence interval. Many data sets are sampled from the probability model with parameters μ and σ^2 fixed. A 95% confidence interval is constructed for μ in each data set. Then 95% of those confidence intervals will include the true value of μ.

Next we turn to the p value in the Neyman–Pearson framework of hypothesis testing, which is even more severely criticized by Bayesians. We use the example of Exercise 1.5, constructed by Lindley and Phillips (1976). Suppose $x = 9$ heads and $r = 3$ tails are observed in $n = 12$ independent tosses of a coin, and we wish to test the null hypothesis H_0: $\theta = 1/2$ against the alternative H_1: $\theta > 1/2$, where θ is the true probability of heads. Suppose the number of tosses n is fixed, so that x has a binomial distribution, with probability

$$f(x|\theta) = \binom{n}{x} \theta^x (1-\theta)^{n-x}. \tag{5.16}$$

The probability distribution under the null hypothesis is shown in Fig. 5.4. Note that the probability of the observed data $x = 9$, which is 0.05371, is not the p value. When there are many possible data outcomes, the probability of observing one particular data set, even if it is the most likely data outcome, can be extremely small. Instead the p value is calculated as the tail probability under the null hypothesis, summing over all data outcomes that are at least as extreme as the observed data:

$$p = P_{\theta=1/2}(x=9) + P_{\theta=1/2}(x=10) + P_{\theta=1/2}(x=11) + P_{\theta=1/2}(x=12) = 0.075, \tag{5.17}$$

where the subscript means that the probability is calculated under the null hypothesis $\theta = 1/2$. One is tempted to interpret the p value as the probability that H_0 is true, but this interpretation is incorrect.

The p values are also criticized for violating the likelihood principle, which says that the likelihood function contains all information in the data about θ and the same inference should be made from two experiments that have the same likelihood (see Subsection 1.4.1). Consider a different experimental design, in which the number of

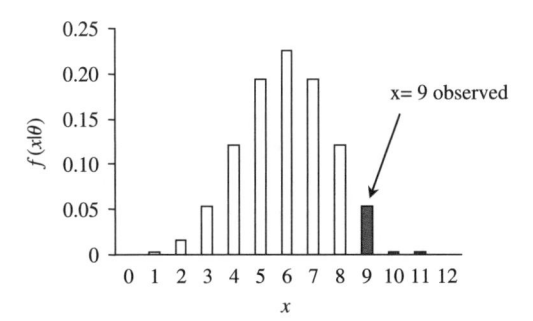

Fig. 5.4 The binomial distribution of x heads out of $n = 12$ tosses of a coin under the null hypothesis that the true probability of heads is $\theta = 1/2$. The observed number of heads is $x = 9$. The p value for testing the null hypothesis $\theta = 1/2$ against the alternative $\theta > 1/2$ sums over data outcomes equal to or more extreme than the observed value, that is, over $x = 9, 10, 11,$ and 12.

tails r was fixed beforehand; in other words, the coin was tossed until $r = 3$ tails were observed, at which point $x = 9$ heads were observed. The data x then have a negative binomial distribution, with probability

$$f(x|\theta) = \binom{r+x-1}{x} \theta^x (1-\theta)^{n-x}. \tag{5.18}$$

If we use this model, the p value becomes

$$p = P_{\theta=1/2}(x = 9) + P_{\theta=1/2}(x = 10) + \ldots = \sum_{j=9}^{\infty} \binom{3+j-1}{j} \left(\frac{1}{2}\right)^j \left(\frac{1}{2}\right)^3$$

$$= 0.0325. \tag{5.19}$$

Thus, at the 5% significance level, we reject H_0 under the negative-binomial model but not under the binomial model. As equations (5.16) and (5.18) differ only by a proportionality constant that is independent of θ, and the likelihood is the same under the two models, the different p values for the two models violate the likelihood principle. It does not appear reasonable that how the experiment is monitored should have any bearing on our inference concerning θ; one would expect that only the *results* of the experiment are relevant. The different p values for the two models are due to the fact that the *sample space*, that is, the space of all data outcomes, differs between the two models. Under the binomial model, the possible data outcomes are $x = 0, 1, \ldots, 9, \ldots, 12$ heads. Under the negative-binomial model, the sample space consists of $x = 0, 1, 2, \ldots, 9, \ldots, 12, \ldots$, heads. The hypothesis-testing approach allows unobserved data outcomes to affect our decision to reject H_0. The small probabilities of x values more extreme than the value actually observed ($x = 9$) are used as evidence against H_0, even though these values did not occur. Jeffreys (1961, p. 385) caricatured the approach as requiring 'that a hypothesis that may be true may be rejected because it has not predicted observable results that have not occurred'!

5.1.3.2 Criticisms of Bayesian methods

All criticisms of Bayesian methods are levied on the prior or the need for it. Bayesians come in two flavours: the objective and the subjective. The *objective* Bayesians consider the prior to be a representation of prior objective information about the parameter. This applies when, for example, a model can be constructed by using knowledge of the biological process to describe uncertainties in the parameter. The approach runs into trouble when no prior information is available about the parameter and the prior is supposed to represent total ignorance. For a continuous parameter, the 'principle of insufficient reason' or 'principle of indifference', largely due to Laplace, assigns a uniform distribution over the range of the parameter. However, such so-called *flat* or *noninformative priors* lead to contradictions. Suppose we want to estimate the size of a square. We know its side is between 1 m and 2 m, so we can let the side have a uniform distribution $U(1, 2)$. On the other hand, the area lies between 1 and 4 m^2, so we can let the area have a uniform distribution $U(1, 4)$. These two priors contradict each other. For example, by the uniform prior for the side, the probability is $1/2$ for the side to be in the interval $(1, 1.5)$ or for the area to be in the interval $(1, 2.25)$. However, by the uniform prior for the area, the probability for the area to be in the interval $(1, 2.25)$ is $(2.25 - 1)/(4 - 1) = 0.3834$, not $1/2$! This contradiction is due to the fact that the prior is not invariant to nonlinear transformations: if x has a uniform distribution, x^2 cannot have a uniform distribution. Similarly, a uniform prior for the probability of different sites p is very different from a uniform prior for sequence distance θ under the JC69 model (see Appendix A). For a discrete parameter that takes m possible values, the principle of indifference assigns probability $1/m$ for each value. This is not as simple as it may seem, as often it is unclear how the parameter values should be partitioned. For example, suppose we are interested in deciding whether a certain event occurred during weekdays or weekends. We can assign prior probabilities $1/2$ for weekdays and $1/2$ for weekends. An alternative is to consider each day of the week as having an equal probability and to assign $5/7$ for weekdays and $2/7$ for weekends. With no information about the parameter, it is unclear which prior is more reasonable. Such difficulties in representing total ignorance have caused the objective Bayesian approach to fall out of favour. Nowadays it is generally accepted that uniform priors are not noninformative and that indeed no prior can represent total ignorance.

The *subjective* Bayesians consider the prior to represent the researcher's *subjective belief* about the parameter before seeing or analysing the data. One cannot really argue against somebody else's subjective beliefs, but 'classical' statisticians reject the notion of subjective probabilities and of letting personal prejudices influence scientific inference. Even though the choice of the likelihood model involves certain subjectivity as well, the model can nevertheless be checked against the data, but no such validation is possible for the prior. However, if prior information is available about the parameter, the use of a prior in the Bayesian framework provides a natural way of incorporating such information, as pointed out by Bayesians.

If one accepts the prior, calculation of the posterior, dictated by the probability calculus, is automatic and self-consistent, free of the criticisms levied on classical statistics discussed above. As mentioned already, Bayesian inference produces direct answers about the parameters that are easy to interpret. Bayesians argue that the concerns about the prior are technical issues in a fundamentally sound theory, while classical statistics is a fundamentally flawed theory, with *ad hoc* fixes occasionally producing sensible results for the wrong reasons.

5.1.3.3 Does it matter?

Thus classical (frequentist) and Bayesian statistics are based on quite different philosophies. To a biologist, an important question is whether the two approaches produce similar answers. This depends on the nature of the problem. Here we consider three kinds of problems.

In so-called *stable-estimation problems* (Savage 1962), a well-formulated model $f(X|\theta)$ is available, and we want to estimate parameters θ from a large data set. The

156 • 5 Bayesian methods

prior will have little effect, and both likelihood and Bayesian estimates will be close to the true parameter value. Furthermore, classical confidence intervals in general match posterior credibility intervals under vague priors. Consider, for example, the uniform prior $f(\theta) = 1/(2c)$, $-c < \theta < c$. (Here use of the uniform prior is not essential as other vague priors lead to the same conclusion.) The posterior is

$$f(\theta|x) = \frac{f(\theta)f(x|\theta)}{\int_{-c}^{c} f(\theta)f(x|\theta)\,d\theta} = \frac{f(x|\theta)}{\int_{-c}^{c} f(x|\theta)\,d\theta}. \tag{5.20}$$

The integral in the denominator gives the area under the curve $f(x|\theta)$ from $-c$ to c. In a large data set, the likelihood $f(x|\theta)$ is highly concentrated in a small region of θ (which should be inside the prior interval as long as the prior is diffuse enough to contain the true θ), outside which it is vanishingly small. The integral is then effectively given by the area in that small region and is insensitive to the precise value of c (Fig. 5.5). Thus the posterior will be insensitive to c or to the prior. A good example is the distance estimation discussed above (Fig. 5.2).

The second kind includes estimation problems in which both the prior and the likelihood exert substantial influence on the posterior. Bayesian inference will then be sensitive to the prior, and classical and Bayesian methods are also likely to produce different answers. The sensitivity of the posterior to the prior can be due to ill-formulated models that are barely identifiable, which lead to strongly correlated parameters, or to paucity of data containing little information about the parameters. Increasing the amount of data will help with the situation.

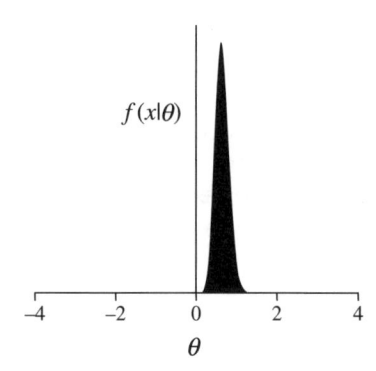

Fig. 5.5 The integral $\int_{-c}^{c} f(x|\theta)\,d\theta$ is the area under the likelihood curve $f(x|\theta)$ over the prior interval $(-c, c)$. When the data set is large, the likelihood $f(x|\theta)$ is highly concentrated in a small interval of θ. Thus as long as c is large enough so that the peak of the likelihood curve is well inside the prior interval $(-c, c)$, the integral will be insensitive to the precise value of c. In the example, the likelihood is concentrated around $\theta = 0.2$ and is vanishingly small outside the interval $(0, 2)$. The areas under the curve over the intervals $(-10, 10)$ or $(-100, 100)$ are virtually the same, and both are essentially identical to the area over the interval $(0, 2)$.

The third kind includes the most difficult problems, of hypothesis testing or model selection when the models involve unknown parameters for which only vague prior information is available. The posterior probabilities of models are very sensitive to the prior for the unknown parameters, and increasing the amount of data does not appear to help with the situation. The best-known case is *Lindley's paradox* (Lindley 1957; see also Jeffreys 1939). Consider the test of a simple null hypothesis $H_0: \theta = 0$ against the composite alternative hypothesis $H_1: \theta \neq 0$ using a random sample $x_1, x_2, \ldots, x_n$ from the normal distribution $N(\theta, 1)$. Let $x = (x_1, x_2, \ldots, x_n)$, and $\bar{x}$ be the sample mean. Suppose the prior is $P(H_0) = P(H_1) = 1/2$, and $\theta \sim N(0, \sigma^2)$ under H_1. The likelihood is given by $\bar{x} \sim N(0, 1/n)$ under H_0 and by $\bar{x} \sim N(\theta, 1/n)$ under H_1. Then the ratio of posterior model probabilities is equal to the ratio of the marginal likelihoods

$$B_{01} = \frac{P(H_0|x)}{P(H_1|x)}$$

$$= \frac{(1/\sqrt{2\pi/n}) \exp\{-(n/2)\bar{x}^2\}}{\int_{-\infty}^{\infty} (1/\sqrt{2\pi\sigma^2}) \exp\{-(\theta^2/(2\sigma^2))\} \times (1/\sqrt{2\pi/n}) \exp\{-(n/2)(\bar{x}-\theta)^2\} d\theta}$$

$$= \sqrt{1 + n\sigma^2} \times \exp\left\{-\frac{n\bar{x}^2}{2[1 + 1/(n\sigma^2)]}\right\}. \tag{5.21}$$

For given data, note that $B_{01} \to \infty$ and $P(H_0|x) \to 1$ if $\sigma^2 \to \infty$. Thus $P(H_0|x)$ can be made as close to 1 as one wishes, by using a sufficiently diffuse prior (that is, by using a large enough σ^2). The difficulty is that when the prior information is weak, one may not be able to decide whether $\sigma^2 = 10$ or 100 is more appropriate, even though the posterior model probability may differ markedly between the two.

Lindley pointed out that in this problem, Bayesian analysis and the traditional significance test can produce drastically different conclusions. The usual hypothesis test is based on $\bar{x}$ having a normal distribution $N(0, 1/n)$ under H_0 and calculates the p value as $\Phi(-\sqrt{n}|\bar{x}|)$, where $\Phi(\cdot)$ is the cumulative density function (cdf) of the standard normal distribution. Suppose that $\sqrt{n}|\bar{x}| = z_{\alpha/2}$, the percentile of the standard normal distribution. Thus we reject H_0 at the significance level α. However, as $n \to \infty$, we see that $B_{01} \to \infty$ and $P(H_0|x) \to 1$. Hence the paradox: while the significance test rejects H_0 decisively at $\alpha = 10^{-10}$, say, the Bayesian method strongly supports H_0 with $P(H_0|x) \approx 1$.

Lindley's paradox, as a manifestation of the sensitivity of posterior model probabilities to the prior, arises whenever we wish to compare different models with unknown parameters but we have only weak prior information about parameters in one or more of the models (e.g. O'Hagan and Forster 2004). For such difficulties to arise, the compared models can have one or more parameters, or one model can be sharp (with no parameters), and the prior can be proper and informative since increasing the size of data while keeping the prior fixed has the same effect.

Bayesian phylogeny reconstruction appears to fall into this third category of difficult problems (Yang and Rannala 2005). See Section 5.6 below for more discussions.

5.2 Prior

If the physical process can be used to model uncertainties in the quantities of interest, it is standard in both classical (likelihood) and Bayesian statistics to treat such quantities as random variables, and derive their conditional (posterior) probability distribution given the data. An example relevant here is the use of the Yule branching process (Edwards 1970) and the birth–death process (Rannala and Yang 1996) to specify the probability distributions of phylogenies. The parameters in the models are the birth and death rates, estimated from the marginal likelihood, which averages over the tree topologies and branch lengths, while the phylogeny is estimated from the conditional (posterior) probability distribution of phylogenies given the data.

Besides use of a biological model, a second approach to specifying the prior is to use past observations of the parameters in similar situations, or to assess prior evidence concerning the parameter. The third approach is to use subjective beliefs of the researcher. When little information is available, uniform distributions are often used as vague priors. For a discrete parameter that can take m possible values, this means assigning probability $1/m$ to each element. For a continuous parameter, this means a uniform distribution over the range of the parameters. Such priors used to be called *noninformative priors* or *flat priors*, but are now referred to as *diffuse* or *vague priors*.

Another class of priors is the *conjugate priors*. Here the prior and the posterior have the same distributional form, and the role of the data or likelihood is to update the parameters in that distribution. Well-known examples include (i) binomial distribution of data $B(n, p)$ with a beta prior for the probability parameter p; (ii) Poisson distribution of data poisson(λ) with a gamma prior for the rate parameter λ; and (iii) normal distribution of data $N(\mu, \sigma^2)$ with a normal prior for the mean μ. In our example of estimating sequence distance under the JC69 model, if we use the probability of different sites p as the distance, we can assign a beta prior beta(α, β). When the data have x differences out of n sites, the posterior distribution of p is beta($\alpha+x, \beta+n-x$). This result also illustrates the information contained in the beta prior: beta(α, β) is equivalent to observing α differences out of $\alpha + \beta$ sites. Conjugate priors are possible only for special combinations of the prior and likelihood. They are convenient as the integrals are tractable analytically, but they may not be realistic models for the problem at hand. Conjugate priors have not found a use in molecular phylogenetics except for the trivial case discussed above, as the problem is typically too complex.

When the prior distribution involves unknown parameters, one can assign priors for them, called *hyper-priors*. Unknown parameters in the hyper-prior can have their own priors. This is known as the *hierarchical Bayesian* or *full Bayesian* approach. Typically one does not go beyond two or three levels, as the effect will become unimportant. For example, the mean μ in the exponential prior in our example of distance calculation under JC69 (equation 5.10) can be assigned a hyper-prior. An alternative is to estimate the hyper-parameters from the marginal likelihood, and use them in posterior probability calculation for parameters of interest. This is known as the *empirical Bayes* approach. For example, μ can be estimated by maximizing $f(x|\mu) = \int f(\theta|\mu) f(x|\theta) \, d\theta$, and the estimate can be used to calculate $f(\theta|x)$ in equation (5.11). The empirical Bayes approach has been widely used in molecular phylogenetics, for example, to estimate evolutionary rates at sites (Yang and Wang 1995; Subsection 4.3.1), to reconstruct ancestral DNA or protein sequences on a phylogeny (Yang et al. 1995a; Koshi and Goldstein 1996a; Section 4.4), to identify amino acid residues under positive selection (Nielsen and Yang 1998, Chapter 8), to infer secondary structure categories of a protein sequence (Goldman et al. 1998), and to construct sequence alignments under models of insertions and deletions (Thorne et al. 1991; Thorne and Kishino 1992).

An important question in real data analysis is whether the posterior is sensitive to the prior. It is always prudent to assess the influence of the prior. If the posterior is dominated by the data, the choice of the prior is inconsequential. When this is not the case, the effect of the prior has to be assessed carefully and reported. Due to advances in computational algorithms (see below), the Bayesian methodology has enabled researchers to fit sophisticated parameter-rich models. As a result, the researcher may be tempted to add parameters that are barely identifiable (Rannala 2002), and the posterior may be unduly influenced by some aspects of the prior even without the knowledge of the researcher. In our example of distance estimation under the JC69 model, problems of identifiability will arise if we attempt to estimate both the substitution rate λ and the time t instead of the distance $\theta = 3\lambda t$ (see Exercise 5.5). It is thus important to understand which aspects of the data provide information about the parameters, which parameters are knowable and which are not, to avoid overloading the model with too many parameters.

5.3 Markov chain Monte Carlo

For most problems, the prior and the likelihood are easy to calculate, but the marginal probability of the data $f(X)$, i.e. the normalizing constant, is hard to calculate. Except for trivial problems such as cases involving conjugate priors, analytical results are unavailable. We have noted above the difficulty in calculating the marginal likelihood $f(x)$ in our simple case of distance estimation (equation 5.11). More complex Bayesian models can involve hundreds or thousands of parameters and high-dimensional integrals. For example, to calculate posterior probabilities for phylogenetic trees, one has to evaluate the marginal probability of data $f(X)$, which is a sum over all possible tree topologies and integration over all branch lengths in those trees and over all parameters in the substitution model. The breakthrough is the development of Markov chain Monte Carlo (MCMC) algorithms, which provide a powerful method for achieving Bayesian computation. Even though MCMC algorithms were also suggested for likelihood calculation, to integrate over random variables in the model (Geyer 1991; Kuhner et al. 1995), they are far less successful in that application (Stephens and Donnelly 2000).

Before introducing MCMC algorithms, we discuss Monte Carlo integration first. The two kinds of algorithms are closely related, and it is important to appreciate their similarities and differences.

5.3.1 Monte Carlo integration

Monte Carlo integration is a simulation method for calculating multidimensional integrals. Suppose we want to compute the expectation of $h(\theta)$ over the density $\pi(\theta)$, with θ possibly multidimensional

$$I = E_\pi(h(\theta)) = \int h(\theta) \pi(\theta) \, d\theta. \tag{5.22}$$

We can draw independent samples $\theta_1, \theta_2, \ldots, \theta_N$ from $\pi(\theta)$. Then

$$\hat{I} = \frac{1}{N} \sum_{i=1}^{N} h(\theta_i) \tag{5.23}$$

is the MLE of I. Since $h(\theta_i)$ are independent and identically distributed, $\hat{I}$ has an asymptotic normal distribution with mean I and variance

$$\text{var}(\hat{I}) = \frac{1}{N^2} \sum_{i=1}^{N} (h(\theta_i) - \hat{I})^2. \tag{5.24}$$

An advantage of Monte Carlo integration is that the variance of the estimate depends on the sample size N, but not the dimension of the integral, unlike numerical integration algorithms whose performance deteriorates rapidly with the increase in dimension.

Monte Carlo integration has serious drawbacks. First, when the function $h(\theta)$ is very different from $\pi(\theta)$, sampling from $\pi(\theta)$ can be very inefficient. Second, the method requires our being able to sample from the distribution $\pi(\theta)$, which may be unfeasible or impossible, especially for high-dimensional problems. Consider applying Monte Carlo integration to calculation of the integral $f(x) = \int f(\theta) f(x|\theta) \, d\theta$ in equation (5.11). It is not easy to sample from the posterior $f(\theta|x)$; the whole point of calculating $f(x)$ is to derive the posterior. Instead one may let $\pi(\theta) = f(\theta)$ and sample from the prior, and then $h(\theta) = f(x|\theta)$ will be the likelihood. While the exponential prior spans the whole positive real line with a slow decay from $1/\mu$ at $\theta = 0$ to 0 at $\theta = \infty$, the likelihood has a spike around $\theta = 0.1$ and is virtually zero outside the narrow interval $(0.06, 0.14)$ (see Fig. 5.2, but note that θ goes from 0 to ∞, so the likelihood curve should look like the spike in Fig. 5.5). Thus most values of θ sampled from the prior will correspond to vanishingly small values of $h(\theta)$ while a few will correspond to high values of $h(\theta)$. The large variation among the $h(\theta_i)$ means that $\text{var}(\hat{I})$ will be large, and a huge number of samples are needed to obtain acceptable estimates. In high dimensions, Monte Carlo integration tends to be even less efficient because random draws from $\pi(\theta)$ are more likely to miss the spike in $h(\theta)$ than in one dimension.

5.3.2 Metropolis–Hastings algorithm

MCMC algorithms generate a dependent sample $\theta_1, \theta_2, \ldots, \theta_N$ from the target density $\pi(\theta)$. Indeed, $\theta_1, \theta_2, \ldots, \theta_N$ form a stationary (discrete-time) Markov chain, with the possible values of θ being the states of the chain. As in the independent samples generated from Monte Carlo integration, the sample mean of $h(\theta)$ provides an unbiased estimate of the integral of equation (5.22):

$$\tilde{I} = \frac{1}{N} \sum_{i=1}^{N} h(\theta_i). \tag{5.25}$$

However, the variance of $\tilde{I}$ is not given by equation (5.24) anymore, as we have to account for the fact that the sample is dependent. Let ρ_k be the coefficient of autocorrelation of $h(\theta_i)$ over the Markov chain at lag k. Then the large-sample variance from the dependent sample is given as

$$\text{var}(\tilde{I}) = \text{var}(\hat{I}) \times [1 + 2(\rho_1 + \rho_2 + \rho_3 + \cdots)] \tag{5.26}$$

(e.g. Diggle 1990, pp. 87–92). In effect, a dependent sample of size N contains as much information as an independent sample of size $N/[1 + 2(\rho_1 + \rho_2 + \cdots)]$. This size is known as the *effective sample size*.

When MCMC is used in Bayesian computation, the target density is the posterior $\pi(\theta) = f(\theta|X)$. While it is in general impossible to generate *independent* samples from the posterior, as required in Monte Carlo integration, MCMC algorithms generate *dependent* samples from the posterior. Here we illustrate the main features of the algorithm of Metropolis *et al.* (1953) by using a simple example of a robot jumping on three boxes (Fig. 5.6a). Parameter θ can take three values: 1, 2, or 3, corresponding to the three boxes, while the heights of the boxes are proportional to the target (posterior)

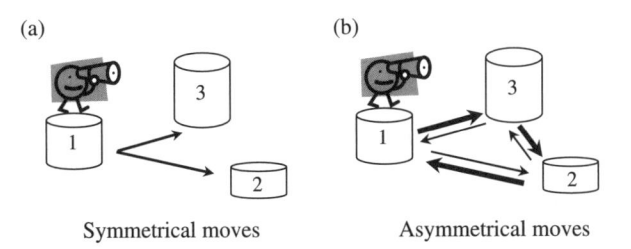

Fig. 5.6 (a) From the current state (say, box 1), a new state is proposed by choosing one out of the two alternative states at random, each with probability 1/2. In the Metropolis algorithm, the proposal density is symmetrical; that is, the probability of going from box 1 to box 2 is the same as the probability of going from box 2 to box 1. (b) In the Metropolis–Hastings algorithm, the proposal density is asymmetrical. The robot has a 'left bias' and proposes the box on the left with probability 2/3 and the box on the right with probability 1/3. Then a proposal ratio is used in calculating the acceptance ratio to correct for the asymmetry of the proposal.

probabilities $\pi(\theta_i) = \pi_i$, for $i = 1, 2, 3$, which we wish to estimate. We give the algorithm before discussing its features

1. Set initial state (say, box $\theta = 1$).
2. Propose a new state θ^*, by choosing one of the two alternative states, each with probability $1/2$.
3. Accept or reject the proposal. If $\pi(\theta^*) > \pi(\theta)$, accept θ^*. Otherwise accept θ^* with probability $\alpha = \pi(\theta^*)/\pi(\theta)$. If the proposal is accepted, set $\theta = \theta^*$. Otherwise set $\theta = \theta$.
4. Print out θ.
5. Go to step 2.

In step 3, one can decide whether to accept or reject the proposal θ^* by generating a random number r from $U(0, 1)$. If $r < \alpha$, accept θ^*, or otherwise reject θ^*. It is easy to see that in this way the proposal is accepted with probability α. Chapter 9 provides more detailed discussions of random numbers and simulation techniques.

Several important features of the algorithm are noteworthy. First, the new state θ^* is proposed through a *proposal density* or *jumping kernel* $q(\theta^*|\theta)$. In the Metropolis algorithm, the proposal density is symmetrical; that is, the probability of proposal from θ to θ^* is equal to the probability of proposal in the opposite direction: $q(\theta^*|\theta) = q(\theta|\theta^*) = 1/2$ for any $\theta \neq \theta^*$. Second, the algorithm generates as output a random sequence of visited states, say

$$1, 3, 3, 3, 2, 1, 1, 3, 2, 3, 3, \ldots.$$

This sequence constitutes a Markov chain; given the current state, the next state to be sampled does not depend on past states. Third, knowledge of the ratio $\pi(\theta^*)/\pi(\theta)$, as opposed to $\pi(\theta)$, is sufficient to implement the algorithm. This is the reason why MCMC algorithms can be used to generate (dependent) samples from the posterior. Since the target distribution is the posterior, $\pi(\theta) = f(\theta|X) = f(\theta)f(X|\theta)/f(X)$, the new state θ^* is accepted with probability

$$\alpha = \min\left(1, \frac{\pi(\theta^*)}{\pi(\theta)}\right) = \min\left(1, \frac{f(\theta^*)f(X|\theta^*)}{f(\theta)f(X|\theta)}\right). \quad (5.27)$$

Here α is called the *acceptance ratio*. Note that, importantly, the normalizing constant $f(X)$ in equation (5.6), which is difficult to compute, cancels in calculation of α.

Fourth, if we let the algorithm run for a long time, the robot will spend more time on a high box than on a low box. Indeed, the proportions of time the robot will spend on boxes 1, 2, and 3 are π_1, π_2, and π_3, respectively, so that $\pi(\theta)$ is the steady-state distribution of the Markov chain. Thus to estimate $\pi(\theta)$, one has only to run the algorithm for a long time and to calculate the frequencies at which the three states are visited in the Markov chain.

The above is a version of the algorithm of Metropolis *et al.* (1953), which generates a Markov chain whose states are the possible values of the parameter θ and whose steady-state distribution is the posterior distribution $\pi(\theta) = f(\theta|X)$.

The algorithm of Metropolis et al. assumes symmetrical proposals. This was extended by Hastings (1970) to allow asymmetrical proposal densities, with $q(\theta^*|\theta) \neq q(\theta|\theta^*)$. The Metropolis–Hastings algorithm thus involves a simple correction in calculation of the acceptance ratio

$$\alpha = \min\left(1, \frac{\pi(\theta^*)}{\pi(\theta)} \times \frac{q(\theta|\theta^*)}{q(\theta^*|\theta)}\right)$$

$$= \min\left(1, \frac{f(\theta^*)}{f(\theta)} \times \frac{f(X|\theta^*)}{f(X|\theta)} \times \frac{q(\theta|\theta^*)}{q(\theta^*|\theta)}\right) \quad (5.28)$$

$$= \min(1, \text{prior ratio} \times \text{likelihood ratio} \times \text{proposal ratio}).$$

By using the *proposal ratio* or the *Hastings ratio*, $q(\theta|\theta^*)/q(\theta^*|\theta)$, the correct target density is recovered, even if the proposal is biased. Consider, for example, calculation of the proposal ratio for $\theta = 1$ and $\theta^* = 2$ in the robot-on-box example when the robot chooses the box on the left with probability 2/3 and the one on the right with probability 1/3 (Fig. 5.6b). We have $q(\theta|\theta^*) = 2/3$, $q(\theta^*|\theta) = 1/3$, so that the proposal ratio is $q(\theta|\theta^*)/q(\theta^*|\theta) = 2$. Similarly the proposal ratio for the move from box 2 to 1 is 1/2. Thus by accepting right boxes more often and left boxes less often, the Markov chain recovers the correct target density even though the robot has a left 'bias' in proposing moves.

For the Markov chain to converge to $\pi(\theta)$, the proposal density $q(\cdot|\cdot)$ has to satisfy certain regularity conditions; it has to specify an irreducible and aperiodic chain. In other words, $q(\cdot|\cdot)$ should allow the chain to reach any state from any other state, and the chain should not have a period. Those conditions are often easily satisfied. In most applications, the prior ratio $f(\theta^*)/f(\theta)$ is easy to calculate. The likelihood ratio $f(X|\theta^*)/f(X|\theta)$ is often easy to calculate as well, even though it may be computationally expensive. The proposal ratio $q(\theta|\theta^*)/q(\theta^*|\theta)$ greatly affects the efficiency of the MCMC algorithm. Therefore a lot of effort is spent on developing good proposal algorithms.

Example. The algorithm for continuous parameters is essentially the same as for discrete parameters, except that the state space of the Markov chain is continuous. As an example, we apply the Metropolis algorithm to estimation of sequence distance under the JC69 model. The reader is invited to write a small program to implement this algorithm (Exercise 5.4). The data are $x = 90$ differences out of $n = 948$ sites. The prior is $f(\theta) = (1/\mu)e^{-\theta/\mu}$, with $\mu = 0.2$. The proposal algorithm uses a sliding window of size w:

1. Initialize: $n = 948$, $x = 90$, $w = 0.01$.
2. Set initial state: $\theta = 0.05$, say.
3. Propose a new state as $\theta^* \sim U(\theta - w/2, \theta + w/2)$. That is, generate a $U(0, 1)$ random number r, and set $\theta^* = \theta - w/2 + wr$. If $\theta^* < 0$, set $\theta^* = -\theta^*$.

4. Calculate the acceptance ratio α, using equation (5.13) to calculate the likelihood $f(x|\theta)$:

$$\alpha = \min\left(1, \frac{f(\theta^*)f(x|\theta^*)}{f(\theta)f(x|\theta)}\right). \tag{5.29}$$

5. Accept or reject the proposal θ^*. Draw $r \sim U(0, 1)$. If $r < \alpha$ set $\theta = \theta^*$. Otherwise set $\theta = \theta$. Print out θ.
6. Go to step 3.

Figures 5.7(a) and (b) show the first 200 iterations of five independent chains, started from different initial values and using different window sizes. Fig. 5.7(c) shows a histogram approximation to the posterior probability density estimated from a long chain, while Fig. 5.7(d) is from a very long chain, which is indistinguishable from the distribution calculated using numerical integration (Fig. 5.2). □

There exist a number of special cases of the general Metropolis–Hastings algorithm. Below we mention some commonly used ones, such as the single-component Metropolis–Hastings algorithm and the Gibbs sampler. We also mention an important generalization, called Metropolis-coupled MCMC or MC^3.

5.3.3 Single-component Metropolis–Hastings algorithm

The advantage of Bayesian inference mostly lies in the ease with which it can deal with sophisticated multiparameter models. In particular, Bayesian 'marginalization' of nuisance parameters (equation 5.9) provides an attractive way of accommodating variation in the data which we cannot ignore but which we are not really interested in. In MCMC algorithms for such multiparameter models, it is often unfeasible or computationally too complicated to update all parameters in θ simultaneously. Instead, it is more convenient to divide θ into components or blocks, of possibly different dimensions, and then update those components one by one. Different proposals are often used to update different components. This is known as *blocking*. Many models have a structure of conditional independence, and blocking often leads to computational efficiency.

A variety of strategies are possible concerning the order of updating the components. One can use a fixed order, or a random permutation of the components. There is no need to update every component in every iteration. One can also select components for updating with fixed probabilities. However, the probabilities should be fixed and not dependent on the current state of the Markov chain, as otherwise the stationary distribution may no longer be the target distribution $\pi(\cdot)$. It is advisable to update highly correlated components more frequently. It is also advantageous to group into one block components that are highly correlated in the posterior density, and update them simultaneously using a proposal density that accounts for the correlation (see Subsection 5.4.3 below).

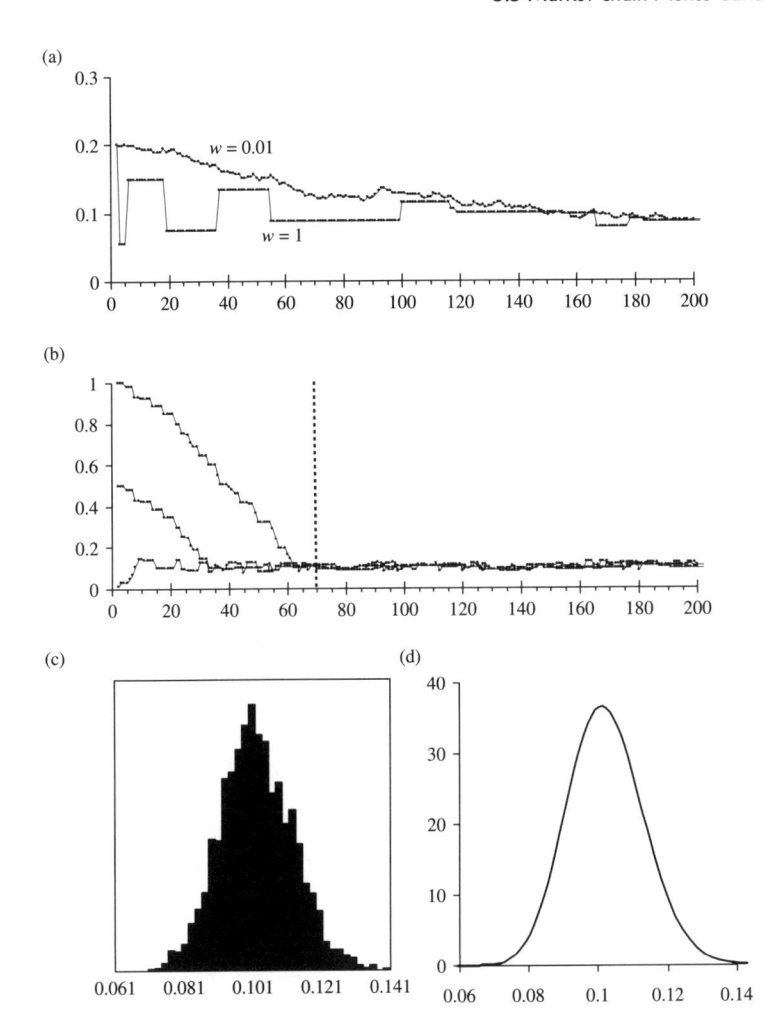

Fig. 5.7 MCMC runs for estimating sequence distance θ under the JC69 model. The data consists of $x = 90$ differences between two sequences out of $n = 948$ sites. (a) Two chains with the window size either too small ($w = 0.01$) or too large ($w = 1$). Both chains started at $\theta = 0.2$. The chain with $w = 0.01$ has an acceptance proportion of 91%, so that almost every proposal is accepted. However, this chain takes tiny baby-steps and mixes poorly. The chain with $w = 1$ has the acceptance proportion 7%, so that most proposals are rejected. The chain often stays at the same state for many iterations without a move. Further experiment shows that the window size $w = 0.1$ leads to an acceptance rate of 35%, and is near optimum (see text). (b) Three chains started from $\theta = 0.01, 0.5$, and 1, with window size 0.1. It appears that after about 70 iterations, the three chains become indistinguishable and have reached stationarity, so that a burn-in of 100 iterations appears sufficient for those chains. (c) Histogram constructed from 10 000 iterations. (d) Posterior density obtained from a long chain of 10 000 000 iterations, sampling every 10 iterations, estimated using a kernel density smoothing algorithm (Silverman 1986).

5.3.4 Gibbs sampler

The *Gibbs sampler* (Gelman and Gelman 1984) is a special case of the single-component Metropolis–Hastings algorithm. The proposal distribution for updating the ith component is the conditional distribution of the ith component given all the other components. This proposal leads to an acceptance ratio of $\alpha = 1$; that is, all proposals are accepted. The Gibbs sampler is the algorithm tested in the seminal paper of Gelfand and Smith (1990). It is widely used in analysis under linear models involving normal prior and posterior densities. However, it has not been used in molecular phylogenetics, where it is in general impossible to obtain the conditional distributions analytically.

5.3.5 Metropolis-coupled MCMC (MCMCMC or MC³)

If the target distribution has multiple peaks, separated by low valleys, the Markov chain may have difficulties in moving from one peak to another. As a result, the chain may get stuck on one peak and the resulting samples will not approximate the posterior density correctly. This is a serious practical concern for phylogeny reconstruction, as multiple local peaks are known to exist in the tree space. A strategy to improve mixing in presence of multiple local peaks is the Metropolis-coupled MCMC or MCMCMC (MC³) algorithm (Geyer 1991), which is similar to the simulated annealing algorithm (Metropolis *et al.* 1953).

In MC³, m chains are run in parallel, with different stationary distributions $\pi_j(\cdot)$, $j = 1, 2, \ldots, m$, where $\pi_1(\cdot) = \pi(\cdot)$ is the target density while $\pi_j(\cdot), j = 2, 3, \ldots, m$ are designed to improve mixing. The first chain is called the *cold chain*, while the other chains are the *hot chains*. Only the cold chain converges to the correct posterior density. For example, one can use incremental heating of the form

$$\pi_j(\theta) \propto \pi(\theta)^{1/[1+\lambda(j-1)]}, \quad \lambda > 0. \tag{5.30}$$

Note that raising the density $\pi(\cdot)$ to the power $1/T$ with $T > 1$ has the effect of flattening out the distribution, making it easier for the algorithm to traverse between peaks across the valleys than in the original distribution (see Fig. 5.8). After each iteration, a swap of states between two randomly chosen chains is proposed through a Metropolis–Hastings step. Let $\theta^{(j)}$ be the current state in chain $j, j = 1, 2, \ldots, m$. A swap between the states of chains i and j is accepted with probability

$$\alpha = \min\left(1, \frac{\pi_i(\theta_j)\pi_j(\theta_i)}{\pi_i(\theta_i)\pi_j(\theta_j)}\right). \tag{5.31}$$

Heuristically, the hot chains will visit the local peaks easily, and swapping states between chains will let the cold chain occasionally jump valleys, leading to better mixing. At the end of the run, output from only the cold chain is used, while outputs from the hot chains are discarded. An obvious disadvantage of the algorithm is that m chains are run but only one chain is used for inference. MC³ is ideally suited to

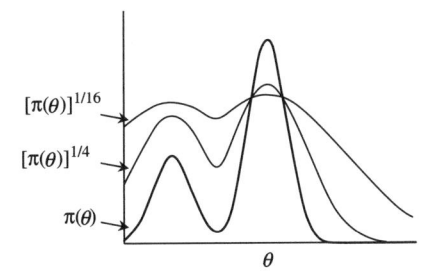

Fig. 5.8 A density $\pi(\theta)$ with two peaks and two 'flattened' densities $[\pi(\theta)]^{1/4}$ and $[\pi(\theta)]^{1/16}$. Note that the two flattened densities are relative, defined up to a normalizing constant (a scale factor to make the density integrate to 1).

implementation on parallel machines or network workstations, since each chain will in general require about the same amount of computation per iteration, and interactions between chains are minimal.

MC^3 may not be very effective when the local peaks are separated by deep valleys. For example, in the data of Fig. 3.15 (see Subsection 3.2.4), two of the fifteen possible trees have higher likelihoods (and thus higher posterior probabilities) than other trees, but the two trees are separated by other trees of much lower likelihoods. In this case, running as many as 10 chains for over 10^8 iterations in MrBayes (Huelsenbeck and Ronquist 2001) did not produce reliable results, as they differ between runs.

5.4 Simple moves and their proposal ratios

The proposal ratio is separate from the prior or the likelihood and is solely dependent on the proposal algorithm. Thus the same proposals can be used in a variety of Bayesian inference problems. As mentioned earlier, the proposal density has only to specify an aperiodic recurrent Markov chain to guarantee convergence of the MCMC algorithm. It is typically easy to construct such chains and to verify that they satisfy those conditions. For a discrete parameter that takes a set of values, calculation of the proposal ratio often amounts to counting the number of candidate values in the source and target. The case of continuous parameters requires more care. This section lists a few commonly used proposals and their proposal ratios. Symbols x and y are sometimes used instead of θ to represent the state of the chain.

Two general results are very useful in deriving proposal ratios, and are presented as two theorems in Appendix A. Theorem 1 specifies the probability density of functions of random variables, which are themselves random variables. Theorem 2 gives the proposal ratio when the proposal is formulated as changes to certain functions of the variables (rather than the variables themselves) in the Markov chain. We will often refer to those two results.

168 • 5 Bayesian methods

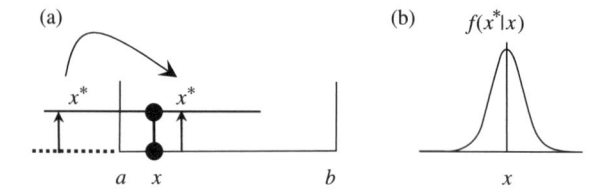

Fig. 5.9 (a) Sliding window using a uniform distribution. The current state is x. A new value x^* is proposed by sampling uniformly from a sliding window of width w centred at the current value. The window width affects the acceptance proportion. If the proposed value is outside the feasible range (a, b), it is reflected back into the interval; for example, if $x^* < a$, it is reset to $x^* = a + (a - x^*) = 2a - x^*$. (b) Sliding window using a normal distribution. The new value x^* is proposed by sampling from a normal distribution $N(x, \sigma^2)$. The variance σ^2 influences the size of the steps taken and plays the same role as the window size w in the proposal of (a).

5.4.1 Sliding window using the uniform proposal

This proposal chooses the new state x^* as a random variable from a uniform distribution around the current state x (Fig. 5.9a):

$$x^* \sim U(x - w/2, x + w/2). \tag{5.32}$$

The proposal ratio is 1 since $q(x^*|x) = q(x|x^*) = 1/w$. If x is constrained in the interval (a, b) and the new proposed state x^* is outside the range, the excess is reflected back into the interval; that is, if $x^* < a$, x^* is reset to $a + (a - x^*) = 2a - x^*$, and if $x^* > b$, x^* is reset to $b - (b - x^*) = 2b - x^*$. The proposal ratio is 1 even with reflection, because if x can reach x^* through reflection, x^* can reach x through reflection as well. Note that it is incorrect to simply set the unfeasible proposed values to a or b. The window size w is a fixed constant, chosen to achieve a reasonable acceptance rate. It should be smaller than the range $b - a$.

5.4.2 Sliding window using normal proposal

This algorithm uses a normal proposal density centred at the current state; that is, x^* has a normal distribution with mean x and variance σ^2, with σ^2 controlling the step size (Fig. 5.9b).

$$x^* \sim N(x, \sigma^2). \tag{5.33}$$

As $q(x^*|x) = (1/\sqrt{2\pi\sigma^2}) \exp -[(x^* - x)^2/(2\sigma^2)] = q(x|x^*)$, the proposal ratio is 1. This proposal also works if x is constrained in the interval (a, b). If x^* is outside the range, the excess is reflected back into the interval, and the proposal ratio remains one. Both with and without reflection, the number of routes from x to x^* is the same as from x^* to x, and the densities are the same in the opposite directions, even if not between the routes. Note that sliding-window algorithms using either uniform or normal jumping kernels are Metropolis algorithms with symmetrical proposals.

How does one choose σ? Suppose the target density is the normal $N(\theta, 1)$, and the proposal is $x^* \sim N(x, \sigma^2)$. A large σ will cause most proposals to fall in unreasonable regions of the parameter space and to be rejected. The chain then stays at the same state for a long time, causing high correlation. A too small σ means that the proposed states will be very close to the current state, and most proposals will be accepted. However, the chain baby-walks in the same region of the parameter space for a long time, leading again to high correlation. Proposals that minimize the autocorrelations are thus optimal. By minimizing the variance of the mean of a normal distribution, var($\tilde{\theta}$) of equation (5.26), Gelman et al. (1996) found analytically the optimal σ to be about 2.4. Thus if the target density is a general normal density $N(\theta, \tau^2)$, the optimal proposal density should be $N(x, \tau^2\sigma^2)$ with $\sigma = 2.4$. As τ is unknown, it is easier to monitor the *acceptance proportion* or *jumping probability*, the proportion at which proposals are accepted, which is slightly below 0.5 at the optimal σ. However, Robert and Casella (2004, p. 317) pointed out that this rule may not always work; the authors constructed an example in which there are multiple peaks in the posterior and the acceptance proportion never goes below 0.8.

5.4.3 Sliding window using the multivariate normal proposal

If the target density is a m-dimensional normal density $N_m(\boldsymbol{\mu}, \mathbf{I})$, where $\mathbf{I}$ is the $m \times m$ identity matrix, one can use the proposal density $q(\mathbf{x}^*|\mathbf{x}) = N_m(\mathbf{x}, \mathbf{I}\sigma^2)$ to propose a move in the m-dimensional space. The proposal ratio is 1. Gelman et al. (1996) derived the optimal scale factor σ to be 2.4, 1.7, 1.4, 1.2, 1, 0.9, 0.7 for $m = 1, 2, 3, 4, 6, 8, 10$, respectively, with an optimal acceptance rate of about 0.5 for $m = 1$, decreasing to about 0.26 for $m > 6$. It is interesting to note that at low dimensions the optimal proposal density is over-dispersed relative to the target density and one should take big steps, while at high dimensions one should use under-dispersed proposal densities and take small steps. In general one should try to achieve an acceptance proportion of about 20–70% for one-dimensional proposals, and 15–40% for multidimensional proposals.

The above result can easily be extended to deal with a general multivariate normal posterior distribution, with a variance–covariance matrix $\mathbf{S}$. Note that a naïve application of the proposal $q(\mathbf{x}^*|\mathbf{x}) = N_m(\mathbf{x}, \mathbf{I}\sigma^2)$ may be highly inefficient for two reasons. First, different variables may have different scales (variances), so that use of one scale factor σ may cause the proposal step to be too small for variables with large variances and too large for variables with small variances. Second, the variables may be strongly correlated, so that ignoring such correlations in the proposal may cause most proposals to be rejected, leading to poor mixing (Fig. 5.10). One approach is to reparametrize the model using $\mathbf{y} = \mathbf{S}^{-1/2}\mathbf{x}$ as parameters, where $\mathbf{S}^{-1/2}$ is the square root of matrix $\mathbf{S}^{-1}$. Note that $\mathbf{y}$ has unit variance, so that the above proposal can be used. The second approach is to propose new states using the transformed variables $\mathbf{y}$, that is, $q(\mathbf{y}^*|\mathbf{y}) = N_m(\mathbf{y}, \mathbf{I}\sigma^2)$, and then derive the proposal ratio in the original

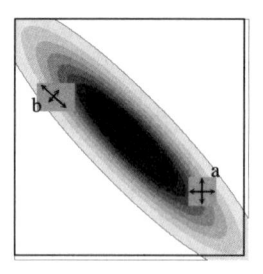

Fig. 5.10 When two parameters are strongly correlated, it is very inefficient to change one variable at a time in the MCMC, since such proposals will have great difficulty in moving along the ridge of the posterior density. Changing both variables but ignoring the correlation (a) is inefficient as well, while accommodating the correlation in the proposal by matching the proposal density to the posterior (b) leads to an efficient algorithm. The correlation structure among the parameters may be estimated by running a short chain before MCMC sampling takes place.

variables **x**. The proposal ratio is 1 according to Theorem 2 in Appendix A. A third approach is to use the proposal $\mathbf{x}^* \sim N_m(\mathbf{x}, \mathbf{S}\sigma^2)$, where σ^2 is chosen according to the above discussion. The three approaches should be equivalent and all of them take care of possible differences in scales and possible correlations among the variables. In real data analysis, **S** is unknown. One can perform short runs of the Markov chain to obtain an estimate $\hat{\mathbf{S}}$ of the variance-covariance matrix in the posterior density, and then use it in the proposal. If **S** is estimated in the same run, samples taken to estimate **S** should be discarded. If the normal distribution is a good approximation to the posterior density, those guidelines should be useful.

5.4.4 Proportional shrinking and expanding

This proposal modifies the parameter by multiplying it with a random variable that is around 1. It is useful when the parameter is always positive or always negative. The proposed value is

$$x^* = x \cdot c = x \cdot e^{\varepsilon(r-1/2)}, \tag{5.34}$$

where $c = e^{\varepsilon(r-1/2)}$ and $r \sim U(0, 1)$, with $\varepsilon > 0$ to be a small fine-tuning parameter, similar to the step length w in the sliding-window proposal. Note that x is shrunk or expanded depending on whether r is $<$ or $> 1/2$. The proposal ratio is c. To see this, derive the proposal density $q(x^*|x)$ through variable transform, considering random variable x^* as a function of random variable r while treating ε and x as fixed. Since $r = 1/2 + \log(x^*/x)/\varepsilon$, and $dr/dx^* = 1/(\varepsilon x^*)$, we have from Theorem 1 in Appendix A

$$q(x^*|x) = f(r(x^*)) \times \left|\frac{dr}{dx^*}\right| = \frac{1}{\varepsilon|x^*|}. \tag{5.35}$$

Similarly $q(x|x^*) = 1/(\varepsilon|x|)$, so the proposal ratio is $q(x|x^*)/q(x^*|x) = c$.

This proposal is useful for shrinking or expanding many variables by the same factor c: $x_i^* = cx_i$, $i = 1, 2, \ldots, m$. If the variables have a fixed order, as in the case of the ages of nodes in a phylogenetic tree (Thorne et al. 1998), the order of the variables will be maintained by this proposal. The proposal is also effective in bringing all variables, such as branch lengths on a phylogeny, into the right scale if all of them are either too large or too small. Although all m variables are altered, the proposal is really in one dimension (along a line in the m-dimensional space). We can derive the proposal ratio using the transform: $y_1 = x_1$, $y_i = x_i/x_1$, $i = 2, 3, \ldots, m$. The proposal changes y_1, but $y_2, \ldots, y_m$ are unaffected. The proposal ratio in the transformed variables is c. The Jacobian of the transform is $|J(\mathbf{y})| = |\partial \mathbf{x}/\partial \mathbf{y}| = y_1^{m-1}$. The proposal ratio in the original variables is $c(y_1^*/y_1)^{m-1} = c^m$, according to Theorem 2 in Appendix A. Another similar proposal multiplies m variables by c and divides n variables by c, with the proposal ratio to be c^{m-n}.

5.5 Monitoring Markov chains and processing output

5.5.1 Validating and diagnosing MCMC algorithms

Developing a correct and efficient MCMC algorithm for a practical application is a challenging task. As remarked by Hastings (1970): 'even the simplest of numerical methods may yield spurious results if insufficient care is taken in their use...The setting is certainly no better for the Markov chain methods and they should be used with appropriate caution.' The power of MCMC enables sophisticated parameter-rich models to be applied to real data analysis, liberating the researcher from the limitations of mathematically tractable but biologically unrealistic models. However, parameter-rich models often cause problems for both inference and computation. There is often a lack of information for estimating the multiple parameters, resulting in nearly flat or ridged likelihood (and thus posterior density) surfaces or strong correlations between parameters. It is usually impossible to independently calculate the posterior probability distribution, making it hard to validate an implementation, that is, to confirm the correctness of the computer program. A Bayesian MCMC program is notably harder to debug than an ML program implementing the same model. In likelihood iteration, the convergence is to a point, and in most optimization algorithms, the log likelihood should always go up, with the gradient approaching zero when the algorithm approaches the MLE. In contrast, a Bayesian MCMC algorithm converges to a statistical distribution, with no statistics having a fixed direction of change.

An MCMC algorithm, even if correctly implemented, can suffer from two problems: slow convergence and poor mixing. The former means that it takes a very long time for the chain to reach stationarity. The latter means that the sampled states are highly correlated over iterations and the chain is inefficient in exploring the parameter space. While it is often obvious that the proposal density $q(\cdot|\cdot)$ satisfies the required regularity conditions so that the MCMC is in theory guaranteed to converge to the target distribution, it is much harder to determine in real data problems whether the chain has reached stationarity. A number of heuristic methods have been suggested to diagnose an MCMC run. Some of them are described below. However, those diagnostics are able to reveal certain problems but are unable to prove the correctness of the algorithm or implementation. Often when the algorithm converges slowly or mixes poorly it is difficult to decide whether this is due to faulty theory, a buggy program, or inefficient but correct algorithms. Currently, to conduct a Bayesian analysis using an MCMC algorithm, one has to run the computer program multiple times, fine-tuning proposal steps and analysing the MCMC samples to diagnose possible problems. It would be nice if such monitoring and diagnoses could be done automatically, ensuring that the chain is run for long enough to achieve reliable inference but not too long to waste computing resources. Currently, such *automatic stopping rules* do not appear feasible (e.g. Robert and Casella 2004). There are always settings which can invalidate any diagnostic criterion, and the stochastic nature inherent in the algorithm prevents any guarantee of performance. In short, diagnostic tools are very useful, but one should bear in mind that they may fail.

In the following we discuss a few strategies for validating and diagnosing MCMC programs. Free software tools are available that implement many more diagnostic tests, such as CODA, which is part of the statistics package R.

1. *Time-series plots* or *trace plots* are a very useful tool for detecting lack of convergence and poor mixing (see, e.g., Figs. 5.7(a) and (b)). One can plot parameters of interest or their functions against the iterations. Note that the chain may appear to have converged with respect to some parameters but not to others, so it is important to monitor many or all parameters.
2. The acceptance rate for each proposal should be neither too high nor too low.
3. Multiple chains run from different starting points should all converge to the same distribution. Gelman and Rubin's (1992) statistic can be used to analyse multiple chains (see below).
4. Another strategy is to run the chain without data, that is, to fix $f(X|\theta) = 1$. The posterior should then be the same as the prior, which may be analytically available for comparison. Theoretical expectations are often available for infinite data as well. Therefore one can simulate larger and larger data sets under a fixed set of parameters and analyse the simulated data under the correct model, to confirm that the Bayesian point estimate becomes closer and closer to the true value. This test relies on the fact that Bayesian estimates are consistent.
5. One can also conduct so-called *Bayesian simulation*, to confirm theoretical expectations. Parameter values are generated by sampling from the prior, and then used to generate data sets under the likelihood model. For a continuous parameter, one can confirm that the $(1-\alpha)100\%$ posterior credibility interval (CI) contains the true parameter value with probability $(1-\alpha)$. We construct the CI for each data set, and examine whether the true parameter value is included in the interval; the proportion of replicates in which the CI includes the true parameter value should equal $(1-\alpha)$. This is called the *hit probability* (Wilson *et al.* 2003). A similar test can be applied to a discrete parameter, such as the tree topology. In this case, each

sequence alignment is generated by sampling the tree topology and branch lengths from the prior and by then evolving sequences on the tree. The Bayesian posterior probabilities of trees or clades should then be the probability that the tree or clade is true. One can bin the posterior probabilities and confirm that among trees with posterior probabilities in the bin 94–96%, say, about 95% of them are the true tree (Huelsenbeck and Rannala 2004; Yang and Rannala 2005).

For a continuous parameter, a more powerful test than the hit probability is the so-called *coverage probability*. Suppose a fixed interval (θ_L, θ_U) covers $(1-\alpha)100\%$ probability density of the prior distribution. Then the posterior coverage of the same fixed interval should on average be $(1-\alpha)100\%$ as well (Rubin and Schenker 1986; Wilson *et al.* 2003). Thus we use Bayesian simulation to generate many data sets, and in each data set calculate the posterior coverage probability of the fixed interval, that is, the mass of posterior density in the interval (θ_L, θ_U). The average of the posterior coverage probabilities over the simulated data sets should equal $(1-\alpha)$.

5.5.2 Potential scale reduction statistic

Gelman and Rubin (1992) suggested a diagnostic statistic called 'estimated potential scale reduction', based on variance-components analysis of samples taken from several chains run using 'over-dispersed' starting points. The rationale is that after convergence, the within-chain variation should be indistinguishable from the between-chain variation, while before convergence, the within-chain variation should be too small and the between-chain variation should be too large. The statistic can be used to monitor any or every parameter of interest or any function of the parameters. Let x be the parameter being monitored, and its variance in the target distribution be τ^2. Suppose there are m chains, each run for n iterations, after the burn-in is discarded. Let x_{ij} be the parameter sampled at the jth iteration from the ith chain. Gelman and Rubin (1992) defined the between-chain variance

$$B = \frac{n}{m-1} \sum_{i=1}^{m} (\bar{x}_{i\cdot} - \bar{x}_{\cdot\cdot})^2, \tag{5.36}$$

and the within-chain variance

$$W = \frac{1}{m(n-1)} \sum_{i=1}^{m} \sum_{j=1}^{n} (x_{ij} - \bar{x}_{i\cdot})^2, \tag{5.37}$$

where $\bar{x}_{i\cdot} = (1/n) \sum_{j=1}^{n} x_{ij}$ is the mean within the ith chain, and $\bar{x}_{\cdot\cdot} = (1/m) \sum_{i=1}^{m} \bar{x}_{i\cdot}$ is the overall mean. If all m chains have reached stationarity and x_{ij} are samples from the same target density, both B and W are unbiased estimates of τ^2, and so is their weighted mean

$$\hat{\tau}^2 = \frac{n-1}{n} W + \frac{1}{n} B. \tag{5.38}$$

If the m chains have not reached stationarity, W will be an underestimate of τ^2, while B will be an overestimate. Gelman and Rubin (1992) showed that in this case $\hat{\tau}^2$ is also an overestimate of τ^2. The *estimated potential scale reduction* is defined as

$$\hat{R} = \frac{\hat{\tau}^2}{W}. \tag{5.39}$$

This should get smaller and approach 1 when the parallel chains reach the same target distribution. In real data problems, values of $\hat{R} < 1.1$ or 1.2 indicate convergence.

5.5.3 Processing output

Before we process the output, the beginning part of the chain before it has converged to the stationary distribution is often discarded as *burn-in*. Often we do not sample every iteration but instead take a sample only for every certain number of iterations. This is known as *thinning* the chain, as the thinned samples have reduced autocorrelations across iterations. In theory, it is always more efficient (producing estimates with smaller variances) to use all samples even if they are correlated. However, MCMC algorithms typically create huge output files, and thinning reduces disk usage and makes the output small enough for further processing.

After the burn-in, samples taken from the MCMC can be summarized in a straightforward manner. The sample mean, median, or mode can be used as a point estimate of the parameter, while the HPD or equal-tail credibility intervals can be constructed from the sample as well. For example, a 95% CI can be constructed by sorting the MCMC output for the variable and then using the 2.5% and 97.5% percentiles. The whole posterior distribution can be estimated from the histogram, after smoothing (Silverman 1986). Two-dimensional joint densities can be estimated as well.

5.6 Bayesian phylogenetics

5.6.1 Brief history

The Bayesian method was introduced to molecular phylogenetics by Rannala and Yang (1996; Yang and Rannala 1997), Mau and Newton (1997), and Li *et al.* (2000). The early studies assumed a constant rate of evolution (the molecular clock) as well as an equal-probability prior for rooted trees, either with or without node ages ordered (labelled histories or rooted trees). Since then, more efficient MCMC algorithms have been implemented in the computer programs BAMBE (Larget and Simon 1999) and MrBayes (Huelsenbeck and Ronquist 2001; Ronquist and Huelsenbeck 2003). The clock constraint is relaxed, enabling phylogenetic inference under more realistic evolutionary models. A number of innovations have been introduced in these programs, adapting tree-perturbation algorithms used in heuristic tree search, such as NNI and SPR (see Subsection 3.2.3), into MCMC proposal algorithms for moving around in the tree space. MrBayes 3 has also incorporated many evolutionary models developed for likelihood inference, and can accommodate heterogeneous data sets from multiple gene loci in a combined analysis. An MC³ algorithm is implemented to overcome multiple local peaks in the tree space. A parallel version of the program makes use of multiple processors on network workstations (Altekar et al. 2004).

5.6.2 General framework

It is straightforward to formulate the problem of phylogeny reconstruction in the general framework of Bayesian inference. Let X be the sequence data. Let θ include all parameters in the substitution model, with a prior distribution $f(\theta)$. Let τ_i be the ith tree topology, $i = 1, 2, \ldots, T_s$, where T_s is the total number of tree topologies for s species. Usually a uniform prior $f(\tau_i) = 1/T_s$ is assumed, although Pickett and Randle (2005) pointed out that this means nonuniform prior probabilities for clades. Let $\mathbf{b}_i$ be the vector of branch lengths on tree τ_i, with prior probability $f(\mathbf{b}_i)$. MrBayes 3 assumes that branch lengths have independent uniform or exponential priors with the parameter (upper bound for the uniform or mean for the exponential) set by the user. The posterior probability of tree τ_i is then

$$P(\tau_i | X) = \frac{\iint f(\theta) f(\tau_i | \theta) f(\mathbf{b}_i | \theta, \tau_i) f(X | \theta, \tau_i, \mathbf{b}_i) \, d\mathbf{b}_i \, d\theta}{\sum_{j=1}^{T_s} \iint f(\theta) f(\tau_j | \theta) f(\mathbf{b}_j | \theta, \tau_j) f(X | \theta, \tau_j, \mathbf{b}_j) \, d\mathbf{b}_j \, d\theta}. \quad (5.40)$$

This is a direct application of equation (5.9), treating τ as the parameter of interest and all other parameters as nuisance parameters. Note that the denominator, the marginal probability of the data $f(X)$, is a sum over all possible tree topologies and, for each tree topology τ_j, an integral over all branch lengths $\mathbf{b}_j$ and substitution parameters θ. This is impossible to calculate numerically except for very small trees. The MCMC algorithm avoids direct calculation of $f(X)$, and achieves the integration over branch lengths $\mathbf{b}_j$ and parameters θ through the Markov chain.

A sketch of an MCMC algorithm may look like the following:

1. Start with a random tree τ, with random branch lengths $\mathbf{b}$, and random substitution parameters θ.
2. In each iteration do the following:
   a. Propose a change to the tree, by using tree rearrangement algorithms (such as NNI or SPR). This step may change branch lengths $\mathbf{b}$ as well.
   b. Propose changes to branch lengths $\mathbf{b}$.
   c. Propose changes to parameters θ.
   d. Every k iterations, sample the chain: save τ, $\mathbf{b}$, θ to disk.
3. At the end of the run, summarize the results.

5.6.3 Summarizing MCMC output

Several procedures have been suggested to summarize the posterior probability distribution of phylogenetic trees. One can take the tree topology with the maximum posterior probability as a point estimate of the true tree. This is called the *MAP tree* (Rannala and Yang 1996), and should be identical or similar to the ML tree under the same model, especially if the data are informative. One can also collect the trees with the highest posterior probabilities into a set of trees with a total probability reaching or exceeding a pre-set threshold, such as 95%. This set constitutes the 95% credibility set of trees (Rannala and Yang 1996; Mau *et al.* 1999).

Posterior probabilities of single (whole) trees can be very low, especially when there are a large number of species on the tree and the data are not highly informative. Thus commonly used procedures summarize shared clades amongst trees visited during the MCMC. For example, one can construct the majority-rule consensus tree and, for each clade on the consensus tree, report the proportion of sampled trees that include the clade (Larget and Simon 1999). Such a proportion is known as the *posterior clade probability*, and is an estimate of the probability that the concerned group is monophyletic (given the data and model). A few practical concerns may be raised concerning the use of clade probabilities. First, as discussed in Subsection 3.1.3, a majority-rule consensus tree may not recognize certain similarities among trees (e.g. the trees of Fig. 3.7), and may thus be a poor summary. It always appears worthwhile to examine posterior probabilities for whole trees. Second, instead of attaching probabilities on the clades on the consensus tree, one may attach probabilities to clades on the MAP tree. This appears more justifiable when the consensus tree and MAP tree differ; using the consensus tree has a logical difficulty similar to constructing a confidence or credibility interval for a parameter that excludes the point estimate. Third, Pickett and Randle (2005) pointed out that with a uniform prior on tree topologies, the probability of a clade depends on both the size of the clade and the number of species on the tree; clades containing very few or very many species have higher prior probabilities than middle-sized clades. This effect can be substantial on large trees, and may lead to spuriously high posterior probabilities for wrong (nonmonophyletic) clades. See an example in Fig. 5.11. The problem stems from using clades to summarize shared features among trees and applies to parsimony and likelihood methods with bootstrapping as well as to Bayesian tree reconstruction.

A phylogenetic tree may best be viewed as a statistical model (see Chapter 6), with the branch lengths on the tree to be parameters in the model. From this viewpoint, the posterior probability for a tree is equivalent to the posterior probability for a model, which is commonly used by Bayesian statisticians. The posterior probability of a clade, however, is an unusual sort of measure as it traverses different models (or trees). Attaching posterior probabilities to clades on the consensus or MAP tree, by examining other trees sampled in the chain, is equivalent to summarizing common features among the better-supported models. While the idea is intuitively appealing, its legitimacy is not obvious.

MrBayes (Huelsenbeck and Ronquist 2001) outputs the posterior means and credibility intervals for branch lengths on the consensus tree. The posterior mean of a branch is calculated by averaging over the sampled trees that share the concerned branch (clade). This procedure does not appear fully justifiable because branch lengths in

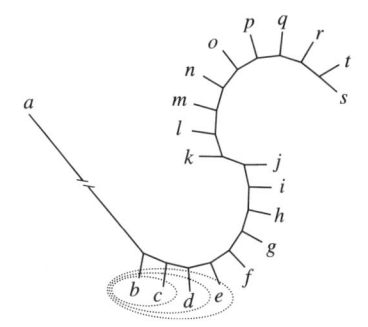

Fig. 5.11 A uniform prior on trees implies biased prior on clades and can lead to spuriously high posterior clade probabilities. Suppose in the true tree for $s = 20$ species, shown here, all branch lengths are 0.2 changes per site except branch a, which has 10 changes per site so that sequence a is nearly random relative to the other sequences. Note that in the true tree, a and b are sister species. If the data are informative, the backbone tree for all species without a will be correctly recovered, but the placement of a on the backbone tree is nearly random and can be on any of the $(2s − 5)$ branches. Thus the resulting $(2s − 5)$ trees will receive nearly equal support. By summing up probabilities across those trees to calculate clade probabilities, the posterior probability for clade (bc) will be about $(2s − 7)/(2s − 5)$, because only two out of the $(2s − 5)$ trees do not include the (bc) clade; these two trees have species a joining either b or c. Similarly clade (bcd), $(bcde)$, $(bcdef)$, etc. will receive support values around $(2s − 9)/(2s − 5)$, $(2s − 11)/(2s − 5)$, and so on. Those clades are wrong as they are not monophyletic, but their posterior probabilities can be close to 1 if s is large. Constructed following Goloboff and Pol (2005).

different trees have different biologically meanings (even if they are for the same shared clade). If branch lengths are of interest, a proper way would be to run another MCMC, sampling branch lengths with the tree topology fixed. An analogy is as follows. Suppose we calculate the posterior probabilities for the normal model $N(\mu, \sigma^2)$ and the gamma model $G(\mu, \sigma^2)$, each with the mean μ and variance σ^2 to be the unknown parameters. If we are interested in the population mean and variance, we should derive their posterior distributions conditional on each model and it is inappropriate to average μ and σ^2 between the two models.

5.6.4 Bayesian versus likelihood

In terms of computational efficiency, stochastic tree search using the program MrBayes (Huelsenbeck and Ronquist 2001; Ronquist and Huelsenbeck 2003) appears to be more efficient than heuristic tree search under likelihood using David Swofford's PAUP program (Swofford 2000). Nevertheless, the running time of the MCMC algorithm is proportional to the number of iterations the algorithm is run for. In general, longer chains are needed to achieve convergence in larger data sets due to the increased number of parameters to be averaged over. However, many users run shorter chains for larger data sets because larger trees require more computation per iteration. As a result, it is not always clear whether the MCMC algorithm has converged in analyses of large data sets. Furthermore, significant improvements to heuristic tree search under likelihood are being made (e.g. Guindon and Gascuel 2003; Vinh and von Haeseler 2004). So it seems that for obtaining a point estimate, likelihood heuristic search using numerical optimization can be faster than Bayesian stochastic search using MCMC. However, no one knows how to use the information in the likelihood tree search to attach a confidence interval or some other measure of the sampling error in the ML tree—as one can use the local curvature or Hessian matrix calculated in a nonlinear programming algorithm to construct a confidence interval for a conventional MLE. As a result, one must currently resort to bootstrapping (Felsenstein 1985a). Bootstrapping under likelihood is an expensive procedure, and appears slower than Bayesian MCMC.

To many, Bayesian inference of molecular phylogenies enjoys a theoretical advantage over ML with bootstrapping. Posterior probability for a tree or clade has an easy interpretation: it is the probability that the tree or clade is correct given the data, model, and prior. In contrast, the interpretation of the bootstrap in phylogenetics has been controversial (see Chapter 6). As a result, posterior probabilities of trees can be used in a straightforward manner in a variety of phylogeny-based evolutionary analyses to accommodate phylogenetic uncertainty; for example, they have been used in comparative analysis to average the results over phylogenies (Huelsenbeck *et al.* 2000b, 2001).

This theoretical advantage is contingent on us accepting the prior and the likelihood model. This is not certain, however. It has been noted that Bayesian posterior probabilities calculated from real data sets are often extremely high. One may observe that while bootstrap support values are published only if they are $> 50\%$ (as otherwise the relationships may not be considered trustworthy), posterior clade probabilities are sometimes reported only if they are $< 100\%$ (as most of them are 100%!). The difference between the two measures of support does not itself suggest anything inappropriate about the Bayesian probabilities, especially given the difficulties in the interpretation of the bootstrap. However, it has been observed that different models may produce conflicting trees when applied to the same data, each with high posterior probabilities. Similarly different genes for the same set of species can produce conflicting trees or clades, again each with high posterior probabilities (e.g. Rokas *et al.* 2005). A number of authors have suggested, based on simulation studies, that posterior probabilities for trees or clades are often misleadingly high (e.g. Suzuki *et al.* 2002; Erixon *et al.* 2003; Alfaro *et al.* 2003).

Bayesian posterior probability for a tree or clade is the probability that the tree or clade is true given the data, the likelihood model and the prior. Thus there can be only three possible reasons for spuriously high clade probabilities: (i) computer program bugs or problems in running the MCMC algorithms, such as lack of convergence and poor mixing, (ii) misspecification of the likelihood (substitution) model, and (iii) misspecification and sensitivity of the prior. First, if the MCMC fails to explore the parameter space properly and only visits an artificially small subset of the space, the posterior probabilities for the visited trees will be too high. This problem may be a serious concern in Bayesian analysis of large data sets, but in principle may be resolved by running longer chains and designing more efficient algorithms. Second, model misspecification, that is, use of an overly-simple substitution model, is found to cause spuriously high posterior probabilities (Buckley 2002; Lemmon and Moriarty 2004; Huelsenbeck and Rannala 2004). The problem can in theory be resolved by implementing more realistic substitution models or taking a model-averaging approach (Huelsenbeck *et al.* 2004). In this regard, use of an overly-complex model is noted to produce accurate posterior probabilities even if the true model is a special case (Huelsenbeck and Rannala 2004). Unlike likelihood, Bayesian inference appears more tolerant of parameter-rich models. Suchard *et al.* (2001) discussed the use of Bayes factors to select the substitution model for Bayesian analysis.

Note that high posterior probabilities were observed in simulated data sets where the substitution model is correct (e.g. Yang and Rannala 2005) and in analyses of small data sets that did not use MCMC (e.g. Rannala and Yang 1996). In those cases, the first two factors do not apply. The third factor, the sensitivity of Bayesian inference to prior specification, is more fundamental and difficult to deal with (see Subsection 5.1.3). Yang and Rannala (2005) assumed independent exponential priors with means μ_0 and μ_1 for internal and external branch lengths, respectively, and noted that the posterior probabilities of trees might be unduly influenced by the prior mean μ_0 on the internal branch lengths. It is easy to see that high posterior probabilities for trees will decrease if μ_0 is small; if $\mu_0 = 0$, all trees and clades will have posterior probabilities near zero. It was observed that in large data sets, the posterior clade probabilities are sensitive to μ_0 only if μ_0 is very small. In an analysis of 40 land plant species, the sensitive region was found to be $(10^{-5}, 10^{-3})$. Such branch lengths seem unrealistically small if we consider estimated internal branch lengths in published trees. However, branch lengths in wrong or poorly supported trees are typically small and often zero. As the prior is specified to represent our prior knowledge of internal branch lengths in all binary trees, the majority of which are wrong or poor trees, a very small μ_0 appears necessary. Yang and Rannala also suggested that μ_0 should be smaller in larger trees with more species.

A similar approach was taken by Lewis *et al.* (2005), who assigned nonzero probabilities to trees with multifurcations. Reversible jump MCMC was used to deal with the different numbers of branch length parameters in the bifurcating and multifurcating trees. This is equivalent to using a mixture-distribution prior for internal branch lengths, with a component of zero and another component from a continuous distribution. The approach of Yang and Rannala uses one continuous prior distribution for internal branch lengths, such as exponential or gamma, and is simpler computationally. While posterior clade probabilities are sensitive to the mean of the prior for internal branch lengths, it is in general unclear how to formulate sensible priors that are acceptable to most biologists. The problem merits further investigation.

180 • *5 Bayesian methods*

5.6.5 A numerical example: phylogeny of apes

We apply the Bayesian approach to the sequences of the 12 mitochondrial proteins from seven ape species, analysed in Subsection 4.2.5. We use MrBayes version 3.0 (Huelsenbeck and Ronquist 2001; Ronquist and Huelsenbeck 2003), with the modification of Yang and Rannala (2005) to allow exponential priors with different means μ_0 and μ_1 for internal versus external branch lengths. The analysis is conducted under the MTMAM model for mitochondrial proteins (Yang *et al.* 1998). We run two chains in the MC^3 algorithm for 10^7 iterations, sampling every 10 iterations. The prior means are initially set at $\mu_0 = \mu_1 = 0.1$ for both internal and external branch lengths. The MAP tree is shown in Fig. 5.12, which is the same as the ML tree of Fig. 4.5 in Chapter 4. The posterior probability is 100% for every node (clade). We then changed the prior mean μ_0 for internal branch lengths, to see its effect on the posterior clade probabilities (Fig. 5.12). Posterior probabilities for all clades are 100% when $\mu_0 = 0.1$ or 10^{-3} but are $< 70\%$ when $\mu_0 = 10^{-5}$. At this stage, it is hard to judge which μ_0 is more reasonable.

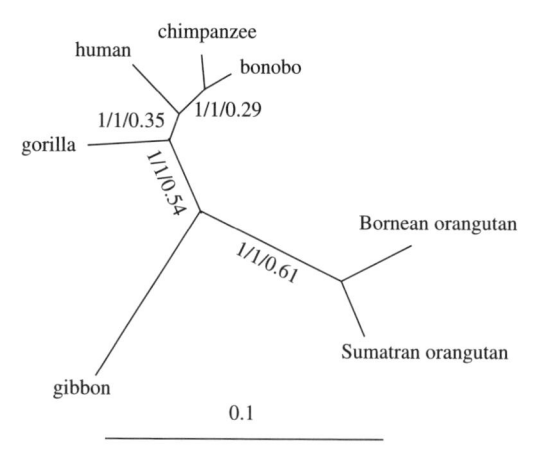

Fig. 5.12 The maximum posterior probability (MAP) tree for the mitochondrial protein sequences from seven ape species. Posterior clade probabilities are calculated under the MTMAM model using MrBayes version 3.0. Independent exponential priors are assumed for branch lengths, with the prior mean for external branch lengths fixed at $\mu_1 = 0.1$, while three values are used for the prior mean μ_0 for internal branch lengths: 10^{-1}, 10^{-3}, and 10^{-5}. The posterior clade probabilities at those three values of μ_0 are shown along the branches in the format '·/ · /·'. The branches are drawn in proportion to the posterior means of the branch lengths calculated using the prior mean $\mu_0 = 0.1$.

5.7 MCMC algorithms under the coalescent model

5.7.1 Overview

MCMC algorithms have been widely used to implement coalescent models of different complexity, to analyse different types of population genetics data such as DNA sequences, microsatellites, and SNPs (single nucleotide polymorphisms). Examples include estimation of mutation rates (e.g. Drummond *et al.* 2002), inference of population demographic processes or gene flow between subdivided populations (e.g. Beerli and Felsenstein 2001; Nielsen and Wakeley 2001; Wilson *et al.* 2003; Hey and Nielsen 2004), and estimation of species divergence times and population sizes (e.g., Yang 2002; Rannala and Yang 2003). Such algorithms have a lot of similarity with phylogenetic algorithms, as both involve sampling in the space of trees (genealogies or phylogenies). Below I provide a flavour of such algorithms by considering estimation of $\theta = 4N\mu$ from one population sample, where N is the long-term (effective) population size and μ is the mutation rate per generation. For other applications under more complex coalescent models see the reviews by Griffiths and Tavaré (1997), Stephens and Donnelly (2000), and Hein *et al.* (2005).

5.7.2 Estimation of θ

The fundamental parameter $\theta = 4N\mu$ measures the amount of genetic variation at a neutral locus maintained in a population with random mating. Traditionally, μ usually refers to the mutation rate per locus, but with molecular sequence data it is more convenient to define it as the rate per site. For population data, the mutation rate is typically assumed to be constant over time and across sites. Suppose we take a random sample of n sequences, and wish to estimate θ. Let X be the data, G be the unknown genealogical tree relating the sequences, and $\mathbf{t} = (t_n, t_{n-1}, \ldots, t_2)$ be the set of $n-1$ coalescent (waiting) times (Fig. 5.13). The data do not contain information to separate the population size N from the mutation rate μ, so we estimate one parameter $\theta = 4N\mu$.

The coalescent model specifies the distribution of the genealogical tree G and the coalescent times $\mathbf{t}$. This may be considered the prior $f(G, \mathbf{t})$. All genealogical trees (called *labelled histories*, or rooted trees with interior nodes ordered according to their ages) have equal probability: that is $f(G) = 1/T_n$, where T_n is the total number of labelled histories for n sequences. We measure time in units of $2N$ generations, and further multiply time by the mutation rate. With this scaling, coalescent times are measured by the expected number of mutations per site, and any two lineages in the sample coalesce at the rate $\theta/2$ (Hudson 1990). The waiting time t_j until the next coalescent event when there are j lineages in the sample has the exponential density

$$f(t_j|\theta) = \frac{j(j-1)}{2} \times \frac{2}{\theta} \exp\left(-\frac{j(j-1)}{2} \times \frac{2}{\theta} t_j\right), \tag{5.41}$$

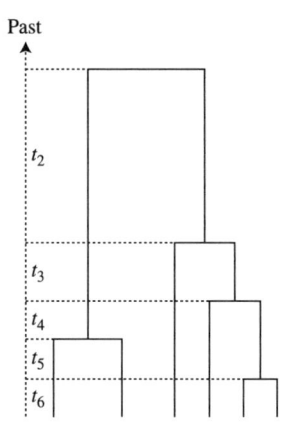

Fig. 5.13 A genealogical tree for six sequences. The coalescent times are defined as the waiting times; that is, t_j is the time when there are j lineages in the sample.

for $j = n, n − 1, \ldots, 2$. The joint density $f(\mathbf{t}|\theta)$ is simply the product of the densities of the $n − 1$ coalescent times. Given the genealogical tree and coalescent times (branch lengths), the likelihood $f(X|\theta, G, \mathbf{t})$ can be calculated using Felsenstein's (1981) pruning algorithm.

Thus with a prior $f(\theta)$, the posterior distribution is given as

$$f(\theta|X) = \frac{\sum_{i=1}^{T_s} \int f(\theta) f(G_i, \mathbf{t}_i|\theta) f(X|\theta, G_i, \mathbf{t}_i)\, \mathrm{d}\mathbf{t}_i}{\sum_{i=1}^{T_s} \iint f(\theta) f(G_i, \mathbf{t}_i|\theta) f(X|\theta, G_i, \mathbf{t}_i)\, \mathrm{d}\mathbf{t}_i\, \mathrm{d}\theta}. \quad (5.42)$$

Here the sum is over T_s possible genealogies and the integral over t is $(n − 1)$-dimensional.

The MCMC algorithm avoids direct calculation of the integrals, and instead achieves the integration and summation through the Markov chain. The algorithm samples θ in proportion to its posterior distribution. The following is a sketch of the MCMC algorithm implemented by Yang and Rannala (2003):

- Start with a random genealogy G, with random coalescent times $\mathbf{t}$ (possibly sampled from the coalescent prior), and a random parameter θ.
- In each iteration do the following:
  - Propose changes to coalescent times $\mathbf{t}$.
  - Propose a change to the genealogy, by using tree-rearrangement algorithms (such as SPR; see Subsection 3.2.3). This step may change coalescent times $\mathbf{t}$ as well.
  - Propose a change to parameter θ.
  - Proposal a change to all coalescent times, using proportional shrinking or expanding

- Every *k* iterations, sample the chain: save θ (and other quantities such as tree height, the time to the most recent common ancestor of the sample).
- At the end of the run, summarize the results.

Example. Estimation of θ from three neutral loci from human populations (Rannala and Yang 2003). The three loci are 61 human sequences from the region 1q24 (about 10 kb) (Yu et al. 2001), 54 human sequences in a region of $\sim$ 6.6 kb at 16q24.3 (Makova et al. 2001), and 63 human sequences of about 10 kb in the region 22q11.2 (Zhao et al. 2000). Since all three loci are noncoding, we assume the same mutation rate to estimate a common θ for all loci. We also estimate the time to the most recent common ancestor t_{MRCA} for the three loci. The data do not contain information to separate the population size N and mutation rate μ, so we estimate one parameter $\theta = 4N\mu$, or we estimate N be fixing μ at 10^{-9} mutations/site/year. We also assume a generation time of $g = 20$ years. The prior on θ is assumed to be a gamma distribution $G(2, 2000)$, with mean 0.001; this corresponds to a prior mean for population size of 12 500 with the 95% prior interval (1 500, 34 800). The likelihood is calculated under the JC69 model. The MCMCcoal program (Rannala and Yang 2003) is used to run the MCMC, using 10 000 iterations as burn-in, followed by 10^6 iterations, sampling every two iterations.

The posterior distribution of parameter θ is shown in Fig. 5.14. The posterior mean and 95% CI are 0.00053 (0.00037, 0.00072), corresponding to a population size of $N = 0.00053/(4g\mu) \approx 6600$ with the 95% CI to be (4600, 9000) for modern humans. The posterior distributions of the time to the most recent common ancestor t_{MRCA} at the three loci are shown in Fig. 5.14. Their posterior means and CIs for the three loci are 0.00036 (0.00020, 0.00060); 0.00069 (0.00040, 0.00110); and 0.00049 (0.00030, 0.00076), respectively. If the mutation rate is $\mu = 10^{-9}$ mutations/site/year, the posterior mean ages of the most recent common ancestors of the samples are 0.36, 0.69, and 0.49 million years. □

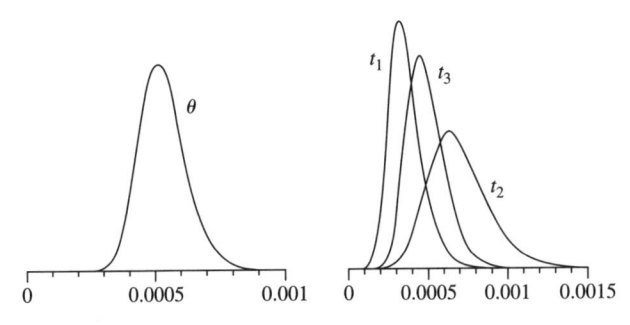

Fig. 5.14 (a) Posterior distribution of θ for modern humans estimated from three neutral loci. (b) Posterior distributions of t_{MRCA} for the three loci, where t_{MRCA} is the time to the most recent common ancestor in the gene genealogy, measured by the expected number of mutations per site.

5.8 Exercises

5.1 (a) In the example of testing for infection in Subsection 5.1.2, suppose that a person tested negative. What is the probability that he has the infection (b) Suppose a person was tested twice and found to be positive both times. What is the probability that he has the infection?

5.2 *Criticism of unbiasedness.* Both likelihood and Bayesian proponents point out that strict adherence to unbiasedness may be unreasonable. For the example of Subsection 5.1.2, consult any statistics textbook to confirm that the expectation of the sample frequency x/n is θ under the binomial model and $\theta(n-1)/n$ under the negative binominal model. Thus the unbiased estimator of θ is $x/n = 9/12$ under the binomial and $x/(n-1) = 9/11$ under the negative binomial. Unbiasedness thus violates the likelihood principle. Another criticism of unbiased estimators is that they are not invariant to reparametrization; if $\hat{\theta}$ is an unbiased estimator of θ, $h(\hat{\theta})$ will not be an unbiased estimator of $h(\theta)$ if h is not a linear function of θ.

*5.3 Suppose the target density is $N(\theta, 1)$, and the MCMC uses the sliding-window proposal with normal proposals, with the jump kernel $x^* \sim N(x, \sigma^2)$. Show that the acceptance proportion (the proportion at which the proposals are accepted) is (Gelman et al. 1996)

$$P_{\text{jump}} = \frac{2}{\pi} \tan^{-1}\left(\frac{2}{\sigma}\right). \tag{5.43}$$

5.4 Write a program to implement the MCMC algorithm of Subsection 5.3.2 to estimate the distance between the human and orangtutan 12s rRNA genes under the JC69 model. Use any programming language of your choice, such as BASIC, Fortran, C/C++, Java, or Mathematica. Investigate how the acceptance proportion changes with the window size w. Also implement the proposal of equation (5.34). (Hint: use the logarithms of the likelihood and prior in the algorithm to avoid numerical problems.)

5.5 Modify the program above to estimate two parameters under the JC69 model: the substation rate $\mu = 3\lambda$ and the time of species divergence T, instead of the distance $\theta = 3\lambda \times 2T$. Consider one time unit as 100 million years, and assign an exponential prior $f(T) = (1/m)e^{-T/m}$ for T with mean $m = 0.15$ (15 million year for human–orangutan divergence) and another exponential prior with mean 1.0 for rate μ (corresponding to a prior mean rate of about 1 substitution per 100 million years). Use two proposal steps, one updating T and another updating μ. Change the prior to examine the sensitivity of the posterior to the prior.

5.6 Modify the program of Exercise 5.4 to estimate the sequence distance under the K80 model. Use the exponential prior $f(\theta) = (1/m)e^{-\theta/m}$ with mean $m = 0.2$ for distance θ and exponential prior with mean 5 for the transition/transversion rate ratio κ. Implement two proposal steps, one for updating θ and another for updating κ. Compare the posterior estimates with the MLEs of Subsection 1.4.2.

6

Comparison of methods and tests on trees

This chapter discusses two problems: the evaluation of statistical properties of tree-reconstruction methods and tests of the significance of estimated phylogenies. As both problems are complex and controversial, I intend this chapter to be a personal appraisal of the large body of complex literature, without any claim for objectivity.

Molecular systematists are now faced with a variety of methods for reconstructing phylogenies and making a choice among them is not always straightforward. For model-based methods there also exist a large collection of models which in principle can be used. Earlier debates concerning the merits and demerits of tree-reconstruction methods mostly centred on philosophical issues, such as whether one can avoid making assumptions about the evolutionary process when inferring phylogenies, or whether one should use model-based methods such as likelihood or 'model-free' methods such as parsimony (e.g. Farris 1983; Felsenstein 1973b, 1983). More recent studies have used computer simulations and 'well-established' phylogenies to assess the performance of different methods. It is common for different studies to reach conflicting conclusions.

Section 6.1 discusses criteria for assessing the statistical properties of tree-reconstruction methods. A summary of simulation studies conducted to evaluate different methods is provided, as well as some recommendations concerning the use of those methods in practical data analysis. Sections 6.2 and 6.3 deal with the likelihood versus parsimony debate from the likelihood and parsimony perspectives, respectively. Even in the statistical framework, tree estimation appears to be much more complex than conventional parameter estimation. These difficulties are discussed in Section 6.2. Section 6.3, on parsimony, summarizes previous attempts to identify the assumptions underlying the method or to provide a statistical justification for it by establishing an equivalence to likelihood under a particular model.

Section 6.4 provides an overview of methods for assessing the reliability of estimated phylogenies. The reconstructed tree, by whatever method, is equivalent to a point estimate. One would like to attach a measure of its reliability, in the same way that a confidence interval or posterior credibility interval provides a measure of the accuracy of a point estimate of a conventional parameter. However, the unconventional nature of the problem creates difficulties with this effort.

6.1 Statistical performance of tree-reconstruction methods

6.1.1 Criteria

In comparing different methods of phylogeny reconstruction, two kinds of error should be distinguished. *Random errors*, also called sampling errors, are due to the finite nature of the data set. In most models used in molecular phylogenetics, the sample size is the number of sites (nucleotides, amino acids, or codons) in the sequence. When the sequence length approaches infinity, sampling errors will decrease and approach zero. *Systematic errors* are due to incorrect assumptions of the method or some other kinds of deficiencies. When the sample size increases, systematic errors will persist and intensify.

One can judge a tree-reconstruction method using a variety of criteria. The computational speed is perhaps the easiest to assess. In general, distance methods are much faster than parsimony, which is in turn faster than likelihood or Bayesian methods. Here we consider the statistical properties of a method.

6.1.1.1 Identifiability

If the probability of the data $f(X|\theta)$ is exactly the same for two parameter values θ_1 and θ_2, that is, $f(X|\theta_1) = f(X|\theta_2)$ for all possible data X, no method will be able to distinguish θ_1 from θ_2 using the observed data. The model is said to be unidentifiable. For example, the model becomes unidentifiable if we attempt to estimate the species divergence time t and the substitution rate r using data of two sequences from two species. The probability of the data or likelihood is exactly the same for $\theta_1 = (t, r)$ and $\theta_2 = (2t, r/2)$, say, and it is impossible to estimate t and r separately. Even if θ is unidentifiable, some functions of θ may be identifiable—in this case, the distance $d = tr$ is. Unidentifiable models are usually due to errors in model formulation and should be avoided.

6.1.1.2 Consistency

An estimation method or estimator $\hat{\theta}$ is said to be (statistically) consistent if it converges to the true parameter value θ when the sample size n increases. Formally, $\hat{\theta}$ is consistent if

$$\lim_{n \to \infty} \Pr(|\hat{\theta} - \theta| < \varepsilon) = 1, \qquad (6.1)$$

for any small number $\varepsilon > 0$. Also $\hat{\theta}$ is said to be *strongly consistent* if

$$\lim_{n \to \infty} \Pr(\hat{\theta} = \theta) = 1. \qquad (6.2)$$

Phylogenetic trees are not regular parameters, but we may use the idea of strong consistency and say that a tree-reconstruction method is consistent if the probability that the estimated tree is the true tree approaches 1 when $n \to \infty$. For model-based methods, the definition of consistency assumes the correctness of the model.

There has been much discussion of consistency since Felsenstein (1978b) demonstrated that parsimony can be inconsistent under certain combinations of branch lengths on a four-species tree (see Subsection 3.4.4 and Exercise 4.3). In a conventional estimation problem, consistency is a weak statistical property, easily satisfied by many good and poor estimators. For example, the usual estimator of the probability p of 'success' from a binomial sample with x successes out of n trials is the sample proportion $\hat{p} = x/n$. This is consistent, but so is an arbitrary poor estimator $\tilde{p} = (x - 1000)/n$, which is even negative if $n < 1000$. Sober (1988) considered likelihood to be a more fundamental criterion than consistency, claiming that consistency is not necessary since real data sets are always finite. Such a position appears untenable as it prefers likelihood for its own sake rather than for its good performance. Statisticians, from both the likelihood/frequentist and Bayesian schools (e.g. Stuart et al. 1999, pp. 3–4; O'Hagan and Forster 2004, pp. 72–74), never appear to have doubted that consistency is a property that any sensible estimator should possess. Fisher (1970, p. 12) considered inconsistent estimators to be 'outside the pale of decent usage'. See Goldman (1990) for more discussions.

6.1.1.3 Efficiency

A consistent estimator with asymptotically the smallest variance is said to be efficient. The variance of a consistent and unbiased estimator cannot be smaller than the Cramér–Rao lower bound; that is, for any unbiased estimator $\hat{\theta}$ of θ, we have (Stuart et al. 1999, pp. 9–14; see also Subsection 1.4.1)

$$\text{var}(\hat{\theta}) \geq 1/I, \qquad (6.3)$$

where $I = -E(d^2 \log(f(X|\theta))/d\theta^2)$ is known as the *expected information* or *Fisher information*. Under quite mild regularity conditions, the MLE has desirable asymptotic properties: that is, when $n \to \infty$, the MLE is consistent, unbiased, and normally distributed, and attains the minimum variance bound of equation (6.3) (e.g. Stuart et al. 1999, Ch. 18).

If t_1 is an efficient estimator and t_2 is another estimator, one can measure the efficiency of t_2 relative to t_1 as $E_{12} = n_2/n_1$, where n_1 and n_2 are the sample sizes required to give both estimators equal variance, i.e. to make them equally precise (Stuart et al. 1999, p. 22). In large samples, the variance is often proportional to the inverse of the sample size, in which case $E_{12} = V_2/V_1$, where V_1 and V_2 are the variances of the two estimators at the same sample size. Suppose we wish to estimate the mean μ of a normal distribution $N(\mu, \sigma^2)$ with variance σ^2 known. The sample mean has variance σ^2/n and is efficient in the sense that no other unbiased estimator can have a smaller variance. The sample median has variance $\pi\sigma^2/(2n)$ for large n. The efficiency of the median relative to the mean is $2/\pi = 0.637$ for large n. Thus the mean achieves the same accuracy as the median with a sample that is 36.3% smaller. The median is nevertheless less sensitive to outliers.

The variance of an estimated tree topology is not a meaningful concept. However, one can measure the relative efficiency of two tree-reconstruction methods by

$$E_{12} = n_2(P)/n_1(P), \qquad (6.4)$$

where $n_1(P)$ and $n_2(P)$ are the sample sizes required for both methods to recover the true tree with the same probability P (Saitou and Nei 1986; Yang 1996a). As $P(n)$, the probability of recovering the correct tree given the sample size n, can be more easily estimated using computer simulation than $n(P)$, an alternative measure may be used:

$$E_{12}^* = (1 - P_2(n))/(1 - P_1(n)). \qquad (6.5)$$

Equations (6.4) and (6.5) are not expected to be identical, but should give the same qualitative conclusion concerning the relative performance of the two methods. Later we will use equation (6.5) to analyse the ML method.

6.1.1.4 Robustness

A model-based method is said to be robust if it still performs well when its assumptions are slightly wrong. Clearly some assumptions matter more than others. Robustness is often examined by computer simulation, in which the data are generated under one model but analysed under another model that is wrong and often too simplistic.

6.1.2 Performance

To judge tree-reconstruction methods, a number of studies have exploited situations in which the true tree is known or believed to be known. The first strategy is the use of laboratory-generated phylogenies. Hillis *et al.* (1992) generated a known phylogeny by 'evolving' the bacteriophage T7 in the laboratory. The restriction-site map of the phage was determined at several time points, and the phage was separated and allowed to diverge into different lineages. Thus both the phylogeny and the ancestral states were known. Parsimony and four distance-based methods were then applied to analyse restriction-site maps of the terminal lineages to infer the evolutionary history. Very impressively, all methods recovered the true phylogeny! Parsimony also reconstructed the ancestral restriction-site maps with $> 98\%$ accuracy.

A second approach uses so-called 'well-established' phylogenies, that is, phylogenetic relationships that are generally accepted based on evidence from fossils, morphology, and previous molecular data. Such phylogenies can be used to assess the performance of tree-reconstruction methods as well as the utilities of different gene loci. For example, Cummings *et al.* (1995), Russo *et al.* (1996), and Zardoya and Meyer (1996) evaluated the performance of different tree-reconstruction methods and of mitochondrial protein-coding genes in recovering mammalian or vertebrate phylogenies. Most modern empirical phylogenetic studies are of this nature, as researchers use a variety of methods to analyse multiple gene loci, and assess the reconstructed phylogenies against previous estimates (e.g. Cao *et al.* 1998; Takezaki and Gojobori 1999; Brinkmann *et al.* 2005).

The third approach is computer simulation. Many replicate data sets are generated under a simulation model and then analysed using various tree-reconstruction methods to estimate the true tree, to see how often the true tree is recovered or what percentage of the true clades are recovered. The shape and size of the tree topology, the evolutionary model and the values of its parameters, and the size of data are all under the control of the investigator and can be varied to see their effects. Chapter 9 should be consulted for discussions of simulation techniques. A criticism of simulation studies is that the models used may not reflect the complexity of sequence evolution in the real world. Another is that a simulation study can examine only a very limited set of parameter combinations, and yet the relative performance of tree-reconstruction methods may depend on the model and tree shape. It is thus unsafe to extrapolate conclusions drawn from simulations in a tiny portion of the parameter space to general situations with real data. Nevertheless, patterns revealed in simulations are repetitively discovered in real data analysis, so simulation will continue to be relevant to comparing and validating phylogenetic methods, especially given that the methods are in general analytically intractable.

Several review articles have been published that summarize previous simulation studies, such as Felsenstein (1988), Huelsenbeck (1995b), and Nei (1996). Conflicting views are often offered concerning the relative performance of different tree-reconstruction methods. The following observations appear to be generally accepted.

1. Methods that assume the molecular clock such as UPGMA perform poorly when the clock is violated, so one should use methods that infer unrooted trees without assuming the clock. UPGMA is nevertheless appropriate for highly similar sequences, such as population data, in which the clock is expected to hold.
2. Parsimony, as well as distance and likelihood methods under simplistic models, is prone to the problem of long-branch attraction, while likelihood under complex and more realistic models are more robust (e.g. Kuhner and Felsenstein 1994; Gaut and Lewis 1995; Yang 1995b; Huelsenbeck 1998).
3. Likelihood methods are often more efficient in recovering the correct tree than parsimony or distance methods (e.g. Saitou and Imanishi 1989; Jin and Nei 1990; Hasegawa and Fujiwara 1993; Kuhner and Felsenstein 1994; Tateno *et al.* 1994; Huelsenbeck 1995b). However, counterexamples have been found, some of which will be discussed in the next section.
4. Distance methods do not perform well when the sequences are highly divergent or contain many alignment gaps, mainly because of difficulties in obtaining reliable distance estimates (e.g. Gascuel 1997; Bruno *et al.* 2000).
5. The level of sequence divergence has a great impact on the performance of tree-reconstruction methods. Highly similar sequences lack information, so that no method can recover the true tree with any confidence. Highly divergent sequences

contain too much noise, as substitutions may have saturated. The amount of information in the data is optimized at intermediate levels of divergence (Goldman 1998). Thus one should ideally sequence fast-evolving genes to study closely related species and slowly evolving genes or proteins to study deep phylogenies. In simulation studies, phylogenetic methods appear to be quite tolerant of multiple substitutions at the same site (Yang 1998b; Bjorklund 1999). However, high divergences are often accompanied by other problems, such as difficulty in alignment and unequal base or amino acid compositions among sequences, which indicate a clear violation of the assumption that the substitution process has been stationary.

6. The shape of the tree as reflected in the relative branch lengths has a huge effect on the success of the reconstruction methods and on their relative performance. 'Hard' trees are characterized by short internal branches and long external branches, with long external branches scattered in different parts of the tree. On such trees, parsimony as well as distance and likelihood methods under simplistic models are prone to errors. 'Easy' trees are characterized by long internal branches relative to external branches. With such trees, every method seems to work fine, and it is even possible for naively simplistic likelihood models or parsimony to outperform likelihood under complex models.

6.2 Likelihood

6.2.1 Contrast with conventional parameter estimation

Yang (1994c; see also Yang 1996a; Yang *et al.* 1995c) argued that the problem of tree reconstruction is not one of statistical parameter estimation but instead is one of model selection. In the former case, the probability distribution of the data, $f(X|\theta)$, is fully specified except for the values of parameters θ. The objective is then to estimate θ. In the latter case, there are several competing data-generating models $f_1(X|\theta_1)$, $f_2(X|\theta_2)$, and $f_3(X|\theta_3)$, each with its own unknown parameters. The objective is then to decide which model is true or closest to the truth. Here, the models correspond to tree topologies while the parameters θ_1, θ_2, or θ_3 in each model correspond to branch lengths in each tree. Tree reconstruction falls into the category of model selection, because the likelihood function and the definition of branch lengths depend on the tree topology (Nei 1987, p. 325). There is a minor difference, however. In a typical model-selection situation, our interest is rarely in the model itself; rather we are interested in inference concerning certain parameters and we have to deal with the model because our inference may be unduly influenced by model misspecification. Selection of the substitution model for tree reconstruction is one such case. In contrast, in phylogenetic tree reconstruction, we assume that one of the trees is true, and our main objective is to identify the true tree.

The distinction between parameter estimation and model selection is not a pedantic one. The mathematical theory of likelihood estimation, such as the consistency and efficiency of MLEs, is developed in the context of parameter estimation and not of model selection. The next two subsections demonstrate that phylogeny reconstruction by ML is consistent but not asymptotically efficient.

6.2.2 Consistency

Early arguments (e.g. Felsenstein 1973b, 1978b; see also Swofford *et al.* 2001) for the consistency of the likelihood method of tree reconstruction (Felsenstein 1981) referred to the proof of Wald (1949) and did not take full account of the fact that the likelihood function changes among tree topologies. Nevertheless, the consistency of ML under commonly used models is easy to establish. Yang (1994c) provided such a proof, assuming that the model is well-formulated so that the different trees are identifiable. The essence of the proof is that when the number of sites approaches infinity, the true tree will predict probabilities of site patterns that match exactly the observed frequencies, thus achieving the maximum possible likelihood for the data; as a result, the true tree will be chosen as the estimate under the ML criterion.

In the model we assume that different sites evolve independently and according to the same stochastic process. The data at different sites are then independently and identically distributed (*i.i.d.*) and can be summarized as the counts of 4^s site patterns for s species: n_i for site pattern i, with the sequence length $n = \sum n_i$. Note that in typical data sets, many of the site patterns are not observed so that $n_i = 0$ for some i. The counts of site patterns are random variables from a multinomial distribution, which has 4^s categories corresponding to the 4^s site patterns. Without loss of generality, suppose tree τ_1 is the true tree, while all other trees are incorrect. The probability of the *i*th category (site pattern) in the multinomial is given by the true tree and true values of parameters $\theta_*^{(1)}$ (branch lengths and substitution parameters), as $p_i = p_i^{(1)}(\theta_*^{(1)})$. Here the superscript $^{(1)}$ indicates that the probabilities and parameters are defined on tree τ_1.

Let $f_i = n_i/n$ be the proportion of sites with the *i*th pattern. Given the data, the log likelihood for any tree cannot exceed the following upper limit

$$\ell_{\max} = n \sum_i f_i \log(f_i). \tag{6.6}$$

Let $p_i^{(k)}(\theta^{(k)})$ be the probability of site pattern i for tree k when its parameters are $\theta^{(k)}$. The maximized log likelihood for tree k is thus

$$\ell_k = n \sum_i f_i \log \{p_i^{(k)}(\hat{\theta}^{(k)})\}, \tag{6.7}$$

where $\hat{\theta}^{(k)}$ are the MLEs on tree τ_k. Then $\ell_k \leq \ell_{\max}$, and the equality holds if and only if $f_i = p_i^{(k)}$ for all i. Similarly

$$(\ell_{\max} - \ell_k)/n = \sum_i f_i \log \left(\frac{f_i}{p_i^{(k)}(\hat{\theta}^{(k)})} \right) \tag{6.8}$$

is known as the *Kullback–Leibler divergence* between the two distributions specified by f_i and $p_i^{(k)}$, which is nonnegative and equals zero if and only if $f_i = p_i^{(k)}$ for all i.

Now estimation of $\theta^{(1)}$ on the true tree τ_1 is a conventional parameter estimation problem, and the standard proof (e.g. Wald 1949) applies. When $n \to \infty$, the data frequencies approach the site-pattern probabilities predicted by the true tree, $f_i \to p_i^{(1)}(\theta_*^{(1)})$, the MLEs approach their true values $\hat{\theta}^{(1)} \to \theta_*^{(1)}$, and the maximum log likelihood for the true tree approaches the maximum possible likelihood $\ell_1 \to \ell_{\max}$. The true tree thus provides a perfect match to the data.

A question is whether it is possible for a wrong tree to attain the same highest log likelihood $\ell_{\max}$, or to provide a perfect match to the site pattern probabilities predicted by tree $\tau_1: p_i^{(1)}(\theta_*^{(1)}) = p_i^{(2)}(\theta^{(2)})$ for all i for a certain tree τ_2. If this occurs, tree τ_1 with parameters $\theta_*^{(1)}$ and tree τ_2 with parameters $\theta^{(2)}$ generate data sets that are identical probabilistically, and the model is said to be unidentifiable. Unidentifiable models are pathological, usually due to conceptual errors in model formulation. Chang (1996a) and Rogers (1997) showed that models commonly used in phylogenetic analysis are indeed identifiable.

6.2.3 Efficiency

In simulation studies, distance methods such as NJ (Saitou and Nei 1987) were noted to recover the true tree with a higher probability when a wrong model was used to calculate sequence distances than when the true model was used (Saitou and Nei 1987; Sourdis and Nei 1988; Tateno *et al.* 1994; Rzhetsky and Sitnikova 1996). In a similar vein, Schoeniger and von Haeseler (1993), Goldstein and Pollock (1994), and Tajima and Takezaki (1994) constructed distance formula under deliberately 'wrong' models for tree reconstruction even though the true model was available. Those results are counterintuitive but may not be surprising since distance methods are not expected to make a full use of the information in the data and no theory predicts that they should have optimal performance.

Similar results, however, were observed when the likelihood method was used. Gaut and Lewis (1995) and Yang (1996a) (see also Siddall 1998; Swofford *et al.* 2001) reported simulation studies where likelihood under the true model had a lower probability of recovering the true tree than parsimony or likelihood under a wrong and simplistic model. Such counterintuitive results may be 'explained away' by suggesting that the data sets may be too small for the asymptotic properties of ML to be applicable.

To see whether small sample sizes are to blame, one can study the asymptotic behaviour of the relative efficiency (measured by equation 6.5) of the methods when the sequence length $n \to \infty$. A few examples from Yang (1997b) are shown in Fig. 6.1. The data sets are generated under the JC69 + Γ_4 model with rates for sites drawn from a gamma distribution with shape parameter $\alpha = 0.2$. The data are analysed using likelihood under JC69 + Γ_4 with α fixed at either 0.2 or ∞; the latter is equivalent to JC69, with one rate for all sites. The two analyses or methods are referred to as *True* and *False*, and both estimate five branch lengths on every tree. Both methods

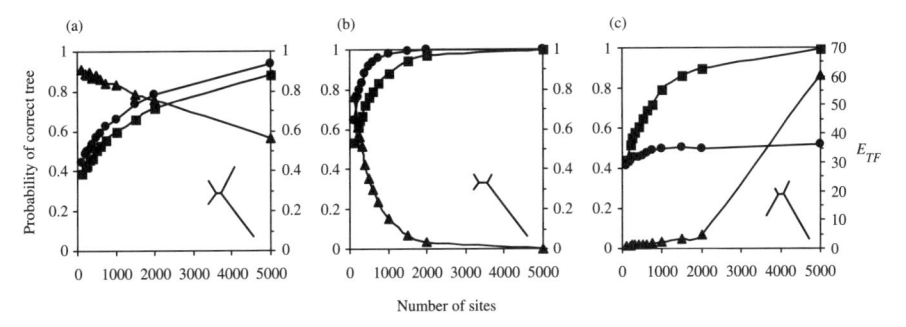

Fig. 6.1 The probability that the ML tree is the true tree plotted against the sequence length. Data sets are simulated under the JC69 + Γ_4 model, with shape parameter $\alpha = 0.2$. The ML analysis assumed either the true JC69 + Γ_4 model with $\alpha = 0.2$ fixed (■, the *True* method) or the false JC69 model, with $\alpha = \infty$ fixed (●, the *False* method). The relative efficiency (▲) of *True* relative to *False* is defined as $E^*_{TF} = (1 - P_F)/(1 - P_T)$. The true trees, shown as insets, have branch lengths as follows: ((a: 0.5, b: 0.5): 0.1, c: 0.6, d: 1.4) in (a); ((a: 0.05, b: 0.05): 0.05, c: 0.05, d: 0.5) in (b); and ((a: 0.1, b: 0.5): 0.1, c: 0.2, d: 1.0) in (c). Both *False* and *True* are consistent for those three trees. Redrawn after Figs. 1B, D, and C of Yang (1997b).

are consistent on the three trees of Fig. 6.1, with the probability of recovering the true tree $P \to 1$ when $n \to \infty$. However, on trees *a* and *b*, *False* approaches this limit faster than *True*, as indicated by the relative efficiency of *True* relative to *False*, $E^*_{TF} = (1 - P_F)/(1 - P_T)$, where P_T and P_F are the probabilities of recovering the true tree by the two methods. Despite the intractability of the likelihood analysis, it appears safe to expect $E^*_{TF} < 1$ when $n \to \infty$ in trees *a* and *b*. In tree *b*, E^*_{TF} apparently approaches zero. Additional simulation shows that increasing the value of α fixed in the *False* method from the true value (0.2) to ∞, so that the model is progressively more wrong, leads to progressively better performance in recovering this tree. Also use of the continuous gamma model instead of the discrete gamma produces similar results. Similar results may be obtained when parsimony is compared with likelihood under the one-rate method (with $\alpha = \infty$) when the data are generated using the one-rate model; on some trees, parsimony outperforms likelihood, with the efficiency of likelihood relative to parsimony, $E_{ML,MP} \to 0$ when $n \to \infty$ (Yang 1996a). In sum, the counterintuitive results are not due to small sample sizes.

A number of authors have suggested that the results might be explained by a 'bias' in parsimony or likelihood under a wrong and simplistic model (e.g. Yang 1996a; Bruno and Halpern 1999; Huelsenbeck 1998; Swofford et al. 2001). Swofford et al. (2001) illustrated this bias by the following analogy. Suppose we want to estimate parameter θ, and one method always returns 0.492 as the estimate, whatever the data. This method cannot be beaten if the true θ is 0.492, and may indeed be a very good method in finite data sets when the true θ is close to 0.492. Similarly, parsimony has the tendency of grouping long branches together irrespective of the true relationships. If the true tree happens to have the long branches together, the inherent bias of parsimony works in its favour, causing the method to outperform likelihood under the true model. This is demonstrated in Fig. 6.2, with the probability of recovering the true tree estimated by simulating 10 000 replicate data sets. The sequence length is 1000 sites. In Fig. 6.2(a), data are generated under the JC69 model

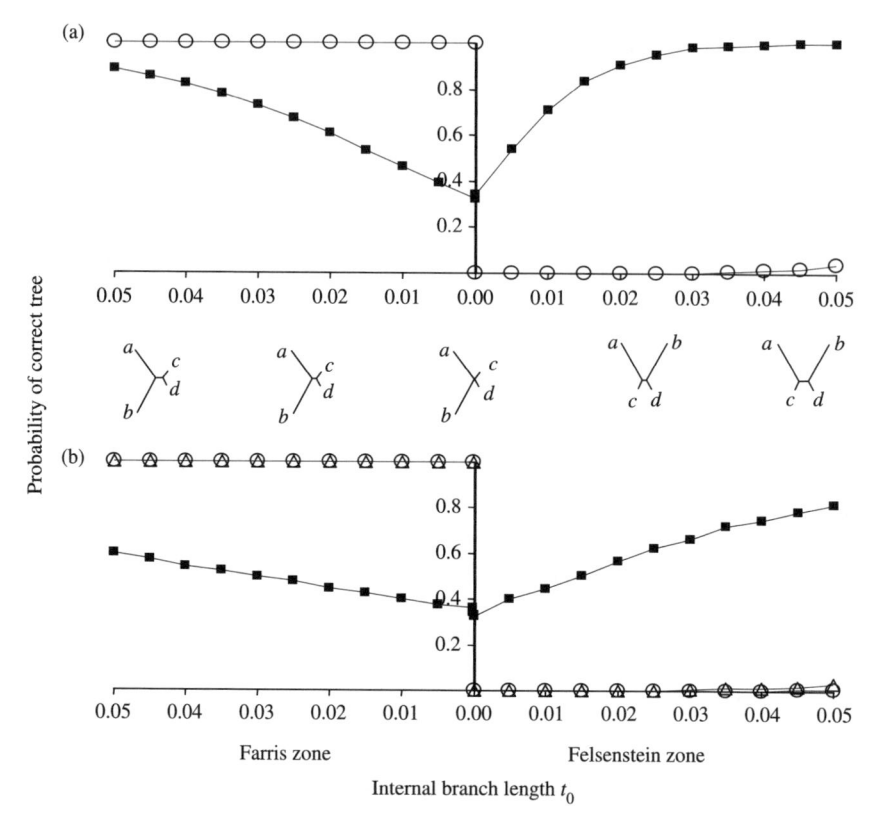

Fig. 6.2 The relative performance of tree-reconstruction methods depends on the tree shape. The external branch lengths, in expected number of substitutions per site, are 0.5 for long branches and 0.05 for short branches. The internal branch length (t_0) varies along the x-axis. On the left is a tree in the 'Farris zone', with its internal branch becoming progressively shorter. Then the topology switches to a tree in the 'Felsenstein zone', with its internal branch becoming increasingly longer. Performance is measured by the probability of recovering the true tree, estimated from 10 000 simulated data sets. The sequence length is 1000 sites. (a) The true model is JC69. Each data set is analysed using likelihood under the true JC69 model (■) and parsimony (○). (b) The true model is JC69 + Γ_4 with shape parameter $\alpha = 0.2$. Each data set is analysed by likelihood under the true JC69 + Γ_4 model, with $\alpha = 0.2$ fixed (■), by likelihood under the false JC69 model (with $\alpha = \infty$) (△), and by parsimony (○). Parsimony in both (a) and (b) and likelihood under the wrong JC69 model in (b) are inconsistent for trees in the Felsenstein zone. At $t_0 = 0$, there is a change in the parameter space and in the definition of the true tree, so that the probability curve for every method is discontinuous. Constructed following Swofford et al. (2001).

and analysed using parsimony as well as ML under JC69. In Fig. 6.2(b), data are generated under JC69 + Γ_4 with $\alpha = 0.2$ and analysed using parsimony as well as ML under both the true model (JC69 + Γ_4 with $\alpha = 0.2$ fixed) and the wrong JC69 model. The results of Fig. 6.2(b) are rather similar to those of Fig. 6.2(a), with likelihood under the wrong JC69 model behaving in the same way as parsimony, so we focus on Fig. 6.2(a). In the right half of the graph, the true tree is $((a, c), b, d)$, with the two long branches a and b separated. Trees of this shape are said to be in the 'Felsenstein zone'. In the left half of the graph, the true tree is $((a, b), c, d)$, with the long branches a and b grouped together. Such trees are said to be in the 'Farris zone'. In both zones, likelihood under the true model recovers the correct tree with reasonable accuracy, with the accuracy improving as the internal branch length t_0 increases. In contrast, parsimony behaves very differently: it recovers the true tree with probability $\sim 100\%$ in the 'Farris zone' but with probability $\sim 0\%$ in the 'Felsenstein zone'; parsimony is indeed inconsistent for trees in the right half of the graphs in Fig. 6.2. If the true tree is a star tree with $t_0 = 0$, parsimony produces the tree $((a, b), c, d)$ with probability $\sim 100\%$, and 0% for the other two trees. Now suppose the true tree is $((a, b), c, d)$, with $t_0 = 10^{-10}$. This branch is so short that not a single change is expected to occur on it at any of the 1000 sites in any of the 10 000 data sets, as the expected total number of changes is $10\,000 \times 1000 \times 10^{-10} = 0.001$. However, parsimony still recovers the true tree in every data set. The site patterns supporting the true tree ($xxyy$) are all generated by convergent evolution or *homoplasies*. Swofford et al. (2001) argued that in this case the evidence for the correct tree is given too much weight by parsimony and is evaluated correctly by likelihood under the correct model. The analyses of Swofford et al. (2001; see also Bruno and Halpern 1999) provide good intuitive explanations for counterintuitive results such as those of Figs. 6.1 and 6.2.

Yet, they are not relevant to the question I posed (Yang 1996a, 1997b): Is the likelihood method of tree reconstruction asymptotically efficient, as is a conventional MLE? I suggest that the answer to this question is 'No'. The asymptotic inefficiency of ML for tree reconstruction is not limited to one set of branch lengths, but applies to a nontrivial region of the parameter space. In the four-species problem, the parameter space is usually defined as a five-dimensional cube, denoted R_+^5, with each of the five branch lengths going from 0 to ∞. Inside R_+^5, one may define a subspace, call it $\aleph$, in which ML is asymptotically less efficient than another method, such as parsimony or ML under a simple and wrong model. For the present, $\aleph$ is not well-characterized; we do not know what shape it takes or even whether it consists of one region or several disconnected regions. However, its existence appears beyond doubt. Now suppose that the biologist always generates data sets from within $\aleph$. If the statistician applies ML for tree reconstruction, ML will be asymptotically less efficient than parsimony over the whole parameter space ($\aleph$). As the problem posed by the biologist is a well-defined statistical inference problem, the option is not open for the statistician to claim that the problem is wrong. The 'bias' argument discussed above lacks force, since in this case the 'bias' is always in the direction of the truth. In the case of conventional parameter estimation, one may demand an estimator that is asymptotically more efficient than

ML over a narrow but nonempty interval, say $\theta \in [0.491, 0.493]$. Such an estimator does not exist (e.g. Cox and Hinkley 1974, p. 292).

While there are reports of real data examples in which a simple and wrong model recovered the phylogeny better than more complex and realistic models (e.g. Posada and Crandall 2001), it should be emphasized that the above discussion is not an endorsement of parsimony or likelihood under simplistic models in real data analysis. The discussion here is more of a philosophical nature. The issue is not that ML often performs more poorly than other methods such as parsimony; it is that this can happen in a certain portion of the parameter space at all. It is important to note that under the true model, ML is always consistent, while parsimony and ML under wrong and simplistic models may be inconsistent. For practical data analysis, the biological process is likely to be far more complex than any of the substitution models we use, and likelihood under complex and realistic models may be necessary to avoid the problem of long-branch attraction. It is prudent to use such complex and realistic models in real data analysis, as recommended by Huelsenbeck (1998) and Swofford et al. (2001).

6.2.4 Robustness

Computer simulations have been conducted to examine the performance of likelihood and other model-based methods when some model assumptions are violated. The results are complex and dependent on a number of factors such as the precise assumptions being violated as well as the shapes of trees assumed in the simulation. Overall the simulations suggest that ML is highly robust to violations of assumptions. ML was found to be more robust than distance methods such as neighbour joining (Fukami-Kobayashi and Tateno 1991; Hasegawa et al. 1991; Tateno et al. 1994; Yang 1994c, 1995b; Kuhner and Felsenstein 1994; Gaut and Lewis 1995; Huelsenbeck 1995a). Certain assumptions have a huge impact on the fit of the model to data, but do not appear to have a great effect on tree reconstruction; the transition/transversion rate difference and unequal base compositions appear to fall into this category (e.g. Huelsenbeck 1995a).

Here I describe two factors that appear to be more important. Both involve some sort of among-site heterogeneity. The first is variation of substitution rates among sites. Chang (1996b) generated data using two rates, so that some sites are fast evolving and others are slowly evolving, and found that likelihood assuming one rate can become inconsistent. This may seem surprising since both sets of sites are generated under the same tree, and yet when they are analysed together, a wrong tree is obtained even from an infinite amount of data. Similar results were found when rates vary according to a gamma distribution (Kuhner and Felsenstein 1994; Tateno et al. 1994; Huelsenbeck 1995a; Yang 1995b). Of course, if the model accommodates variable rates among sites, ML is consistent.

A somewhat similar case was constructed by Kolaczkowski and Thornton (2004). Data sets were generated using two trees, or more precisely two sets of branch lengths for the same four-species tree $((a, b), (c, d))$, shown in Fig. 6.3. Half of the sites

6.2 Likelihood · 197

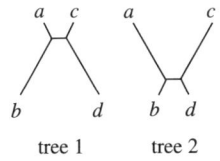

```
   a  c      a      c
    \/        \    /
    /\         \  /
   /  \         \/
  b    d        /\
              b    d
  tree 1     tree 2
```

Fig. 6.3 If the data are generated using two trees (or more precisely, two sets of branch lengths on the same tree topology), it is possible for parsimony to outperform likelihood assuming one tree.

evolved on tree 1 with branches a and c short and b and d long, while the other half evolved on tree 2 with branches a and c long and b and d short. Such data may arise when different sites in a gene sequence evolve at different rates but the rates for sites also drift over time (possibly due to changing functional constraints), a process called *heterotachy*. Likelihood under the homogeneous model, assuming one set of branch lengths for all sites, was found to perform much worse than parsimony. Note that each of the two trees is hard to recover, and parsimony can be inconsistent on any of them if the internal branch is short enough (Felsenstein 1978b). However, the probabilities of observing the three parsimony-informative site patterns, $xxyy$, $xyyx$, $xyxy$, are the same on both trees and remain the same if the data are any mixture of sites generated on the two trees. Thus the performance of parsimony is the same whether one tree or two trees are used to generate the data. In contrast, the use of two trees violates the likelihood model, causing ML to become inconsistent and to perform even worse than parsimony. Previously, Chang (1996b) found that a distance method assuming a homogeneous model in distance calculation is inconsistent under such a mixture model, and his explanation appears to apply to ML as well. In data sets generated under the mixture model, species a and c have an intermediate distance since they are very close on tree 1 but far away on tree 2. Similarly, species b and d have an intermediate distance since they are close on tree 2 but far away on tree 1. However, every other pair of species have a large distance since they are far away on both trees. The topology compatible with these requirements is $((a, c), b, d)$, which is different from the true topology.

Kolaczkowski and Thornton (2004) suggested that the performance difference between parsimony and likelihood for the trees of Fig. 6.3 also applied to other schemes of branch-length heterogeneity. This claim of generality has been challenged by a number of studies, which examined other combinations of branch lengths and found that likelihood under the homogeneous model performed much better than parsimony (e.g. Spencer *et al.* 2005; Gadagkar and Kumar 2005; Gaucher and Miyamoto 2005; Philippe *et al.* 2005; Lockhart *et al.* 2006). Of course, if two sets of branch lengths are assumed in the model, likelihood will perform well in the test of Kolaczkowski and Thornton since the model is then correct (Spencer *et al.* 2005). Such a mixture model, however, contains many parameters, especially on large trees, and does not appear to be useful for practical data analysis.

6.3 Parsimony

Parsimony, also known as *cladistics*, was initially developed to analyse discrete morphological data. When molecular sequences became available, and indeed became the predominant form of data, the method is applied to molecules as well, with each position (nucleotide or amino acid) considered a character. Such application provoked a long-standing debate about whether phylogeny reconstruction should be viewed as a statistical problem and whether parsimony or likelihood should be the method of choice. See Felsenstein (2001b, 2004) and Albert (2005) for recent reviews of that debate.

Parsimony does not make explicit assumptions about the evolutionary process. Some authors argue that parsimony makes no assumptions at all and that, furthermore, phylogenies should ideally be inferred without invoking any assumptions about the evolutionary process (Wiley 1981). Others point out that it is impossible to make any inference without a model; that a lack of explicit assumptions does not mean that the method is 'assumption-free' as the assumptions may be merely implicit; that the requirement for explicit specification of the assumed model is a strength rather than weakness of the model-based approach since then the fit of the model to data can be evaluated and improved (e.g. Felsenstein 1973b).

The latter position is taken in this book. Given this position, it is meaningful to ask what assumptions parsimony makes concerning the evolutionary process, and how parsimony can be justified. In this section I provide an overview of studies that attempt to establish an equivalence between parsimony and likelihood under particular models, and of arguments put forward to justify parsimony.

6.3.1 Equivalence with misbehaved likelihood models

A number of authors have sought to establish an equivalence between parsimony and likelihood under a particular model. Equivalence here means that the most parsimonious tree and the ML tree under the said model are identical in every possible data set. One can easily show that the two methods produce the same tree in particular data sets, but such results are not useful. Our focus is on the tree topology, so that additional inferences such as reconstruction of ancestral states or estimation of branch lengths, if provided by the method, are ignored. An established equivalence between parsimony and likelihood may serve two purposes. First, it may indicate that the likelihood model is the evolutionary model assumed by parsimony, thus helping us to understand the implicit assumptions made by parsimony. Second, it may provide a statistical (likelihood) justification for parsimony; if likelihood has nice statistical properties under the model, parsimony will share these properties.

However, both objectives are somewhat compromised. First, as Sober (2004) argued, an equivalence between parsimony and likelihood under a particular model means that the said likelihood model is sufficient for the equivalence but may not be necessary. In other words, it is possible for parsimony to be equivalent to likelihood under some other evolutionary models. Second, likelihood comes in many flavours and some likelihood methods are known not to work. Establishing an equivalence with a misbehaved likelihood method provides little justification for parsimony. Indeed there are about a dozen likelihood methods proposed in the literature, such as profile, integrated, marginal, conditional, partial, relative, estimated, empirical, hierarchical, penalized, and so on. Most of them have been proposed to deal with nuisance parameters. (Profile and integrated likelihoods are described in Subsection 1.4.3. See Goldman (1990) for a careful discussion of some of those methods.) The well-known asymptotic properties of MLEs, such as consistency and asymptotic efficiency, only hold under certain regularity conditions. One such condition is that the number of parameters in the model should not increase without bound with increase in the sample size. Unfortunately, most likelihood models used to establish an equivalence with parsimony are such pathological ones with infinitely many parameters, under which likelihood is known to misbehave.

Here I provide a brief review of studies that attempt to establish an equivalence between parsimony and likelihood under such infinite-parameters models. They are rarely used in real data analysis. The next subsection considers equivalence with well-behaved likelihood models, which are in common use.

Felsenstein (1973b, 2004, pp. 97–102) considered a Markov model of character evolution in which a set of branch lengths are applied to all sites in the sequence, but every site has a rate. The model is very similar to models of variable rates across sites discussed in Subsection 4.3.1, except that the site rates are separate parameters so that the number of rate parameters increases without bound with the increase in sequence length. Felsenstein showed that the parsimony and likelihood trees coincide when the substitution rates for sites approach zero. The result, though short of a rigorous proof, provides a mathematical justification for the intuitive expectation that parsimony and likelihood are highly similar at low sequence divergences. The proof, however, requires that the MLE of the rate for any site $\hat{r}$ approaches zero (with rates at different sites approaching zero proportionally), which is slightly different from letting the rate parameters r approach zero and showing that likelihood and parsimony trees coincide in every data set. Even if the true rate r is small, data sets are still possible in which $\hat{r}$ is not very small, in which case parsimony and likelihood may not be equivalent.

Farris (1973) considered a stochastic model of character evolution and attempted to derive MLEs of not only the ancestral sequences at the interior nodes of the tree but also the time points of changes, that is, a complete specification of the history of sequence evolution through time. He argued that likelihood and parsimony gave the same estimates. Goldman (1990) used a symmetrical substitution model to estimate the tree topology, branch lengths, and the ancestral states at the interior nodes of the tree. All branches on the tree are assumed to have the same length, so that the model lacks a time structure (Thompson 1975). Goldman showed that likelihood under the model produces the same tree as parsimony. He argued that estimation of ancestral character states makes the likelihood method inconsistent, and equivalence with such a likelihood method may provide an explanation for the inconsistency of parsimony. This likelihood model is similar to the 'maximum parsimony likelihood' of Barry and Hartigan (1987b) in that both estimate ancestral states together with model parameters such as branch lengths or branch transition probabilities. Strictly speaking, the likelihood formulations in Farris (1973), Barry and Hartigan (1987b), and Goldman (1990) appear to need more justification. Given the assumed models of character evolution, the ancestral states are random variables, with fully specified distributions. They should be averaged over in the definition of likelihood, as in Felsenstein (1973b, 1981), and not treated as parameters. The 'likelihood function' Goldman (1990, equation 6) used is not $f(\mathbf{X}|\tau, t, \mathbf{Y})$ but $f(\mathbf{X}, \mathbf{Y}|\tau, t)$, where $\mathbf{X}$ are sequences at the tips of the tree, $\mathbf{Y}$ the ancestral states, τ the tree topology, and t the branch lengths. Although $f(\mathbf{X}, \mathbf{Y}|\tau, t)$ may be given a penalized or hierarchical likelihood interpretation (Silverman 1986, pp. 110–119; Lee and Nelder 1996), the approach is not likelihood in the usual sense of the word. The same cricism applies to the 'maximum parsimony likelihood' of Barry and Hartigan (1987b, pp. 200–201).

Tuffley and Steel (1997) presented an insightful analysis of Felsenstein's (1981) likelihood method applied to data sets consisting of one single character (site). The analysis is an elegant application of both combinatorics (for parsimony) and probability theory (for likelihood), and is explained in Section 6.5. The principal result is that under a model of equal rate between any two characters, the maximized likelihood on any tree is given by $(1/c)^{1+l}$, where c is the number of character states ($c = 4$ for the JC69 model) and l is the character length or the minimum number of changes according to parsimony. Thus tree 1 is shorter than tree 2 if and only if it has a higher likelihood than tree 2; parsimony and likelihood under the model will always produce the same tree on one character.

Tuffley and Steel's theory does not apply to multiple characters if the same set of branch lengths is applied to all characters, as in Felsenstein's (1981) formulation. However, if a separate set of branch lengths is assumed for every character, the likelihood analysis will effectively be separate analyses of the characters. The tree with the highest likelihood, calculated by multiplying probabilities across characters, is then always the minimum-length or most-parsimonious tree, with the character lengths summed over characters. Tuffley and Steel called this model the *no common mechanism model*. Note that the model assumes the same relative rate between states, although it allows the amount of evolution to vary across sites and branches.

As stressed by Tuffley and Steel (1997; see also Steel and Penny 2000), the equivalence under the no-common mechanism model does not establish equivalence between parsimony and likelihood under commonly implemented likelihood models, nor does it provide a justification for parsimony under such models. The theory is of a philosophical nature. Application of likelihood to models involving infinitely many parameters is detested by statisticians (e.g. Stein 1956; Kalbfleisch and Sprott 1970; see also Felsenstein and Sober 1986; Goldman 1990), and the no-common mechanism model is not an acceptable way of dealing with among-site heterogeneity in the likelihood framework. If one is concerned about certain aspects of the evolutionary process being heterogeneous among sites, one may let such quantities be drawn from a parametric or nonparametric distribution, in the same way that models of random rates for sites are constructed. One may then use likelihood to estimate the parameters in such a 'super-process'. Such an approach accommodates the among-site heterogeneity while keeping the number of parameters under control. Interestingly, my analysis of such a model with branch lengths varying at random among sites simply leads to Felsenstein's (1981) likelihood, effectively with a redefinition of branch lengths.

6.3.2 Equivalence with well-behaved likelihood models

There are many likelihood models in common use in molecular phylogenetics, as discussed in Chapters 1, 2, and 4. These have the essential features that the model is identifiable and the number of parameters does not increase without bound with the increase of sequence length. Establishing an equivalence between parsimony and such a likelihood model will go a long way to justifying parsimony. Unfortunately, likelihood analysis under such models is most often intractable.

The only likelihood model that is tractable involves three rooted trees for three species, with binary characters evolving under the molecular clock. The three rooted trees are $\tau_1 = ((1,2),3)$, $\tau_2 = ((2,3),1)$, and $\tau_3 = ((3,1),2)$, shown in Fig. 6.4, where t_{i0} and t_{i1} are the two branch lengths in tree τ_i, for $i = 1, 2, 3$. There are only four site patterns, *xxx*, *xxy*, *yxx*, and *xyx*, where x and y are any two distinct characters. Let n_0, n_1, n_2, n_3 be the counts of sites with those patterns, with $n = n_0 + n_1 + n_2 + n_3$. We may also represent the data by the frequencies of the site patterns: $f_i = n_i/n$, $i = 0, 1, 2, 3$. If evolution proceeds at a constant rate, one may expect pattern *xxy* to occur with higher probability than patterns *yxx* and *xyx* if τ_1 is the true tree, since the former pattern is generated by a change over a longer time period while either of the latter patterns is generated by a change in a shorter time period. Similarly, patterns *yxx* and *xyx* 'support' trees τ_2 and τ_3, respectively. Thus one should select tree τ_1, τ_2, or τ_3 as the estimate of the true tree if n_1, n_2, or n_3 is the greatest among the three, respectively. We will consider this as a parsimony argument, but note that this differs from the parsimony method in common use, which minimizes the number of changes and does not distinguish among the three rooted trees. Sober (1988) discussed this

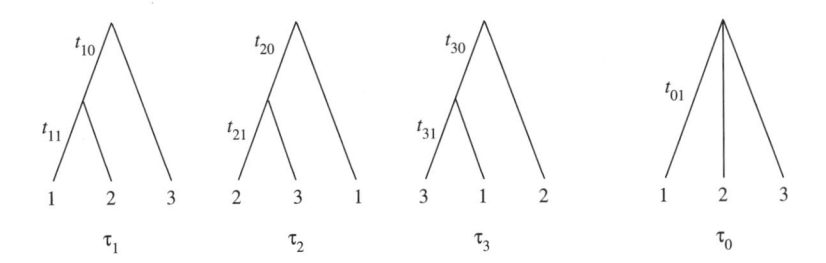

Fig. 6.4 The three rooted trees for three species: τ_1, τ_2, and τ_3. Branch lengths t_{i0} and t_{i1} in each tree τ_i ($i = 1, 2, 3$) are measured by the expected number of character changes per site. The star tree τ_0 is also shown with its branch length t_{01}.

model extensively, but his likelihood analysis is limited to one character and does not appear to be valid as it fails to accommodate the fact that the estimated branch lengths differ across the three trees.

The likelihood solution is illustrated in Fig. 6.5 (Yang 2000a, Table 4). The sample space, i.e. the space of all possible data sets, represented by (f_1, f_2, f_3), is a tetrahedron. This is divided into four regions, corresponding to the four trees; if the data fall into

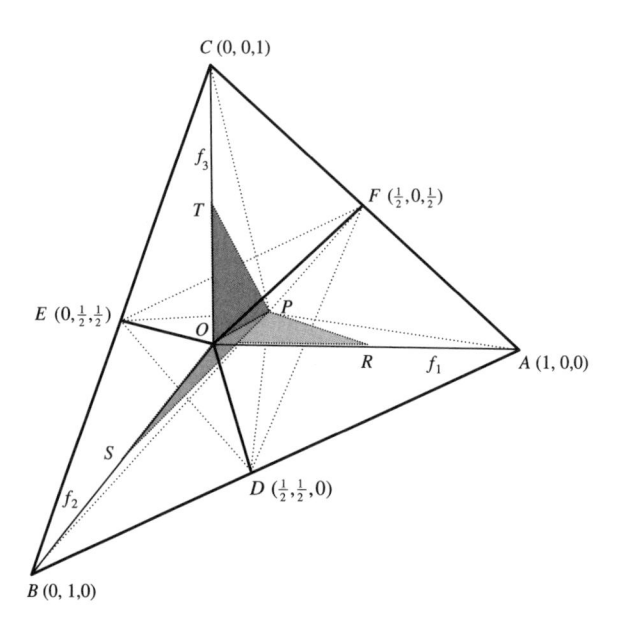

Fig. 6.5 Tree reconstruction corresponds to partitioning of the sample space, i.e. the space of all possible data sets. Here binary characters evolving under a clock with symmetrical substitution rates are used to reconstruct the rooted trees for three species of Fig. 6.4. Each data set is represented as (f_1, f_2, f_3), the frequencies of three site patterns xxy, yxx, and xyx, with the frequency of the constant pattern xxx to be $f_0 = 1 - f_1 - f_2 - f_3$. The sample space is thus the tetrahedron $OABC$. The origin is at $O(0, 0, 0)$, while OA, OB, and OC are the three axes of the coordinate system, representing f_1, f_2, and f_3, respectively. Point $P(1/4, 1/4, 1/4)$ is inside the tetrahedron. The sample space is partitioned into four regions, corresponding to the four trees; tree τ_i is the ML tree if and only if the data fall within region i ($i = 0, 1, 2, 3$). The region for τ_0 is the line segment OP plus the tetrahedron $PDEF$. In this region, the three binary trees have the same likelihood as the star tree, so τ_0 is taken as the ML tree. The region for τ_1 is a contiguous block $OPFAD$, consisting of three tetrahedrons $OPAD$, $OPAF$, and $PDAF$. The regions for τ_2 and τ_3 are $OPDBE$ and $OPECF$, respectively. The parameter (probability) space for each tree is superimposed onto the sample space. For τ_0, this is line segment OP, corresponding to $0 \leq f_1 = f_2 = f_3 < 1/4$ or $0 \leq t_{01} < \infty$. For τ_1, τ_2, and τ_3, the parameter space is triangle OPR, OPS, or OPT, respectively, where the coordinates of points R, S, and T are $R(1/2, 0, 0)$, $S(0, 1/2, 0)$, and $T(0, 0, 1/2)$. Phylogeny reconstruction may also be viewed as projecting the observed data point onto the three parameter planes, with the distance measured by the Kullback–Leibler divergence. Redrawn after Yang (2000a).

any region, the corresponding tree will be the ML tree. Parsimony and likelihood produce the same tree as the best estimate except in the tetrahedron *PDEF*, in which the three binary trees have the same likelihood as the star tree, so τ_0 is taken as the estimate. In this region, the sequences are more divergent than random sequences; for example, if $f_1 > (f_2, f_3)$ and $f_2 + f_3 \geq 1/2$, the parsimony tree is τ_1 but the ML tree is τ_0. Given that such extreme data sets are rare, we may consider likelihood and parsimony to be equivalent under the model.

Under this model, the least-squares tree based on pairwise distances is the same as the ML or parsimony tree (Yang 2000a). Similarly, if one assigns priors on branch lengths t_0 and t_1 and define the likelihood for tree τ_i as

$$L_i = \int \int f(n_0, n_1, n_2, n_3 | \tau_i, t_{i0}, t_{i1}) f(t_{i0}) f(t_{i1}) \, dt_{i0} \, dt_{i1}, \qquad (6.9)$$

the *maximum integrated likelihood tree* is τ_1, τ_2, or τ_3 if n_1, n_2, or n_3 is the greatest, respectively (Yang and Rannala 2005). In addition, if the three binary trees are assigned the same prior probability, the maximum posterior probability (MAP) tree will also be the ML tree by the Bayesian approach. Thus in terms of point estimate of the tree topology, all methods agree with each other under this model.

A slight extension of the model is to JC69 for four character states (Saitou 1988; Yang 2000a). There is then an additional site pattern *xyz*, but this pattern does not affect tree selection under either parsimony or likelihood (again except for rare data sets with extremely high divergences). The full likelihood solution does not appear tractable (see Exercise 4.2), but it is easy to show that if the ML tree is a binary tree, it must correspond to the greatest of n_1, n_2, and n_3 (Yang 2000a, pp. 115–116). The conditions under which the ML tree is the star tree are not well characterized. Again one may disregard rare extreme data sets of very divergent sequences and consider parsimony and likelihood under the JC69 model to be equivalent.

More complex models and larger trees with more species are difficult to analyse. However, a number of authors have studied cases in which parsimony is inconsistent. In such cases likelihood and parsimony must not be equivalent since likelihood is always consistent. Hendy and Penny (1989) showed that with four species and binary characters evolving under the clock, parsimony is always consistent in recovering the (unrooted) tree, although ML and parsimony do not appear to be equivalent under the model. With five or more species evolving under the clock, it is known that parsimony can be inconsistent in estimating the unrooted trees (Hendy and Penny 1989; Zharkikh and Li 1993; Takezaki and Nei 1994). Thus it is not equivalent to likelihood. For estimating unrooted trees without the clock assumption, parsimony can be inconsistent for trees of four or more species and is not equivalent to likelihood (Felsenstein 1978b; DeBry 1992). Those studies often used small trees with four to six species. The cases for much larger trees are not known. However, it appears easier to identify cases of inconsistency of parsimony on large trees than on small trees (Kim 1996; Huelsenbeck and Lander 2003), suggesting that likelihood and parsimony are in general not equivalent on large trees.

Some authors (e.g. Yang 1996a) have suggested that parsimony is closer to a simplistic likelihood model, such as JC69, than to a more complex model, such as HKY85 + Γ. Under the assumption of equal rates between any two character states, as in JC69, and under the additional assumption that all branch lengths on the tree are equal (Goldman 1990), likelihood will produce exactly the same ancestral reconstructions as parsimony. However, it is unclear whether the ML tree is the same as the parsimony tree.

6.3.3 Assumptions and Justifications

6.3.3.1 Ockham's razor and maximum parsimony

The *principle of parsimony* is an important general principle in the generation and testing of scientific hypotheses. Also known as Ockham's razor, it states that one should not increase, beyond what is necessary, the number of entities required to explain anything. This principle, used sharply by William Ockham (died *c.* 1349), assumes that simpler explanations are inherently better than complicated ones. In statistical model building, models with fewer parameters are preferred to models with more parameters if they fit the data nearly equally well. LRT, AIC, and BIC are all mathematical exemplifications of this principle; see Subsection 4.7.1.

The parsimony (minimum-step) method of tree reconstruction is often claimed to be based on the parsimony principle in science. The number of character changes on the tree is taken as the number of *ad hoc* assumptions that one has to invoke to explain the data. This correspondence appears superficial, in term rather than in content. For example, both the parsimony and minimum-evolution methods minimize the amount of evolution to select the tree, the only difference being that parsimony uses the number of changes without correction for multiple hits while minimum evolution uses the number of changes after correction for multiple hits (see Sections 3.3 and 3.4). It is not reasonable to claim that parsimony, because of its failure to correct for multiple hits, enjoys a philosophical justification that minimum evolution lacks. One may ask how parsimony should proceed and be justified if the number of changes at every site is known and given.

6.3.3.2 Is parsimony a nonparametric method?

Some authors suggest that parsimony is a nonparametric method (e.g. Sanderson and Kim 2000; Holmes 2003; Kolaczkowski and Thornton 2004). This claim appears to be a misconception. In statistics, some methods make no or weak assumptions about the distribution of the data, so that they may apply even if the distributional assumptions of the parametric methods are violated. For example, for two normally distributed variables, we usually calculate Karl Pearson's product–moment correlation coefficient r and use a t statistic to test its significance. However, this parametric method may not work well if the variables are proportions or ranks, in which case the normal assumption is violated. One may then use Spearman's rank correlation coefficient r_s, which does not rely on the normal assumption and is a nonparametric measure.

Nonparametric methods make fewer or less stringent assumptions about the data-generating model. However, they are not assumption free. They often make the same assumptions of randomness and independence of data samples as parametric methods. A nonparametric method works under situations where the parametric method works, perhaps with a slight loss of power. Parsimony for tree reconstruction is known to break down over a range of parameter values under simple parametric models (Felsenstein 1978b). This failure disqualifies it as a nonparametric method. As pointed out by Spencer *et al.* (2005), simply not requiring a parametric model is not a sufficient criterion for a satisfactory nonparametric method. A useful nonparametric model should perform well over a wide range of possible evolutionary models, but parsimony does not have this property.

6.3.3.3 Inconsistency of parsimony

As mentioned in Chapter 3, parsimony can be inconsistent over some portions of the parameter space; in particular, the method tends to suffer from the problem of long-branch attraction. Felsenstein (1973b, 1978b) conjectured that when the amount of evolution is small and the rate of evolution is more or less constant among lineages, parsimony may be consistent. Later studies have shown that even the existence of a molecular clock combined with a small amount of evolution does not guarantee consistency, and the situation seems to become worse when there are more sequences in the data (Hendy and Penny 1989; Zharkikh and Li 1993; Takezaki and Nei 1994). Huelsenbeck and Lander (2003) generated random trees using a model of cladogenesis, and found that parsimony is often inconsistent.

Farris (1973, 1977; see also Sober in Felsenstein and Sober 1986) dismissed the argument of Felsenstein (1978b) by claiming that the model Felsenstein used is overly simplistic and thus irrelevant to the performance of parsimony in real data analysis. This response is off the mark. Sober (1988) pointed out that the failure of parsimony under the simple model shows that parsimony assumes that evolution does not obey this admittedly simple model, contrary to the claim that parsimony makes no assumptions at all about the evolutionary process. Furthermore, the model is unrealistic because it assumes that evolution occurs independently among sites and lineages, at the same rate across sites and between character states, and so on. A more complex and realistic model relaxing those restrictions will include the simplistic models as special cases: for example, the rates may be allowed to differ between character states but they do not have to be. The inconsistency of parsimony under a simplistic model means inconsistency under the more complex and realistic model as well. Furthermore, as pointed out by Kim (1996), the conditions that lead to inconsistency of parsimony are much more general than outlined by Felsenstein (1978b); further relaxation of the assumptions simply exacerbates the problem.

6.3.3.4 Assumptions of parsimony

While many authors agree that parsimony makes assumptions about the evolutionary process, it has been difficult to identify them. It appears to be safe to suggest that parsimony assumes independence of the evolutionary process across characters (sites in the sequence) and among evolutionary lineages (branches on the tree). However, much more cannot be said without generating some controversy. Parsimony assigns equal weights to different types of changes, which appears to suggest that it assumes equal rates of change between characters; weighted parsimony is designed to relax this assumption. Similarly, parsimony assigns equal weights to character lengths at different sites, which suggests that it assumes the same (stochastic) evolutionary process at different sites; successive weighting (Farris 1969) attempts to modify this assumption by down-weighting sites with more changes.

Felsenstein (1973a, 1978a) argued that parsimony makes the assumption of low rates; see Subsection 6.3.1 above. As Farris (1977; see also Kluge and Farris 1969) pointed out, the minimum-change criterion does not mean that the method works only if changes are rare. Similarly, the likelihood method maximizes the likelihood (the probability of the data) to estimate parameters, but it is not true that the method works only if the likelihood is high. In large data sets, the likelihood is vanishingly small, but it is valid to compare likelihood for different hypotheses. The effect of the amount of evolution on the performance of parsimony appears to depend on the tree shape. If the tree is hard and parsimony is inconsistent, reducing the evolutionary rate indeed helps to remedy the problem. However, when the tree is easy, parsimony can outperform likelihood, with greater superiority at higher evolutionary rates (e.g. Yang 1996a, Fig. 6).

One aspect of the similarity between parsimony and likelihood may be worth emphasizing. In the Markov-chain models used in likelihood analysis, the transition-probability matrix $P(t) = e^{Qt}$ has the property that the diagonal element is greater than the off-diagonals in the same column:

$$p_{jj}(t) > p_{ij}(t), \qquad (6.10)$$

for any fixed j and any $i \neq j$; here $p_{ij}(t)$ is the transition probability from i to j over time t. As a result, an identity tends to have a higher probability than a difference if both can fit the data; ancestral reconstructions and phylogenetic trees requiring fewer changes tend to have higher likelihoods, irrespective of the sequence divergence level. Sober (1988) discussed this implication and referred to equation (6.10) as 'backward inequality'. This observation appears to provide an intuitive justification for the view that parsimony may provide a reasonable approximation to likelihood (Cavalli-Sforza and Edwards 1967). Perhaps one should be content to consider parsimony as a heuristic method of tree reconstruction that often works well under simple conditions, rather than seeking a rigorous statistical justification for it. Heuristic methods often involve *ad hoc* treatments of the data that cannot be justified rigorously under any model. As long as its limitations are borne in mind, parsimony is a simple and useful method.

6.4 Testing hypotheses concerning trees

A phylogenetic tree reconstructed using likelihood, parsimony, or distance methods may be viewed as a point estimate. It is desirable to attach a measure of reliability on it. However, the tree represents a complex structure that is quite different from a conventional parameter, making it difficult to apply conventional procedures for constructing confidence intervals or performing significance tests. This section describes several commonly used methods for evaluating the reliability of the estimated tree: bootstrap, test of internal branch lengths, and likelihood-based tests of Kishino and Hasegawa (1989) and Shimodaira and Hasegawa (1999). The Bayesian method provides measures of accuracy in the form of posterior probabilities for trees or clades. These are discussed in Chapter 5 and will not be considered further here.

6.4.1 Bootstrap

6.4.1.1 The bootstrap method

Bootstrap is perhaps the most commonly used method for assessing uncertainties in estimated phylogenies. It was introduced by Felsenstein (1985a) as a more-or-less straightforward application of the bootstrap technique in statistics developed by Efron (1979; Efron and Tibshirani 1993). Bootstrap can be used in combination with any tree-reconstruction method. However, as Felsenstein (1985a) pointed out, the tree-reconstruction method must be consistent; otherwise high support values may be generated by bootstrap for incorrect clades. Here we describe the bootstrap procedure and discuss some of the difficulties in its interpretation.

The method generates a number of pseudo data sets, called *bootstrap samples*, by resampling sites from the original data set with replacement (Fig. 6.6). Each bootstrap sample has the same number of sites as the original data. Each site in the bootstrap sample is chosen at random from the original data set, so that some sites in the original data may be sampled multiple times while others may not be sampled at all. For example, in Fig. 6.6, site 1 is included twice in the bootstrap sample, while sites 6 and 10 are not sampled. Each bootstrap data set is then analysed in the same way as the original data set to reconstruct the phylogenetic tree. This process generates a set of trees from the bootstrap samples, which can then be summarized. For example, for every clade in the tree from the original data set, we can determine the proportion of bootstrap trees that include the clade (Fig. 6.7). This proportion is known as the *bootstrap support* or *bootstrap proportion* for the clade. Some authors use the bootstrap trees to construct a majority-rule consensus tree and attach support values for clades in the consensus tree. This procedure appears less sound logically if the consensus tree differs from the tree estimated from the original data set. See Sitnikova *et al.* (1995) and Whelan *et al.* (2001) for similar comments.

6.4.1.2 The RELL approximation

The bootstrap method involves intensive computation when used with the likelihood method of tree reconstruction, as in theory each bootstrap data sample has to be analysed in the same way as the original data, which may involve a full-scale tree search. To analyse 100 bootstrap samples, the computation will be 100 times that needed to analyse the original data set. For the likelihood method, an approximation suggested

```
Original    Site         1  2  3  4  5  6  7  8  9  10
alignment   human        N  E  N  L  F  A  S  F  I  A
            chimpanzee   N  E  N  L  F  A  S  F  A  A
            bonobo       N  E  N  L  F  A  S  F  A  A
            gorilla      N  E  N  L  F  A  S  F  I  A
            orangutan    N  E  D  L  F  T  P  F  T  T
            Sumatran     N  E  S  L  F  T  P  F  I  T
            gibbon       N  E  N  L  F  T  S  F  A  T


Bootstrap   Site         2  4  1  9  5  8  9  1  3  7
sample      human        E  L  N  I  F  F  I  N  N  S
            chimpanzee   E  L  N  A  F  F  A  N  N  S
            bonobo       E  L  N  A  F  F  A  N  N  S
            gorilla      E  L  N  I  F  F  I  N  N  S
            orangutan    E  L  N  T  F  F  T  N  D  P
            Sumatran     E  L  N  I  F  F  I  N  S  P
            gibbon       E  L  N  A  F  F  A  N  N  S
```

Fig. 6.6 Construction of a bootstrap sample, by sampling sites in the original sequence alignment with replacement.

by Kishino and Hasegawa (1989) is to use the MLEs of branch lengths and substitution parameters from the original data to calculate the log likelihood values in each bootstrap sample; in theory one should re-estimate those parameters by likelihood iteration for every tree in every bootstrap sample. Resampling sites in the alignment is then equivalent to resampling the logarithms of the probabilities of data at the sites in the original data set. The method is thus called the RELL bootstrap (for Resampling Estimated Log Likelihoods). Hasegawa and Kishino's (1994) computer simulation suggests that the RELL bootstrap provides a good approximation to Felsenstein's real bootstrap. The RELL method is convenient for evaluating a fixed set of trees; one simply evaluates all trees in the set using the original data set, stores the logarithms of site probabilities for every tree, and then resamples them to calculate log likelihood values for trees in each bootstrap sample. The method is not applicable if heuristic tree search has to be performed for every bootstrap sample, since different trees are visited during the tree search in different bootstrap samples.

6.4.1.3 Interpretations of bootstrap support values

How bootstrap support values should be interpreted is not very clear. Intuitively, if the different sites have consistent phylogenetic signals, there will be little conflict among the bootstrap samples, resulting in high bootstrap support values for all or most clades. If the data lack information or different sites contain conflicting signals, there will be more variation among the bootstrap samples, resulting in lower bootstrap proportions for most clades. Thus higher bootstrap values indicate stronger support for the concerned clade. In the literature, at least three interpretations have been offered (see, e.g., Berry and Gascuel 1996).

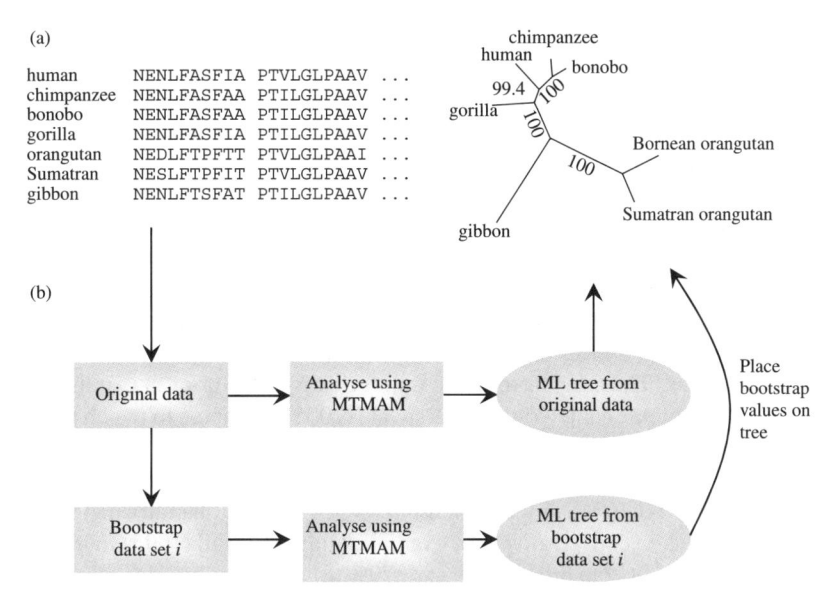

Fig. 6.7 The bootstrap method is used to calculate support values for clades on the ML tree. The mitochondrial proteins of seven ape species are analysed under the MTMAM model. From the original data, an exhaustive tree search is performed, identifying the ML tree, as shown. Bootstrap is then used to attach support values (in percentages) for clades on the ML tree. A number of (say, 1000) bootstrap data sets are generated by resampling sites in the original alignment (Fig. 6.6). Each bootstrap data set is analysed in the same way as the original data set, yielding an ML tree for each bootstrap sample. The proportion of bootstrap trees that contain a clade on the original ML tree is calculated and placed on the original ML tree. Constructed following Whelan et al. (2001).

The first is *repeatability*. A clade with bootstrap proportion P in the original data set is expected to be in the estimated tree with probability P if many new data sets are generated from the same data-generating process and if the same tree-reconstruction method is used to analyse the data sets (Felsenstein 1985a). The rationale for re-sampling the original data by bootstrap is that the distribution of the bootstrap samples around the observed data set is a good approximation of the unknown distribution of observed data from the data-generating process (Efron 1979; Efron et al. 1996). However, the simulations of Hillis and Bull (1993) suggest that the bootstrap proportion varies so much among replicate data sets that it has little value as a measure of repeatability.

A second interpretation is the frequentist *type-I error rate* or *false positive rate*, using a multifurcating tree as the null hypothesis (Felsenstein and Kishino 1993). If we generate many data samples under the multifurcating tree in which the concerned clade (with bootstrap proportion P from the original data set) is absent, then the clade will be in the estimated tree with probability $< 1 - P$. A rigorous justification for this interpretation is yet to be found, although Felsenstein and Kishino (1993) constructed an argument using an analogy to model selection concerning parameters of a normal distribution.

A third interpretation is *accuracy*; the bootstrap proportion is the probability that the tree or clade is true. This is a Bayesian interpretation and appears to be the one that most empirical phylogeneticists use or would like to use, perhaps in the same way that frequentist confidence levels and p values are often given Bayesian misinterpretations. The interpretation is meaningful only if we assign prior probabilities to trees, branches, and other parameters. Efron et al. (1996) argued that bootstrap proportions could be interpreted as posterior probabilities under an uninformative prior. The concerned prior assumes that the 4^s site patterns (for s species) have equal probabilities. When translated into a prior on branch lengths, this assigns a point mass of infinity on branch lengths. Thus, far from being uninformative, this prior is extreme to the extent of being unreasonable. Yang and Rannala (2005) attempted to match the bootstrap proportions with posterior probabilities by adjusting the prior, and concluded that the goal was in general unachievable. Based on simulation studies, many authors suggested that the bootstrap proportion, if interpreted as the probability that the clade is correct, tends to be conservative. For example, Hillis and Bull (1993) found that in their simulations bootstrap proportions of $\geq 70\%$ usually correspond to a probability of $\geq 95\%$ that the corresponding clade is true. However, a number of authors have pointed out that this result is not true in general; bootstrap proportions can be either too conservative or too liberal (e.g. Efron et al. 1996; Yang and Rannala 2005).

A few refinements to the bootstrap procedure of Felsenstein (1985a) have been suggested, including the complete-and-partial bootstrap of Zharkikh and Li (1995) and the modified method of Efron et al. (1996). Those methods involve more intensive computation and have not been widely used in real data analysis.

6.4.2 Interior branch test

Another procedure for evaluating the significance of the estimated tree is to test whether the length of an interior branch is significantly greater than zero. This is known at the *interior branch test*, and appears to have been first suggested by Felsenstein (1981) as a test of the reliability of the ML tree. One can calculate the variance covariance matrix of the branch lengths on the ML tree by using the local curvature of the log likelihood surface, relying on the asymptotic normal approximation to the MLEs. If any interior branch length is not significantly different from zero, alternative branching patterns in that portion of the tree are not rejected. Alternatively one can construct a likelihood ratio test, calculating the log likelihood either with or without constraining the interior branch length to zero (Felsenstein 1988). The χ_1^2 distribution may be used to evaluate the test statistic (twice the log likelihood difference between the two models), although in theory it is more appropriate to use a 50 : 50 mixture of 0 and χ_1^2, as the branch length must be ≥ 0 so that the value 0 is at the boundary of the parameter space (Self and Liang 1987; Gaut and Lewis 1995; Yang 1996b). The LRT is expected to be more reliable than the normal approximation for testing the positivity of an interior branch length.

The same test may be applied if a distance-based method is used for tree reconstruction. Nei *et al.* (1985) tested whether the estimated interior branch length in the UPGMA tree is significantly greater than zero by calculating the standard error of the estimated branch length and then applying a normal approximation. Li and Gouy (1991) and Rzhetsky and Nei (1992) discussed the use of this test when the tree is inferred without assuming the clock, for example, by using the neighbour-joining method (Saitou and Nei 1987).

The interior branch test involves some difficulties. First, the hypothesis is not specified *a priori* and is instead derived from the data since the ML or NJ tree is unknown until after the data are analysed (Sitnikova *et al.* 1995). Second, if one applies the same test to all interior branch lengths, multiple hypotheses are being tested using the same data set, and some sort of correction for multiple testing is called for (e.g. Li 1989). Third, the rationale of the test is not so straightforward; it is not obvious how testing an interior branch length on a tree that may be wrong should inform us of the reliability of the tree. With an infinite amount of data, one should expect all interior branch lengths in the true tree to be positive, but it is uncertain that interior branch lengths in a wrong tree should all be zero. In this regard, there appear to be qualitative differences between ML and least-squares estimates of interior branch lengths in wrong trees. For the ML method, Yang (1994c) found the MLEs of interior branch lengths in wrong trees when $n \to \infty$ to be often strictly positive. Thus the positivity of the interior branch length does not appear to have anything to do with the significance of the ML tree. Results of computer simulation appear to support this observation (Tateno *et al.* 1994; Gaut and Lewis 1995). For distance-based methods, Yang (1994c) found that in a few cases of small trees, all least-squares estimates of interior branch lengths on wrong trees are zero in infinite data sets (under the constraint that they are nonnegative). Sitnikova *et al.* (1995) allowed branch lengths to be negative and showed that the *expectations* of estimated interior branch lengths in a wrong tree in finite data sets can be positive, but pointed out that a tree with negative expected interior branch lengths must be wrong. The simulations of Sitnikova *et al.* (1995) suggest that the test of interior branch lengths provides some sort of measure of accuracy for trees estimated using distance methods.

6.4.3 Kishino-Hasegawa test and modifications

Kishino and Hasegawa (1989) suggested an approximate test, known as the K-H test, for comparing two candidate phylogenetic trees in the likelihood framework. The log likelihood difference between the two trees is used as the test statistic and its approximate standard error is calculated by applying a normal approximation to the estimated log likelihoods. Suppose the two trees are 1 and 2. The test statistic is $\Delta = \ell_1 - \ell_2$. The log likelihood for tree 1 is

$$\ell_1 = \sum_{h=1}^{n} \log f_1(\mathbf{x}_h | \hat{\theta}_1), \qquad (6.11)$$

where $\hat{\theta}_1$ are the MLEs of branch lengths and other parameters for tree 1. The log likelihood ℓ_2 for tree 2 is defined similarly. Now let

$$d_h = \log(f_1(\mathbf{x}_h|\hat{\theta}_1)) - \log(f_2(\mathbf{x}_h|\hat{\theta}_2)) \qquad (6.12)$$

be the difference between the two trees in the logarithm of the probability of data at site h, calculated at the MLEs. The mean is $\bar{d} = \Delta/n$. Kishino and Hasegawa (1989) suggest that the differences d_h are approximately independently and identically distributed (i.i.d.), so that the variance of $\bar{d}$ and thus of Δ can be estimated by the sample variance.

$$\text{var}(\Delta) = \frac{n}{n-1} \sum_{h=1}^{n} (d_h - \bar{d})^2. \qquad (6.13)$$

The two trees are significantly different if Δ is greater than twice (or 1.96 times) its standard error, $[\text{var}(\Delta)]^{1/2}$ (not one standard error as sometimes stated in the literature).

The K-H test is only valid if the two compared models or trees are specified beforehand. In molecular phylogenetics, the more common practice is to estimate the ML tree from the data and then test every other tree against the ML tree. As pointed out emphatically by Shimodaira and Hasegawa (1999) and Goldman et al. (2000), such a use of the K-H test is not valid, tending to cause false rejections of non-ML trees. In this case, the test is said to suffer from a *selection bias*, since the ML tree to be tested is selected or identified from the data, which are then used to conduct the test. The problem is of the same nature as multiple comparisons or multiple testing. Shimodaira and Hasegawa (1999) developed a test, known as the S-H test, that corrects for the selection bias. To ensure correct overall type-I error rates under all true models (true trees and parameter values), the test is constructed by considering the worst-case scenario or the least-favourable condition, that is, under the null hypothesis $E(\ell_1)/n = E(\ell_2)/n = \ldots$, with the likelihood defined for infinite data and the expectations taken over the true model. The S-H test is very conservative. Later, Shimodaira (2002) developed an approximately unbiased test, or AU test, which is less conservative. The unbiasedness here simply means that the test is less conservative than the S-H test; the power (or the probability of rejecting the null when the null is false) of an unbiased test is higher than α, the significance level of the test. The AU test controls the overall type-I error rate in most but not all cases. The AU test is implemented in the CONSEL program (Shimodaira and Hasegawa 2001).

As pointed out by Shimodaira and Hasegawa (1999), the same idea underlying the K-H test was independently suggested in the statistics literature by Linhart (1988) and Vuong (1989). The null hypothesis underlying the K-H and S-H tests is not entirely clear. In Vuong's (1989) version, the two models compared are both wrong and the null hypothesis is explicitly stated as $E(\ell_1)/n = E(\ell_2)/n$. Vuong also contrasted the test with Cox's (1961, 1962) test, in which one of the two compared models is assumed to be true. In the context of tree comparison, one tree should be true so that the null hypothesis $E(\ell_1)/n = E(\ell_2)/n = \ldots$ appears unreasonable. Alternatively, one may consider one of the possible trees to be true in the null hypothesis, but the least favourable condition $E(\ell_1)/n = E(\ell_2)/n = \ldots$ is simply used to ensure that the test controls the overall type-I error rate under all possible model (tree) and parameter combinations.

6.4.4 Indexes used in parsimony analysis

In parsimony analysis, several indexes have been suggested to measure clade support levels or, more vaguely, phylogenetic signal. They do not have a straightforward statistical interpretation, but are mentioned here as they are commonly reported in parsimony analysis.

6.4.4.1 Decay index

The *decay index*, also called *branch support* or *Bremer support*, is the difference between the tree length of the globally most parsimonious tree and the tree length achievable among trees that do not have a particular clade (Bremer 1988). It is the 'cost', in terms of the extra number of changes incurred by removing a particular clade on the most parsimonious tree. In general, Bremer support does not have an immediate statistical interpretation; its statistical significance depends on other factors such as the size of the tree, the tree length or overall sequence divergence, and the evolutionary model (e.g. Cavender 1978; Felsenstein 1985c; Lee 2000; DeBry 2001; Wilkinson *et al.* 2003).

6.4.4.2 Winning-sites test

For every site, one can score which of the two compared trees has the shorter character length and thus 'wins'. One can represent the outcome by '+' or '−', with '0' representing a tie. A binomial distribution may be used to test whether the numbers of '+' and '−' deviate significantly from 1 : 1. A test of this sort was discussed by Templeton (1983) for restriction site data. Note its similarity to the K-H test for the likelihood method. In theory, the binomial is not the correct distribution, since the total number of '+' and '−' is not fixed. Instead the trinomial distribution $M_3(n, p_+, p_-, p_0)$ may be used to test the null hypothesis $p_+ = p_-$, with all sites in the sequence used, including constant and parsimony-noninformative sites. To extend the test to more than two trees, one has to account for the problem of multiple comparisons.

6.4.4.3 Consistency index (CI) and retention index (RI)

The consistency index for a character measures the 'fit' of the character to a tree. It is defined as m/s, the minimum possible number of changes on any tree (m) divided by the step length of the current tree (s) (Kluge and Farris 1969; Maddison and Maddison 1982). The consistency index (CI) for the whole data set for a tree is defined as $\sum m_i / \sum s_i$, with the sum taken over characters (sites). If the characters are perfectly congruent with each other and with the tree, the CI will be 1. If there is a

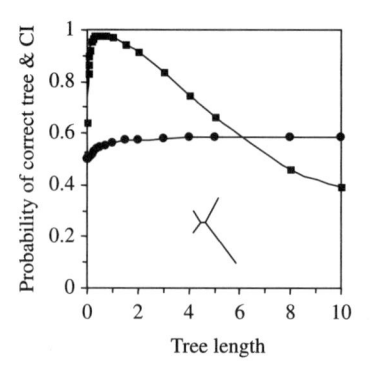

Fig. 6.8 The probability of recovering the correct tree by parsimony (P_c, ■) and the consistency index of the most parsimonious tree (CI, •), plotted against the true tree length. Both P_c and CI range from 0 to 1, so the same y-axis is used for both. Data are simulated using the tree topology of Fig. 6.1a: $((a: 0.5, b: 0.5): 0.1, c: 0.6, d: 1.4)$, but all five branch lengths are multiplied by a constant so that the tree length (the expected number of changes per site on the tree) is as specified.

lot of convergent evolution or *homoplasy*, the CI will be close to zero. The retention index (RI) for a data set is defined as $\sum (M_i - s_i)/\sum (M_i - m_i)$, where M_i is the maximum conceivable number of steps for character i on any tree (Farris 1989). Like the CI, the RI also ranges from 0 to 1, with 0 meaning a lot of homoplasies and 1 meaning perfect congruence among characters and between characters and the tree. For molecular data, homoplasy is not a good indicator of phylogenetic information content in the data set even for parsimony, and indexes such as CI and RI are not very useful despite their pleasant-sounding names. For example, Fig. 6.8 shows a simulation study on a four-species tree. The probability that parsimony recovers the true tree increases very rapidly with the increase of sequence divergence when the sequences are highly similar, peaks at an optimal divergence level, and then drops slowly with further increase in sequence divergence. The consistency index for the most parsimonious tree, however, increases very slowly over the whole range of the sequence divergence.

6.4.5 Example: phylogeny of apes

We apply bootstrap to attach support values for clades on the ML tree inferred in Subsection 4.2.5 for the mitochondrial protein sequences. The ML tree inferred under the MTMAM model, referred to as τ_1, is shown in Fig. 6.7. The second best tree (τ_2) groups human and gorilla together, and the third tree (τ_3) groups the two chimpanzees with gorilla. These two trees are worse than the ML tree τ_1 by 35.0 and 38.2 log likelihood units, respectively (Table 6.1).

We generate 1000 bootstrap data sets. Each data set is analysed using an exhaustive tree search under the MTMAM model, and the resulting ML tree is recorded. Thus

Table 6.1 Tests of trees for the mitochondrial protein data set

Tree	ℓ_i	$\Delta\ell_i - \ell_{\max}$	SE	K-H	S-H	Bootstrap	RELL
τ_1: ((human, chimps), gorilla)	$-14,558.6$	0	0	NA	NA	0.994	0.987
τ_2: ((human, gorilla), chimps)	$-14,593.6$	-35.0	16.0	0.014	0.781	0.003	0.010
τ_3: ((chimps, gorilla), human)	$-14,596.8$	-38.2	15.5	0.007	0.754	0.003	0.003

p values are shown for the K-H and S-H tests, while bootstrap proportions are shown for the bootstrap (Felsenstein 1985a) and RELL approximate bootstrap (Kishino and Hasegawa 1989).

1000 bootstrap trees are generated from the 1000 bootstrap data sets. We then find the percentage of the bootstrap trees that include every clade on τ_1, and present those as bootstrap support values for clades on τ_1. The support value for one clade is 99.4%, while all others are 100% (see Fig. 6.7). Similarly, the bootstrap proportions for trees τ_1, τ_2, and τ_3 are 99.4%, 0.3%, and 0.3%, respectively, and are $\sim 0\%$ for all other trees. As discussed in Subsection 5.6.5 in Chapter 5, the Bayesian analysis under the same model produced 100% for every clade when the prior mean for internal branch lengths is $\mu_0 > 10^{-3}$ but can be much smaller if μ_0 is very small, say $\mu_0 = 10^{-5}$.

The RELL approximation to the bootstrap uses the parameter estimates obtained from the original data set to calculate the log likelihood values for the bootstrap samples. Application of this approach to evaluate all 945 trees leads to approximate bootstrap proportions 98.6%, 1.0%, and 0.3% for trees τ_1, τ_2, and τ_3, respectively, and 0% for all other trees. Thus the RELL bootstrap provides a good approximation to the bootstrap in this analysis.

Use of the K-H test leads to rejection of all trees at the 1% level except for τ_2, for which the p value is 1.4%. As discussed above, the K-H test fails to correct for multiple comparisons and tends to be overconfident. The S-H test is far more conservative; at the 1% level, it fails to reject 27 trees when compared with the ML tree (Table 6.1).

*6.5 Appendix: Tuffley and Steel's likelihood analysis of one character

This section describes Tuffley and Steel's (1997) likelihood analysis of data sets consisting of one single character under a symmetrical substitution model. The reader is invited to consult the original paper for a mathematically rigorous proof.

The model assumes the same rate between any two of c character states. The transition probability is

$$p_{ij}(t) = \begin{cases} \frac{1}{c} + \frac{c-1}{c} \exp\left(-\frac{c}{c-1}t\right), & \text{if } i = j, \\ \frac{1}{c} - \frac{1}{c} \exp\left(-\frac{c}{c-1}t\right), & \text{if } i \neq j. \end{cases} \quad (6.14)$$

Note that when t goes from 0 to ∞, $p_{ii}(t)$ decreases monotonically from 1 to $1/c$, while $p_{ij}(t)$, $i \neq j$, increases from 0 to $1/c$ (see Fig. 1.3 in Chapter 1). If $c = 4$, the model becomes JC69 (Jukes and Cantor 1969). We use JC69 and the four-species tree

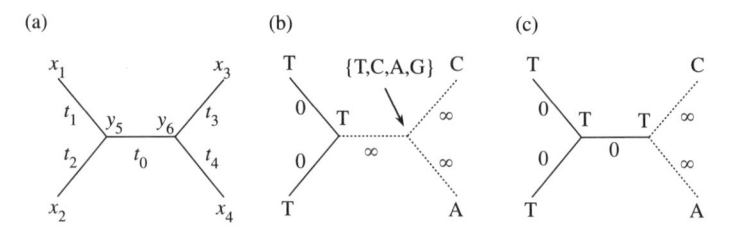

Fig. 6.9 Illustration of Tuffley and Steel's (1997) likelihood analysis of one character for the case of $c = 4$ character states (nucleotides) on a tree of four species. (a) The data, $\mathbf{x} = x_1x_2x_3x_4$, are the nucleotides observed at the tips, while $\mathbf{y} = y_5y_6$ are the unknown ancestral states. The parameters are $\mathbf{t} = (t_1, t_2, t_0, t_3, t_4)$, where each branch length t is the expected number of changes per site. The optimal branch lengths are either 0 or ∞, represented by solid and dotted lines, respectively, in (b) and (c). (b) Calculation of likelihood for one data set $\mathbf{x} = $ TTCA for a given set of branch lengths $\mathbf{t} = (0, 0, \infty, \infty, \infty)$. The character length (minimum number of changes) on the tree is $l = 2$. A set of branch lengths $\mathbf{t}$ is viable if the data can occur given $\mathbf{t}$ or if at least one reconstruction is compatible with $\mathbf{t}$. Shown here is such a set, with $k = 3$ infinity branch lengths and $m = 4$ compatible reconstructions: $y_5y_6 = $ TT, TC, TA, TG. (c) If $\mathbf{t}$ is viable and includes only $k = l$ infinity branch lengths, only one reconstruction is compatible with $\mathbf{t}$, and it is the (or a) most parsimonious reconstruction. Here $\mathbf{t} = (0, 0, 0, \infty, \infty)$ includes two infinity branch lengths, and only the most parsimonious reconstruction $y_5y_6 = $ TT is compatible with it.

of Fig. 6.9(a) for illustration. The reader may replace c by 4 in the discussion below. Let $\mathbf{x}$ be the data, the states observed at the tips of the tree, and $\mathbf{y}$ be the unknown states at the ancestral nodes, called a reconstruction. In Fig. 6.9(a), $\mathbf{x} = x_1x_2x_3x_4$ and $\mathbf{y} = y_5y_6$. Let $\mathbf{t}$ be the set of branch lengths on the unrooted tree. The objective is to maximize the likelihood $f(\mathbf{x}|\mathbf{t})$ by adjusting $\mathbf{t}$. Tuffley and Steel's principal result is that the maximized likelihood is

$$\max_{\mathbf{t}} f(\mathbf{x}|\mathbf{t}) = (1/c)^{1+l}, \quad (6.15)$$

where l is the character length or the minimum number of changes. Thus tree 1 is shorter than tree 2 if and only if tree 1 has a higher likelihood than tree 2; parsimony and likelihood will always produce the same tree.

The proof of equation (6.15) is in several stages. First, we establish that the MLEs of branch lengths are 0, ∞, or indefinite. Readers dismayed at such poor estimates may note that the likelihood method rarely performs well in data sets of size 1; for example, the MLE of the probability of heads from one flip of a coin is either 0 or 1 (depending on whether the coin lands on tail or head). The likelihood is a summation over all possible reconstructions:

$$f(\mathbf{x}|\mathbf{t}) = \sum_{\mathbf{y}} f(\mathbf{x}, \mathbf{y}|\mathbf{t}). \quad (6.16)$$

6.5 Tuffley and Steel's likelihood analysis • 217

For the example tree of Fig. 6.9(a), the sum is over the 16 possible ancestral reconstructions, and each term in the sum is given as

$$f(\mathbf{x}, \mathbf{y}|\mathbf{t}) = (1/c) p_{y_5 x_1}(t_1) p_{y_5 x_2}(t_2) p_{y_5 y_6}(t_0) p_{y_6 x_3}(t_3) p_{y_6 x_4}(t_4). \quad (6.17)$$

Here the 'root' is placed at node 5, so that $f(\mathbf{x}, \mathbf{y}|\mathbf{t})$ is equal to $1/c$, the probability of observing y_5 at node 5, times five transition probabilities along the five branches. See Subsection 4.2.1 and equation (4.2) for a detailed discussion of likelihood calculation on a tree.

Now consider the likelihood $f(\mathbf{x}|\mathbf{t})$ as a function of any one branch length t. To be concrete, let us consider t_0 in Fig. 6.9(a), but it should be obvious that the argument applies to any branch length on any tree. Note that $f(\mathbf{x}, \mathbf{y}|\mathbf{t})$ of equation (6.17) includes only one term that is a function of t_0; this term is either $p_{ij}(t_0)$ or $p_{ii}(t_0) = 1 - (c-1) \cdot p_{ij}(t_0)$. The transition probabilities for the other branches (of lengths t_1, t_2, t_3, t_4) are independent of t_0 and are constants when we study the change of likelihood $f(\mathbf{x}|\mathbf{t})$ with the change of t_0. Let $q = p_{ij}(t_0)$ so that $p_{ii}(t_0) = 1 - (c-1)q$. Note that t_0 and q consitute a one-to-one mapping, with $0 \le t_0 \le \infty$ and $0 \le q \le 1/c$. With this notation, it is easy to see that $f(\mathbf{x}, \mathbf{y}|\mathbf{t})$ and thus $f(\mathbf{x}|\mathbf{t})$ are linear functions of q. The maximum of $f(\mathbf{x}|\mathbf{t})$ thus occurs at the boundary of q, either 0 or $1/c$; in other words, the MLE of t_0 is either 0 or ∞ (Fig. 6.10). There is a third possibility, in which the likelihood $f(\mathbf{x}|\mathbf{t})$ is independent of q or t_0 so that the MLE of t can take any value (Fig. 6.10). Below we assume that the MLE is either 0 or ∞ but will consider the third possibility later.

Now that the optimal branch lengths are either 0 or ∞, we consider $\mathbf{t}$ consisting of 0 and ∞ only. The transition probability then takes only three values: 0 or 1 if $t = 0$, or $1/c$ if $t = \infty$ (see equation 6.14). If a branch length is 0, no change can occur on that branch. A reconstruction $\mathbf{y}$ that does not assign changes to zero-length branches in $\mathbf{t}$ is said to be compatible with $\mathbf{t}$. For example, given $\mathbf{t} = (t_1, t_2, t_0, t_3, t_4) = (0, 0, \infty, \infty, \infty)$ in Fig. 6.9(b), reconstruction $\mathbf{y} = y_5 y_6 = \mathrm{AC}$ is not compatible, while $\mathbf{y} = \mathrm{TT}$ is. Furthermore, we say $\mathbf{t}$ is viable if there exists

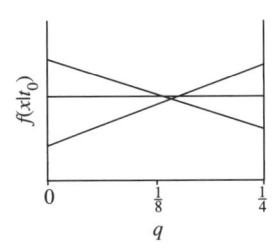

Fig. 6.10 The likelihood $f(\mathbf{x}|\mathbf{t})$ is a linear function of the transition probability over the internal branch in the tree of Fig. 6.9(a): $q = p_{ij}(t_0)$. Depending on whether the straight line for likelihood decreases or increases with q, the likelihood is maximized when $q = 0$ (corresponding to $t_0 = 0$) or $q = 1/c$ (corresponding to $t_0 = \infty$). If the likelihood line is horizontal, any value of q or t_0 achieves the same likelihood.

at least one reconstruction that is compatible with **t**. For example **t** = (0, 0, 0, 0, 0) is not viable for the data of Fig. 6.9(b), since it is impossible to find a compatible reconstruction; in other words, $f(\mathbf{x}|\mathbf{t}) = 0$ and the data are impossible for the given **t**. Let k be the number of infinity branch lengths in **t**. We must have $k \geq l$ for **t** to be viable. Branch lengths that are not viable can be ignored. For any viable **t**, note that any incompatible reconstruction makes no contribution to the likelihood. Every compatible reconstruction **y** makes the same contribution to the likelihood, irrespective of how **y** assigns changes to branches; this contribution is $f(\mathbf{x}, \mathbf{y}|\mathbf{t}) = (1/c)^{1+k}$, which is equal to the prior probability $1/c$ of the state at the root, multiplied by the transition probabilities, each equal to $1/c$, along the k infinity branches. Thus the likelihood is

$$f(\mathbf{x}|\mathbf{t}) = m(1/c)^{1+k}, \tag{6.18}$$

where m is the number of reconstructions compatible with **t**.

We now establish that for any viable **t**, with k infinity branch lengths, the number of compatible reconstructions $m \leq c^{k-l}$, so that the likelihood, according to equation (6.18), is $\leq (1/c)^{1+l}$, as claimed in equation (6.15). When we assign character states to ancestral nodes to produce reconstructions compatible with **t**, we must avoid changes on branches of zero length. Suppose we draw branches of zero length in solid lines and those of infinite length in dotted lines (Fig. 6.11). Then nodes connected by solid (zero) branches must have the same state. A set of such nodes may be called a *connected component* of the tree; they are enclosed by circles in Fig. 6.11. If one of the nodes in the set is a tip, the set is called an *external connected component*; all nodes in the set must have the state of the tip. If the set consists of internal nodes only, it is called an *internal connected component*. An internal component can have any of

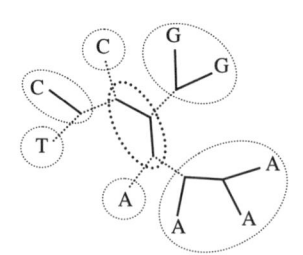

Fig. 6.11 Illustration of the concept of internal connected components, used for counting the number of reconstructions compatible with a viable set of branch lengths **t**: with s internal components, there are c^s compatible reconstructions. Branches of zero and infinity lengths are shown in solid and dotted lines, respectively. A connected component (circled) is the set of nodes connected by solid zero-length branches; all nodes in the set must have the same state. If a tip is in the set, all nodes in the set must have the state of the tip. A connected component that contains internal nodes only is called an internal connected component, and can have any of the c states. There is only one internal connected component in the example, as shown by the dotted circle, so that $s = 1$. Note that there are $k = 6$ infinity branch lengths, and that the character length is $l = 3$.

the c states; the resulting reconstructions are all compatible with **t**. Thus if **t** induces s internal components on the tree, there will be $m = c^s$ reconstructions compatible with **t**. In the example of Fig. 6.9(b), only one node ($s = 1$) is not connected to any tip by a solid branch. Then there are $m = c^s = 4^1 = 4$ compatible reconstructions: TT, TC, TA, TG. The branch lengths of Fig. 6.11 induces one internal connected component as well, with $c^s = 4$ compatible reconstructions.

We have now to show that $s \leq k - l$ to complete the proof that $m = c^s \leq c^{k-l}$. Tuffley and Steel pointed out that among the reconstructions compatible with **t**, some have at most $s - k$ changes. Since this number must not be smaller than the character length l (which is the global minimum over all possible **t**), we have $s - k \geq l$, or $s \leq k - l$, as required. The argument is as follows. If every dotted (infinity) branch links two different states, there will be k changes to start with (since there are k infinity branches). Now we can reassign character states to internal connected components to reduce the number of changes; we let an internal component take the state of an external component next to it or of another internal component with assigned state. The process is repeated until all internal components are assigned states. This way, for each of the s internal components, the number of changes is reduced at least by one. Thus the minimum number of changes among reconstructions compatible with **t** is at most $k - s$. As said above, this number must not be smaller than l, so we have $k - s \geq l$, or $s \leq k - l$. In the example of Fig. 6.11, there are $k = 6$ infinity branches and $s = 1$ internal component. The external components next to the internal component have states A, C, or G. By assigning A, C, or G to the internal component (that is, to the three internal nodes in the component), one can have at most $k - s = 5$ changes. Indeed the number of changes achieved when the internal component is assigned A, C, or G is 4, 4, or 5, respectively. All of them are greater than the character length $l = 3$.

To summarize, we have shown that any viable **t** with k infinity branch lengths induces at most $k - l$ internal connected components or at most $m = c^{k-l}$ compatible reconstructions, so that the likelihood cannot exceed $(1/c)^{1+l}$, according to equation (6.18).

It is easy to see that this likelihood is achievable. Suppose **y** is a (or the) most parsimonious reconstruction. We can use **y** to construct **t** as follows: if **y** assigns a change to a branch, set the branch length to ∞; otherwise set it to 0. In Fig. 6.9(c), TT is a most parsimonious reconstruction, and $\mathbf{t} = (0, 0, 0, \infty, \infty)$ is the corresponding set of branch lengths. It is clear that **t** constructed this way is viable. Furthermore, **y** is the only reconstruction compatible with **t**, because from the above, the number of internal connected components is $s \leq k - l = 0$, so all internal nodes are connected to the tips by zero-length branches. Thus $m = 1$. Equation (6.18) then gives the likelihood for **t** as $(1/c)^{1+l} = (1/4)^3$.

If two reconstructions $\mathbf{y}_1$ and $\mathbf{y}_2$ are equally most parsimonious, one can use them to construct two sets of branch lengths $\mathbf{t}_1$ and $\mathbf{t}_2$, both having the same highest likelihood. This is the situation of two local optima on the likelihood surface, discussed by Steel (1994a). Indeed, there are at least as many local optima as the number of equally most parsimonious reconstructions.

Table 6.2 Likelihood solutions in data sets of one character under the JC69 model on a four-species tree

Data (x)	$\hat{t}_1$	$\hat{t}_2$	$\hat{t}_0$	$\hat{t}_3$	$\hat{t}_4$	$f(x\|\hat{t})$	$y_5 y_6$
xxxx	0	0	0	0	0	$1/4$	xx
xxxy	0	0	0	0	∞	$(1/4)^2$	xx
xxyx	0	0	0	∞	0	$(1/4)^2$	xx
xyxx	0	∞	0	0	0	$(1/4)^2$	xx
yxxx	∞	0	0	0	0	$(1/4)^2$	xx
xxyy	0	0	∞	0	0	$(1/4)^2$	xy
xyyx	0	∞	0	∞	0	$(1/4)^3$	xx
	∞	0	0	0	∞		yy
xyxy	0	∞	0	0	∞	$(1/4)^3$	xx
	∞	0	0	∞	0		yy
xxyz	0	0	0	∞	∞	$(1/4)^3$	xx
	0	0	∞	0	∞		xy
	0	0	∞	∞	0		xz
	0	0	∞	∞	∞		xw
			...				...
xyzw	0	∞	0	∞	∞	$(1/4)^4$	xx
	∞	0	0	∞	∞		yy
	0	∞	∞	∞	0		xz
			...				...

Each data set, represented by $x = x_1 x_2 x_3 x_4$, consists of the nucleotide states observed at one site in the four species of Fig. 6.9(a). In the table, x, y, z, and w are any distinct nucleotides. Branch lengths t_1, t_2, t_0, t_3, t_4 are defined in Fig. 6.9(a). The table shows the MLEs of branch lengths and the maximized likelihood $f(x|\hat{t})$ for each data set. The optimal ancestral states ($y_5 y_6$) are shown as well. Multiple local peaks exist for some data sets, such as xyyx, xyxy, xxyz, and xyzw.

The case of four species is detailed in Table 6.2. Here we note that it is possible for the likelihood $f(x|t)$ to be independent of branch lengths t. This is the situation of horizontal likelihood line in Fig. 6.10. Consider the data TTCA on the tree of Fig. 6.9(b). When two of the three dotted branches have length ∞, the other dotted branch can have any length between 0 and ∞, and the likelihood stays at the same maximum. If the data are TCAG (or xyzw as in Table 6.2), the likelihood is maximized as long as each pair of tips is separated by at least one infinity branch. Thus parsimony and likelihood may not produce identical reconstructions of ancestral states, although they always agree on the tree topology. Tuffley and Steel show that for binary characters ($c = 2$), the two methods indeed produce the same ancestral reconstructions, even though this may not be the case for $c \geq 3$.

Part III

Advanced topics

7

Molecular clock and estimation of species divergence times

7.1 Overview

The hypothesis of a *molecular clock* asserts that the rate of DNA or protein sequence evolution is constant over time or among evolutionary lineages. In the early 1960s, when protein sequences became available, it was observed that the differences between proteins from different species, such as haemoglobin (Zuckerkandl and Pauling 1962), cytochrome *c* (Margoliash 1963), and fibrinopeptides (Doolittle and Blomback 1964), were roughly proportional to the divergence times between the species. The observations led to the proposal of the *molecular evolutionary clock* by Zuckerkandl and Pauling (1965).

    A few clarifications are in order. First, the clock was envisaged as a stochastic clock due to the random nature of the amino acid or nucleotide-substitution process. It 'ticks' at random intervals rather than regularly as a wristwatch normally does. Under the Markov models of nucleotide or amino acid substitution, substitution events ('ticks') arrive at time intervals that are exponentially distributed. Second, it was noted from the beginning that different proteins or regions of a protein evolve at drastically different rates, so that the clock hypothesis allows for such rate differences among proteins; the different proteins are said to have their own clocks, which tick at different rates. Third, rate constancy may not hold globally for all species and usually applies to a group of species. For example, we may say that the clock holds for a particular gene within primates.

    As soon as it was proposed, the molecular-clock hypothesis had an immediate and tremendous impact on the burgeoning field of molecular evolution and has been a focus of controversy throughout the four decades of its history. First, the utility of the molecular clock was well recognized. If proteins evolve at roughly constant rates, they can be used to reconstruct phylogenetic trees and to estimate divergence times among species (Zuckerkandl and Pauling 1965). Second, the reliability of the clock and its implications for the mechanism of molecular evolution were a focus of immediate controversy, entwined in the neutralist–selectionist debate. See Section 8.2 for a brief review of the latter debate. At the time, the neo-Darwinian theory of evolution was generally accepted by evolutionary biologists, according to which the evolutionary process is dominated by natural selection. A constant rate of evolution among species as different as elephants and mice was incompatible with that theory, as species living in different habits, with different life histories, generation times, etc. must be under very different regimes of selection. When the *neutral theory of molecular evolution* was proposed (Kimura 1968; King and Jukes 1969), the observed clock-like behaviour of molecular evolution became 'perhaps the strongest evidence for the theory' (Kimura and Ohta 1971). This theory emphasizes random fixation of neutral or nearly neutral mutations with relative fitness close to zero. The rate of molecular evolution is then equal to the neutral mutation rate, independent of factors such as environmental changes and population sizes. If the mutation rate is similar and the function of the protein remains the same among a group of species so that the same proportion of mutations are neutral, a constant evolutionary rate is expected by the theory (Ohta and Kimura 1971). Rate differences among proteins are explained by the presupposition that different proteins are under different functional constraints, with different fractions of neutral mutations. Nevertheless, the neutral theory is not the only one compatible with clock-like evolution and neither does the neutral theory always predict a molecular clock. Controversies also exist concerning whether the neutral theory predicts rate constancy over generations or over calendar time, or whether the clock should apply to silent DNA changes or to protein evolution (Kimura 1983; Li and Tanimura 1987; Gillespie 1991; Li 1997).

Since the 1980s, DNA sequences have accumulated rapidly and have been used to conduct extensive tests of the clock and to estimate evolutionary rates in different groups of organisms. Wu and Li (1985) and Britten (1986) noted that primates have lower rates than rodents, and humans have lower rates than apes and monkeys, characterized as *primate slowdown* and *hominoid slowdown* (Li and Tanimura 1987). Two major factors that are proposed to account for between-species rate differences are generation time, with a shorter generation time associated with more germ-line cell divisions per calendar year and thus a higher substitution rate (Laird *et al.* 1969; Wilson *et al.* 1977; Wu and Li 1985; Li *et al.* 1987), and DNA repair mechanism, with a less reliable repair mechanism associated with higher mutation or substitution rate (Britten 1986). Martin and Palumbi (1993) found that substitution rates in both nuclear and mitochondrial genes are negatively related to body size, with high rates in rodents, intermediate rates in primates, and slow rates in whales. Body size is not expected to affect substitution rate directly, but is highly correlated to a number of physiological and life-history variables, notably generation time and metabolic rate. Species with small body sizes tend to have shorter generation times and higher metabolic rates. The negative correlation between substitution rate and body size has been supported in some studies (e.g. Bromham *et al.* 1996; Bromham 2002) but questioned in others (e.g. Slowinski and Arbogast 1999). The disagreements do not appear to have been resolved.

Application of the clock to estimate divergence times began with Zuckerkandl and Pauling (1962), who used an approximate clock to date duplication events among α, β, γ, and δ globins of the haemoglobin family. The molecular clock has since been used widely to date species divergences, and has produced a steady stream of controversies, mostly because molecular dates are often at odds with the fossil records. The conflict is most evident concerning several major events in evolution. The first is the origin of the major animal forms. Fossil forms of metazoan phyla appear as an 'explosion' about 500–600 million years ago (MYA) in the early Cambrian (Knoll and Carroll 1999), but most molecular estimates have been much older, sometimes twice as old (e.g. Wray *et al.* 1996). Another is the origins and divergences of modern mammals and birds following the demise of the dinosaurs about 65 MYA at the Cretaceous–Tertiary boundary (the K-T boundary). Molecules again generated much older dates than expected from fossils (e.g. Hedges *et al.* 1996). Part of the discrepancy is due to the incompleteness of fossil data; fossils represent the time when species developed diagnostic morphological characters and were fossilized, while molecules represent the time when the species became genetically isolated, so fossil dates have to be younger than molecular dates (Foote *et al.* 1999; Tavaré *et al.* 2002). Part of the discrepancy appears to be due to inaccuracies and deficiencies in molecular time estimation. Despite the sometimes acrimonious controversies, the interactions between molecules and fossils have been a driving force in this research area, as they prompted reinterpretations of fossils, critical evaluations of molecular dating techniques, and the development of more advanced analytical methods.

A number of nice reviews have been published recently on the molecular clock and its use in the estimation of divergence times. See, for example, Morgan (1998) and Kumar (2005) for the history of the molecular clock; Bromham and Penny (2003) for a discussion of the clock in relation to theories of molecular evolution; Cooper and Penny (1997), Cooper and Fortey (1998), Smith and Peterson (2002), and Hasegawa *et al.* (2003) for assessments of the conflicts between molecules and fossils concerning the divergences of mammals at the K-T boundary and the origins of the major animal phyla around the Cambrian. My focus in this chapter is on statistical methods for testing the clock hypothesis, and on likelihood and Bayesian methods for dating species divergence events under global and local clock models. In such an analysis, fossils are used to calibrate the clock, that is, to translate sequence distances into absolute geological times and substitution rates. A similar situation concerns viral genes, which evolve so fast that changes are observed over the years. Then one can use the dates at which the sequences are determined to calibrate the clock and to estimate divergence times, using essentially the same techniques as discussed here. See Rambaut (2000) and Drummond *et al.* (2002) for such analyses.

7.2 Tests of the molecular clock

7.2.1 Relative-rate tests

The simplest test of the clock hypothesis examines whether two species A and B evolve at the same rate by using a third outgroup species C (Fig. 7.1). This has become known as the *relative-rate test*, even though almost all tests of the clock compare relative rather than absolute rates. The relative-rate test has been discussed by a number of authors, starting with Sarich and Wilson (1973; see also Sarich and Wilson 1967). If the clock hypothesis is true, the distances from ancestral node O to species A and B should be equal: $d_{OA} = d_{OB}$ or $a = b$ (Fig. 7.1). Equivalently

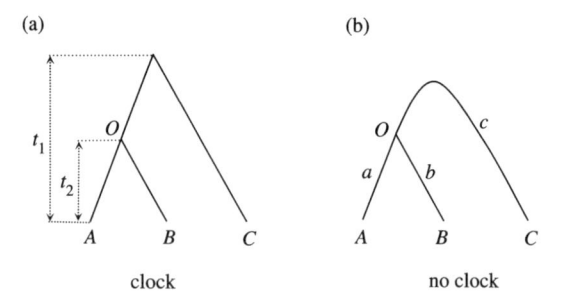

Fig. 7.1 Rooted and unrooted trees used to explain the relative-rate test. (a) Under the clock, the parameters are the ages of two ancestral nodes t_1 and t_2, measured by the expected number of substitutions per site. (b) Without the clock, the parameters are the three branch lengths a, b, and c, also measured by the expected number of substitutions per site. The clock model is a special case of the nonclock model with the constraint $a = b$; that is, the nonclock model reduces to the clock model when $a = b = t_2$ and $(a + c)/2 = t_1$.

one can formulate the clock hypothesis as $d_{AC} = d_{BC}$. Sarich and Wilson (1973) did not describe how to decide on the significance of the difference. Fitch (1976) used the number of differences to measure the distance between species, and calculated $a = d_{AB} + d_{AC} - d_{BC}$ and $b = d_{AB} + d_{BC} - d_{AC}$ as the numbers of changes along branches OA and OB. The clock was then tested by comparing $X^2 = (a-b)^2/(a+b)$ against χ_1^2. This is equivalent to using the binomial distribution $B(a + b, 1/2)$ to test whether the observed proportion $a/(a+b)$ deviates significantly from $1/2$, with $[a/(a+b) - 1/2]/\sqrt{(1/2) \times (1/2)/(a+b)} = (a-b)/\sqrt{a+b}$ compared against the standard normal distribution. This approach fails to correct for multiple hits, and furthermore, the χ^2 and binomial approximations may not be reliable if $a+b$ is small. Wu and Li (1985) corrected for multiple hits under the K80 model (Kimura 1980), and calculated $d = d_{AC} - d_{BC}$ and its standard error. Then $d/\text{SE}(d)$ was compared against the standard normal distribution. Of course, calculation of the distances and their standard errors relies on a substitution model. A version of the test that is not sensitive to the substitution model is proposed by Tajima (1993). This compares the counts (say, m_1 and m_2) of two site patterns xyy and xyx, where x and y are any two distinct nucleotides. Tajima suggested comparison of $(m_1 - m_2)^2/(m_1 + m_2)$ against χ_1^2, like Fitch's test above. The relative-rate test can also be conducted in a likelihood framework (Muse and Weir 1992; see also Felsenstein 1981). One calculates ℓ_0 and ℓ_1, the log likelihood values with and without constraining $a = b$, respectively (Fig. 7.1b). Then $2\Delta\ell = 2(\ell_1 - \ell_0)$ is compared against χ_1^2.

7.2.2 Likelihood ratio test

The likelihood ratio test of the clock (Felsenstein 1981) applies to a tree of any size. Under the clock (H_0), there are $s-1$ parameters corresponding to the ages of the $s-1$ internal nodes on the rooted tree with s species, measured by the expected number of

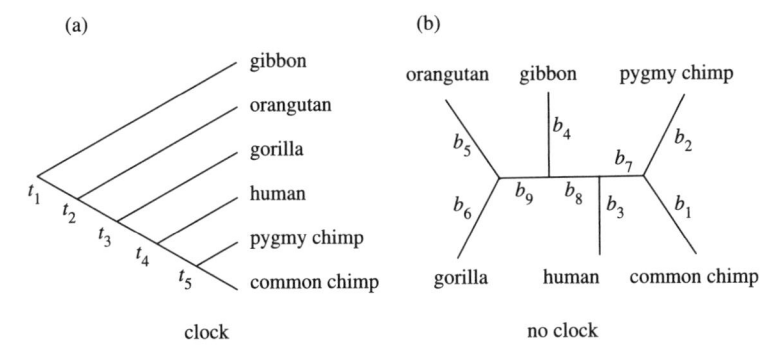

![Figure 7.2: rooted tree (a) with taxa gibbon, orangutan, gorilla, human, pygmy chimp, common chimp and internal node times t_1–t_5; unrooted tree (b) with taxa orangutan, gibbon, pygmy chimp, gorilla, human, common chimp and branches b_1–b_9.]

Fig. 7.2 Rooted and unrooted trees for six primate species for the data of Horai et al. (1995), used to conduct the likelihood ratio test of the molecular clock. (a) Under the clock model, the parameters are the ages of the ancestral nodes t_1-t_5, measured by the expected number of substitutions per site. (b) Without the clock, the parameters are the nine branch lengths b_1-b_9, measured by the expected number of substitutions per site. The two models can be compared using a likelihood ratio test to decide whether the clock holds.

changes per site. The more general nonclock model (H_1) allows every branch to have its own rate. Because time and rate are confounded, this model involves $2s - 3$ free parameters, corresponding to the branch lengths in the unrooted tree. In the example of Fig. 7.2 with $s = 6$ species, there are $s - 1 = 5$ parameters under the clock (t_1-t_5), and $2s - 3 = 9$ parameters without the clock (b_1-b_9). The clock model is equivalent to the nonclock model by applying $s - 2$ equality constraints. In the example, the four constraints may be $b_1 = b_2$, $b_3 = b_7 + b_1$, $b_4 = b_8 + b_3$, and $b_5 = b_9 + b_4$. The inequality constraint $b_6 > b_5$ does not reduce the number of parameters. Let ℓ_0 and ℓ_1 be the log likelihood values under the clock and nonclock models, respectively. Then $2\Delta\ell = 2(\ell_1 - \ell_0)$ is compared with the χ^2 distribution with d.f. $= s - 2$ to decide whether the clock is rejected.

Example. Test of the molecular clock in the mitochondrial 12s rRNA genes of human, common chimpanzee, pygmy chimpanzee, gorilla, orangutan, and gibbon. The rooted and unrooted trees are shown in Fig. 7.2. See Horai et al. (1995) for the GenBank accession numbers of the sequences. There are 957 sites in the alignment, with the average base compositions to be 0.216 (T), 0.263 (C), 0.330 (A), and 0.192 (G). We use the K80 model to conduct the test. The log likelihood values are $\ell_0 = -2345.10$ under the clock and $\ell_1 = -2335.80$ without the clock. Comparison of $2\Delta\ell = 2(\ell_1 - \ell_0) = 18.60$ with the χ^2_4 distribution indicates significant difference, with $P < 0.001$. The clock is thus rejected for those data. □

7.2.3 Limitations of the clock tests

A few limitations of the clock tests may be noted here. First, none of the tests discussed above examines whether the rate is constant over time. Instead they test a weaker hypothesis that all tips of the tree are equidistant from the root, with distance measured by the number of substitutions. For example, if the evolutionary rate has been accelerating (or decelerating) over time in all lineages, the tree will look clock-like, although the rate is not constant. Similarly the relative-rate test using three species may detect a rate difference between ingroup species A and B, but will never detect any rate difference between the outgroup species C and the ingroup species. Second, the tests of molecular clock cannot distinguish a constant rate from an average variable rate *within* a lineage, although the latter may be a more sensible explanation than the former when the clock is rejected and the rate is variable across lineages. Finally, failure to reject the clock may simply be due to lack of information in the data or lack of power of the test. In particular, the relative-rate test applied to only three species often has little power (Bromham *et al.* 2000). These observations suggest that failure to reject the clock may not be taken as strong evidence for a constant rate.

7.2.4 Index of dispersion

Many early studies tested the molecular clock using the *index of dispersion R*, defined as the variance to mean ratio of the number of substitutions among lineages (Ohta and Kimura 1971; Langley and Fitch 1974; Kimura 1983; Gillespie 1984, 1986b, 1991). When the rate is constant and the species are related to each other through a star tree, the number of substitutions on each lineage should follow a Poisson distribution, according to which the mean should equal the variance (Ohta and Kimura 1971). An R value significantly greater than 1, a phenomenon known as the *over-dispersed clock*, means that the neutral expectation of the clock hypothesis is violated. The dispersion index is used more as an diagnosis of the relative importance of mutation and selection and as a test of the neutral theory than as a test of rate constancy over time. A dispersion index greater than 1 can be taken as evidence for selection, but it can also indicate nonselective factors, such as generation times, mutation rates, efficiency of DNA repair, and metabolic rate effect. Kimura (1983, 1987) and Gillespie (1986a, 1991) used gene sequences from different orders of mammals and found that R is often greater than 1. However, the authors' use of the star phylogeny for mammals appears to have led to overestimates of R (Goldman 1994). Ohta (1995) used three lineages only (primates, artiodactyls, and rodents) to avoid the problem of phylogeny, but the analysis may be prone to sampling errors as the variance is calculated using only three lineages. The dispersion index appears somewhat out of date (but see Cutler 2000), as the hypotheses can often be tested more rigorously using likelihood ratio tests, with more realistic models used to estimate evolutionary rates (e.g. Yang and Nielsen 1998).

7.3 Likelihood estimation of divergence times

7.3.1 Global-clock model

The molecular-clock assumption provides a simple yet powerful way of dating evolutionary events. Under the clock, the expected distance between sequences increases

7.3 Likelihood estimation of divergence times • 229

linearly with their time of divergence. When external information about the geological ages of one or more nodes on the phylogeny is available (typically based on the fossil record), sequence distances or branch lengths can be converted into absolute geological times. This is known as *calibration of the molecular clock*. Both distance and likelihood methods can be used to estimate the distances from the internal nodes to the present time. The assumed substitution model may be important, as a simplistic model may not correct for multiple hits properly and may underestimate distances. Often the underestimation is more serious for large distances than for small ones, and the nonproportional underestimation may generate systematic biases in divergence time estimation. Nevertheless, there is no difficulty in using any of the substitution models discussed in Chapters 1, 2, and 4, so the effect of the substitution model can easily be assessed. Instead my focus in this chapter is on assumptions about the rates, incorporation of fossil uncertainties, assessment of errors in time estimates, etc.

Similarly the rooted tree topology is assumed to be known in molecular-clock dating. Uncertainties in the tree may have a major impact on estimation of divergence times, but the effects are expected to be complex, depending on whether the conflicts are around short internal branches and on the locations of the nodes to be dated in relation to calibration nodes and to the unresolved nodes. When the tree topology involves uncertainties, it appears inappropriate to use a consensus tree with poorly supported nodes collapsed into polytomies. A polytomy in a consensus tree indicates unresolved relationships rather than the best estimate of the true relationship. It is better to use a binary tree that is most likely to be correct. The use of several binary trees may provide an indication of the robustness of time estimation to uncertainties in the tree topology.

There are $s - 1$ ancestral nodes in a rooted tree of s species. Suppose that the ages of c ancestral nodes are known without error, determined from fossil data. The model then involves $s - c$ parameters: the substitution rate μ and the ages of $s - 1 - c$ nodes that are not calibration points. For example, the tree of Fig. 7.3 has $s = 5$ species, with four interior node ages: t_1, t_2, t_3, and t_4. Suppose node ages t_2 and t_3 are fixed according to fossil records. Then three parameters are estimated under the model: μ, t_1, and t_4. Given rate μ and the times, each branch length is just the

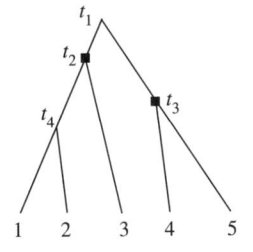

Fig. 7.3 A phylogenetic tree for five species, used to explain likelihood and Bayesian methods for divergence time estimation. Fossil calibrations are available for nodes 2 and 3.

product of the rate and the time duration of the branch, so that the likelihood can be calculated using standard algorithms (see Chapter 4). Times and rates are then estimated by maximizing the likelihood. The time parameters have to satisfy the constraints that any node should not be older than its mother node; in the tree of Fig. 7.3, $t_1 > \max(t_2, t_3)$, and $0 < t_4 < t_2$. Numerical optimization of the likelihood function has to be performed under such constraints, which can be achieved by using constrained optimization or through variable transformations (Yang and Yoder 2003).

7.3.2 Local-clock models

The molecular clock often holds in closely related species, for example within the hominoids or even primates, but is most often rejected in distant comparisons, for example among different orders of mammals (Yoder and Yang 2000; Hasegawa et al. 2003; Springer et al. 2003). Given that sequences provide information about distance but not time and rate separately, one may expect divergence time estimation to be sensitive to assumptions about the rate. This is indeed found to be the case in many studies (e.g. Takezaki et al. 1995; Rambaut and Bromham 1998; Yoder and Yang 2000; Aris-Brosou and Yang 2002).

One approach to dealing with violation of the clock is to remove some species so that the clock approximately holds for the remaining species. This may be useful if one or two lineages with grossly different rates can be identified and removed (Takezaki et al. 1995), but is difficult to use if the rate variation is more complex. The approach of filtering data may also suffer from a lack of power of the test. Another approach is to take explicit account of among-lineage rate variation when estimating divergence times. This has been the focus of much recent research, with both likelihood and Bayesian methodologies employed. In this section we describe the likelihood approach, including heuristic rate-smoothing algorithms. The Bayesian approach is discussed in Section 7.4.

In the likelihood method, one may assign different rates to branches on the tree and then estimate, by ML, both the divergence times and the branch rates. The first application of such *local-clock models* appears to be Kishino and Hasegawa (1990), who estimated divergence times within hominoids under models with different transition and transversion rates among lineages. The likelihood is calculated by using a multivariate normal approximation to the observed numbers of transitional and transversional differences between sequences. Rambaut and Bromham (1998) discussed a likelihood *quartet-dating* procedure, which applies to the tree of four species: $((a, b), (c, d))$, shown in Fig. 7.4, with a rate for the left part of the tree and another rate for the right part of the tree. This was extended by Yoder and Yang (2000) to trees of any size, so that an arbitrary number of rates can be used with an arbitrary assignment of branches to rates. The implementation is very similar to that of the global-clock model discussed above. The only difference is that under a local-clock model with k branch rates, one estimates $k - 1$ extra rate parameters.

A serious drawback of such local-clock models is their arbitrariness in the number of rates to assume and in the assignment of rates to branches. Note that some assignments

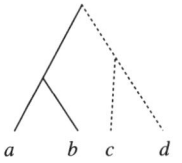

Fig. 7.4 The tree of four species used in the quartet-dating method of Rambaut and Bromham (1998). Two substitution rates are assumed for branches on the left and on the right of the tree, respectively.

are not feasible as they cause the model to become unidentifiable. This approach may be straightforward to use if extrinsic reasons are available for assigning branches to rates; for example, two groups of species may be expected to be evolving at different rates and thus assigned different rates. However, in general too much arbitrariness may be involved. Yang (2004) suggested the use of a rate-smoothing algorithm to assist automatic assignment of branches to rates, but the procedure becomes rather complicated.

7.3.3 Heuristic rate-smoothing methods

Sanderson (1997) described a heuristic rate-smoothing approach to estimating divergence times under local-clock models without *a priori* assignment of branches to rates. The approach follows Gillespie's (1991) idea that the rate of evolution may itself evolve, so that the rate is autocorrelated across lineages on the phylogeny, with closely related lineages sharing similar rates. The method minimizes changes in rate across branches, thus smoothing rates and allowing joint estimation of rates and times. Let the branch leading to node k be referred to as branch k, which has rate r_k. Let **t** be the times and **r** the rates. Sanderson (1997) used parsimony or likelihood to estimate branch lengths on the tree (b_k) and then fitted times and rates to them, by minimizing

$$W(\mathbf{t}, \mathbf{r}) = \sum_k (r_k - r_{\mathrm{anc}(k)})^2, \qquad (7.1)$$

subject to the constraints

$$r_k T_k = b_k. \qquad (7.2)$$

Here $\mathrm{anc}(k)$ is the node ancestral to node k, and T_k is the time duration of branch k, which is $t_{\mathrm{anc}(k)}$ if node k is a tip and $t_{\mathrm{anc}(k)} - t_k$ if node k is an interior node. The summation in equation (7.1) is over all nodes except the root. For the two daughter nodes of the root (let them be 1 and 2), which do not have an ancestral branch, the sum of squared rate differences may be replaced by $(r_1 - \bar{r})^2 + (r_2 - \bar{r})^2$, where $\bar{r} = (r_1 + r_2)/2$. The approach insists on a perfect fit to the estimated branch lengths while minimizing changes in rate between the ancestral and descendant branches. Sampling errors in the estimated branch lengths are thus ignored.

An improved version was suggested by Sanderson (2002), which maximizes the following 'log likelihood function'

$$\ell(\mathbf{t}, \mathbf{r}, \lambda | X) = \log\{f(X|\mathbf{t}, \mathbf{r})\} - \lambda \sum_k (r_k - r_{\text{anc}(k)})^2. \quad (7.3)$$

The second term is the same as in equation (7.1) and penalizes rate changes across branches. The first term is the log likelihood of the data, given $\mathbf{t}$ and $\mathbf{r}$. This could be calculated using Felsenstein's (1981) pruning algorithm. To reduce computation, Sanderson (2002) used a Poisson approximation to the number of changes for the whole sequence along each branch estimated by parsimony or likelihood. The probability of the data is then given by multiplying Poisson probabilities across branches. The approximation ignores correlations between estimated branch lengths. Furthermore, the substitution process is Poisson only under simple models like JC69 and K80. Under more complex models such as HKY85 or GTR, different nucleotides have different rates and the process is not Poisson.

The smoothing parameter λ determines the relative importance of the two terms in equation (7.3) or how much rate variation should be tolerated. If $\lambda \to 0$, the rates are entirely free to vary, leading to a perfect fit to the branch lengths. If $\lambda \to \infty$, no rate change is possible and the model should reduce to the clock. There is no rigorous criterion by which to choose λ. Sanderson (2002) used a cross-validation approach to estimate it from the data. For any value of λ, one sequentially removes small subsets of the data, estimates parameters $\mathbf{r}$ and $\mathbf{t}$ from the remaining data, and then uses the estimates to predict the removed data. The value of λ that gives the overall best prediction is used as the estimate. Sanderson used the so-called 'leave-one-out' cross-validation and removed in turn the branch length (the number of changes) for every terminal branch. This approach is computationally expensive, if the data consist of multiple genes, which are expected to have different rates and different λs.

A few modifications were introduced by Yang (2004), who suggested maximizing the following 'log likelihood function'

$$\ell(\mathbf{t}, \mathbf{r}, \sigma^2; X) = \log\{f(X|\mathbf{t}, \mathbf{r})\} + \log\{f(\mathbf{r}|\mathbf{t}, \sigma^2)\} + \log\{f(\sigma^2)\}. \quad (7.4)$$

The first term is the log likelihood. Yang (2004) used a normal approximation to the MLEs of branch lengths estimated without the clock, following Thorne et al. (1998) (see below). The second term, $f(\mathbf{r}|\mathbf{t})$, is a prior density for rates. This is specified using the geometric Brownian motion model of rate drift of Thorne et al. (1998). The model is illustrated in Fig. 7.5 and will be discussed in more detail below in Section 7.4. Here it is sufficient to note that given the ancestral rate $r_{\text{anc}(k)}$, the current rate r has a log-normal distribution with density

$$f(r_k | r_{\text{anc}(k)}) = \frac{\exp\{-(1/(2t\sigma^2))[\log(r_k/r_{\text{anc}(k)}) + (t\sigma^2/2)]^2\}}{r_k \sqrt{2\pi t \sigma^2}}, \quad 0 < r_k < \infty. \quad (7.5)$$

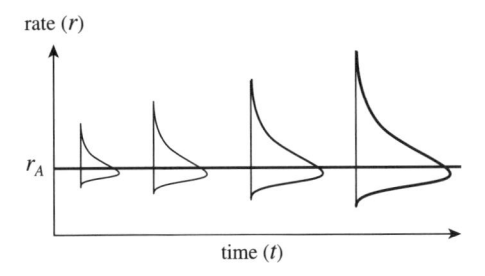

Fig. 7.5 The geometric Brownian motion model of rate drift. The mean rate stays the same as the ancestral rate r_A, but the variance parameter $t\sigma^2$ increases linearly with time t.

Here parameter σ^2 specifies how clock-like the tree is, with a large σ^2 meaning highly variable rates and serious violation of the clock. The density $f(\mathbf{r}|\mathbf{t})$ of all rates is given by multiplying the log normal densities across branches. Because the density of equation (7.5) is high when r_k is close to $r_{\text{anc}(k)}$ and low when the two rates are far apart, maximizing $f(\mathbf{r}|\mathbf{t})$ has the effect of minimizing rate changes over branches, similar to minimizing the sum of squared rate differences in equation (7.3). Compared with Sanderson's approach, this has the advantage of automatically taking account of different scales of rates and branch lengths, so that there is no need for cross-validation. If the data contain sequences from multiple genes, one simply multiplies the likelihood $f(X|\mathbf{t},\mathbf{r})$ and the rate prior $f(\mathbf{r}|\mathbf{t})$ across loci. Lastly, the third term in equation (7.4) applies an exponential prior with a small mean (0.001) to penalize large values of σ^2.

The rate-smoothing algorithms (Sanderson 1997, 2002; Yang 2004) are *ad hoc*, and suffer from several problems. First, the 'log likelihood functions' of equations (7.3) and (7.4) are not log likelihood in the usual sense of the word since the rates $\mathbf{r}$ are unobservable random variables in the model and should ideally be integrated out in the likelihood calculation. While such 'penalized likelihood' is used by statisticians, for example, in smoothing empirical frequency densities (Silverman 1986, pp. 110–119), their statistical properties are uncertain. Second, the reliability of the Poisson or normal approximations to the likelihood is unknown. Exact calculation on sequence alignment does not appear feasible computationally because of the large number of rates and the high dimension of the optimization problem. Yang (2004) used rate smoothing to assign branches into rate classes only; divergence times are then estimated by ML from the original sequence alignment, together with the branch rates.

7.3.4 Dating primate divergences

We apply the likelihood global- and local-clock models to analyse the data set of Steiper *et al.* (2004) to estimate the time of divergence between hominoids (apes and humans) and cercopithecoids (Old World monkeys) (Fig. 7.6). The data consist of

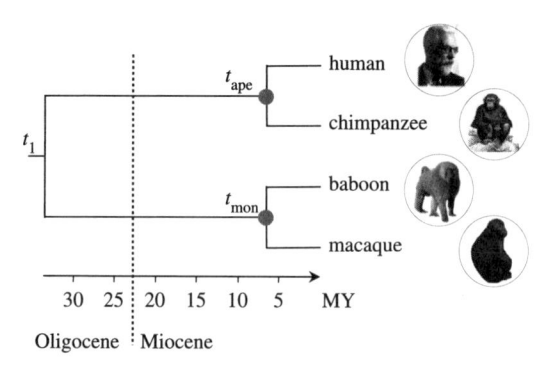

234 • *7 Molecular clock and species divergence times*

Fig. 7.6 The phylogenetic tree of four primate species for the data of Steiper *et al.* (2004). The branch lengths are drawn in proportion to the posterior means of divergence times estimated under the Bayesian local-clock model (clock 3 in Table 7.2).

five genomic contigs (each of 12–64 kbp) from four species: human (*Homo sapiens*), chimpanzee (*Pan troglodytes*), baboon (*Papio anubis*), and rhesus macaque (*Macaca mulatta*). See Steiper *et al.* (2004) for the GenBank accession numbers. The five contigs are concatenated into one data set; in the analysis of Yang and Rannala (2006), accommodating the differences among the contigs, such as differences in substitution rate, in base compositions, and in the extent of rate variation among sites, produced very similar results. The JC69 and HKY + Γ_5 models are used for comparison. Under the HKY85 + Γ_5 model, the MLEs of parameters are $\hat{\kappa} = 4.4$ and $\hat{\alpha} = 0.68$, while the base compositions are estimated using their observed values: 0.327 (T), 0.177 (C), 0.312 (A), and 0.184 (G). HKY85 + Γ_5 has four more parameters than JC69 and the log likelihood difference between the two models is > 7700; JC69 is easily rejected by the data.

Two fossil calibrations are used. The human-chimpanzee divergence was assumed to be between 6 and 8 MYA (Brunet *et al.* 2002) and the divergence of baboon and macaque is assumed to be between 5 and 7 MYA (Delson *et al.* 2000). See Steiper *et al.* (2004) and Raaum *et al.* (2005) for reviews of relevant fossil data. In the likelihood analysis, we fix those two calibration points at $t_{ape} = 7$ MY and $t_{mon} = 6$ MY. The likelihood ratio test does not reject the clock (Yang and Rannala 2006). Under the clock, two parameters are involved in the model: t_1 and r. The estimates are $\hat{t}_1 = 33$–34 MY for the age of the root and $\hat{r} = 6.6 \times 10^{-10}$ substitutions per site per year under both JC69 and HKY85 + Γ_5 models. A local-clock model assumes that the apes and monkeys have different rates. This is the quartet-dating approach of Rambaut and Bromham (1998). Estimates of t_1 are largely the same as under the clock, while the rate is lower in the apes than in the monkeys: $\hat{r}_{ape} = 5.4$ and $\hat{r}_{mon} = 8.0$. Note that the likelihood analysis ignores uncertainties in the fossil calibrations and grossly overestimates the confidence in the point estimates. Steiper *et al.* (2004) used four combinations of the fossil calibrations, with $t_{ape} = 6$ or 8 MY and $t_{mon} = 5$ or 7 MY, and used the range of estimates of t_1 as an assessment of the effect of fossil uncertainties. At any rate, the estimates of divergence time t_1 between hominoids and cercopithecoids are close to those of Steiper *et al.*

*7.3.5 Uncertainties in fossils

7.3.5.1 Complexity of specifying fossil calibration information

It appears sensible to use a statistical distribution to describe the age of a fossil calibration node for use in molecular-clock dating. The distribution then characterizes our best assessment of the fossil records. However, use of fossils to specify calibration information is a complicated process (Hedges and Kumar 2004; Yang and Rannala 2006). First, determining the date of a fossil is prone to errors, such as experimental errors in radiometric dating and assignment of the fossil to the wrong stratum. Second, placing the fossil correctly on the phylogeny can be very complex. A fossil may be clearly ancestral to a clade but by how much the fossil species precedes the common ancestor of the clade may be hard to determine. Misinterpretations of character state changes may also cause a fossil to be assigned to an incorrect lineage. A fossil presumed to be ancestral may in fact represent an extinct side branch and is not directly relevant to the age of the concerned clade. Here I happily leave such difficult tasks to the palaeontologists (Benton *et al.* 2000) and focus only on the question of how to incorporate such information, when it is already specified, in molecular-clock dating.

7.3.5.2 Problems with naïve likelihood implementations

Methods discussed above, including the global- and local-clock likelihood models and rate-smoothing algorithms, can use one or more fossil calibrations, but assume that the ages of calibration nodes are known constants, without error. The problem of ignoring uncertainties in fossil calibrations is well-appreciated (Graur and Martin 2004; Hedges and Kumar 2004). If a single fossil calibration is used and the fossil age represents the minimum age of the clade, the estimated ages for all other nodes will also be interpreted as minimum ages. However, with this interpretation, it will not be appropriate to use multiple calibrations simultaneously, because the minimum node ages may not be compatible with each other or with the sequence data. Suppose that the true ages of two nodes are 100 MY and 200 MY, and fossils place their minimum ages at 50 MY and 130 MY. Used as calibrations, those minimum ages will be interpreted by current dating methods as true node ages, and will imply, under the clock, a ratio of 1 : 2.6 for the distances from the two nodes to the present time, instead of the correct ratio 1 : 2. Time estimates will be systematically biased as a result.

Sanderson (1997) argued for the use of lower and upper bounds on node ages as a way of incorporating fossil date uncertainties, with constrained optimization algorithms used to obtain estimates of times and rates in his penalized likelihood method (equation 7.3). This approach does not appear to be valid, as it causes the model to become unidentifiable. Consider a single fossil calibration with the node age t_C constrained to be in the interval (t_L, t_U). Suppose for a certain t_C inside the interval, the estimates of other node ages and of rates are $\hat{\mathbf{t}}$ and $\hat{\mathbf{r}}$, with the smoothing parameter $\hat{\lambda}$. If $t'_C = 2t_C$ is also in the interval, then $\hat{\mathbf{t}}' = 2\hat{\mathbf{t}}$, $\hat{\mathbf{r}}' = \hat{\mathbf{r}}/2$, and $\hat{\lambda}' = 4\hat{\lambda}$ will fit the likelihood (equation 7.3) or the cross-validation criterion equally well; in other words, by making all divergence times twice as old and all rates half as small, one achieves exactly the same fit and there is no way to distinguish between the two or many other sets of parameter values. If multiple fossil calibrations are used simultaneously and if they are in conflict with each other or with the molecules, the optimization algorithm may get stuck at the borders of the intervals, but the uncertainties in the fossil node ages are not properly accommodated. Thorne and Kishino (2005) also pointed out a problem with the standard use of the nonparametric bootstrap method to construct confidence intervals for estimated divergence times in Sanderson's approach. If one resamples sites in the alignment to generate bootstrap data sets, and analyse each bootstrap sample using the same fossil calibrations, the method will fail to account for uncertainties in the fossils and lead to misleadingly narrow confidence intervals. This problem is obvious when one considers data sets of extremely long sequences. When the amount of data approaches infinity there will be no difference among time estimates from the bootstrap samples, and the constructed confidence intervals will have zero width. The correct intervals, however, should have positive width to reflect fossil uncertainties.

7.3.5.3 Approaches for incorporating fossil uncertainties in likelihood analysis

Unlike the Bayesian approach, in which fossil uncertainties are naturally incorporated into the prior on divergence times and automatically accommodated in the posterior distribution (see next section), likelihood methods for dealing with fossil uncertainties have yet to be developed. The estimation problem is unconventional, as the model is not fully identifiable and the errors in the estimates do not decrease to zero with the increase of sequence data. Here I discuss a few possible approaches, focusing on conceptual issues and dealing with only simple cases for which the answers can be judged intuitively. Although the ideas appear to apply to general cases, computationally feasible algorithms have yet to be developed.

Given that time estimates will always involve uncertainties even with an infinite amount of sequence data, it appears sensible to use a probability density to represent such uncertainties, in the form of $f(\mathbf{t}|x)$. However, without a prior on $\mathbf{t}$ (as in Bayesian inference), the conditional density of times $\mathbf{t}$ given data x is not a meaningful concept (see Subsection 5.1.3). Fisher's (1930b) concept of *fiducial probability* seems to provide a useful framework, in that a distribution on parameters $\mathbf{t}$ is used to represent our state of knowledge about $\mathbf{t}$. To illustrate the concept, suppose a random sample of size n is taken from a normal distribution $N(\theta, 1)$, with unknown θ. The sample mean $\bar{x}$ has the density $f(\bar{x}|\theta) = (1/\sqrt{2\pi/n}) \exp\{n(\bar{x} - \theta)^2/2\}$. This is also the likelihood for θ. Fisher suggested that this be viewed as a density of θ given data x, and called it the fiducial distribution

$$f(\theta|x) = \frac{1}{\sqrt{2\pi/n}} \exp\left\{\frac{n}{2}(\theta - \bar{x})^2\right\}. \tag{7.6}$$

Note that $f(\theta|x)$ is not a frequency density in the usual sense. Rather it is an expression of the credence we place on possible values of the parameter, similar to the degree of belief in Bayesian statistics. The difference from the Bayesian posterior density is that no prior is assumed. Savage (1962) characterized fiducial inference as 'an attempt to make the Bayesian omelette without breaking the Bayesian eggs'. This is a controversial concept, and is sometimes described as Fisher's great failure. As Wallace (1980) commented, 'Of all R.A. Fisher's creations, fiducial inference was perhaps the most ambitious, yet least accepted'. It appears to have been regenerated recently in the name of *generalized confidence intervals* (Weerahandi 1993, 2004).

Below I provide a tentative application of fiducial inference to the estimation of divergence time in the likelihood framework. I will also discuss a few other promising methods for incorporating fossil uncertainties.

After this chapter was written, it was realized that my tentative use of Fisher's fiducial statistics below does not match exactly Fisher's original definition (Fisher 1930b). For the simple case of observation x and one parameter θ, Fisher defined $f(\theta|x) = -\partial F(x|\theta)/\partial \theta$ as the fiducial density of θ, where $F(x|\theta)$ is the cumulative density function. This does not seem tractable in the present case of estimation of divergence times, and what I considered as fiducial distribution below is in fact a rescaled likelihood, equivalent to the posterior density using the improper uniform prior on branch lengths. The discussion is retained to stimulate further research.

7.3.5.4 Two species, infinite data under the clock

A simple case is the estimation of rate r when the divergence time t between two species involves uncertainties. Let $g(t)$ be a probability density for t; this may represent our best interpretation of the fossil records in relation to the age of the root (Fig. 7.7). Let x be the data. We first assume that the sequences are infinitely long, so that the branch length $b(= tr)$ is determined completely. If nothing is known *a priori* about r, the fiducial distribution of r may be given by a simple variable transform, $r = b/t$, as

$$f(r|x) = g(b/r) \times b/r^2 \qquad (7.7)$$

(see Theorem 1 in Appendix A). Here data x is the branch length b, which is a constant. Fig. 7.8(b) shows this density when $b = 0.5$, and t has a uniform or gamma distribution

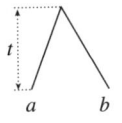

Fig. 7.7 The tree for two species which diverged time t ago, for estimating the substitution rate r under the clock. Time t is not known completely and its uncertainties are described by a probability density $g(t)$, perhaps based on the fossil records. The sequence data provide information about the branch length $b = tr$.

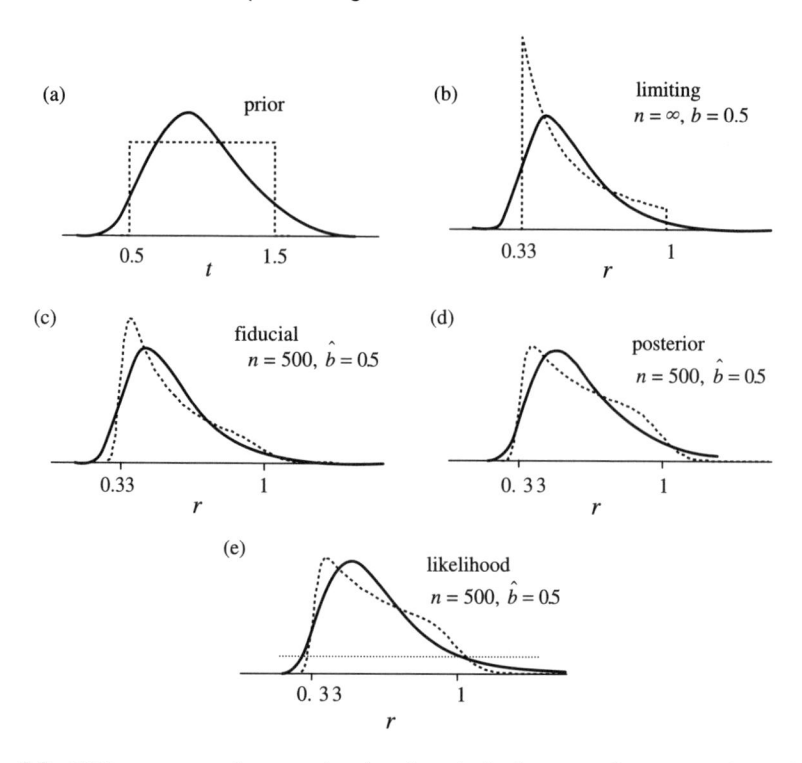

Fig. 7.8 Different approaches to estimating the substitution rate r from comparison of two sequences, when the divergence time t involves uncertainties described by a distribution $g(t)$ (Fig. 7.7). (a) Two distributions for t are shown, mimicking our state of knowledge of t from the fossil record. The first (dotted line) has $t \sim U(0.5, 1.5)$; that is, $g(t) = 1$ if $0.5 < t < 1.5$ and $= 0$ otherwise. The second (solid line) has $t \sim G(10, 10)$, a gamma distribution with mean 1. If one time unit is 100 MY, the mean age will be 100 MY in both distributions. (b) The distributions of r when the sequences are infinitely long so that the branch length, $b = tr = 0.5$ substitutions per site (or one change per site between the two sequences), is known without error. With a divergence time of 100 MY, this distance corresponds to a rate of 0.5×10^{-8} substitutions per site per year. The densities are calculated using equation (7.7). (c) The fiducial distribution of r, given by equation (7.12), when the data consist of $x = 276$ differences out of $n = 500$ sites. (d) The posterior densities (equation 7.13), when an improper uniform prior is specified for r; that is, $f(r) = 1$ for $0 < r < \infty$. (e) The (integrated) likelihood function, given by equation 7.15. The horizontal lines represent likelihood values equal to $1/e^{1.92}$ times the maxima; they delineate the approximate 95% likelihood (confidence) intervals. Integrals are calculated numerically using Mathematica. See Table 7.1 for a summary of the results.

with mean 1; that is, $t \sim U(0.5, 1)$ or $t \sim G(10, 10)$, as shown in Fig. 7.8(a). Since r is a parameter, $f(r|x)$ does not have a frequentist interpretation; $r dr$ cannot be interpreted as the expected frequency that r, sampled from a population, falls in the interval $(r, r + dr)$.

7.3.5.5 Two species, finite data under the clock

Next, we consider a finite data set, consisting of x differences between two sequences of n sites. Assume the JC69 model. The likelihood given branch length $b = tr$ is

$$\phi(x|b) = \left(\tfrac{3}{4} - \tfrac{3}{4}e^{-8b/3}\right)^x \left(\tfrac{1}{4} + \tfrac{3}{4}e^{-8b/3}\right)^{n-x}. \tag{7.8}$$

In large data sets, this should be close to a normal density. One should expect r to involve more uncertainty in finite data than in infinite data. It appears reasonable to define the fiducial distribution of b as

$$f(b|x) = C_x \phi(x|b), \tag{7.9}$$

where $C_x = 1/\int_0^\infty \phi(x|b)\,db$ is a normalizing constant. As uncertainties in b are independent of uncertainties in t, the joint distribution of b and t becomes

$$f(b,t|x) = C_x \phi(x|b)g(t). \tag{7.10}$$

To derive the distribution of r, change variables from (b,t) to (t,r), so that

$$f(t,r|x) = C_x \phi(x|tr)g(t)t. \tag{7.11}$$

Then the fiducial distribution of r may be obtained by integrating out t:

$$f(r|x) = \int_0^\infty f(t,r|x)\,dt$$

$$= \frac{\int_0^\infty \left(\tfrac{3}{4} - \tfrac{3}{4}e^{-8tr/3}\right)^x \left(\tfrac{1}{4} + \tfrac{3}{4}e^{-8tr/3}\right)^{n-x} g(t)t\,dt}{\int_0^\infty \left(\tfrac{3}{4} - \tfrac{3}{4}e^{-8b/3}\right)^x \left(\tfrac{1}{4} + \tfrac{3}{4}e^{-8b/3}\right)^{n-x}\,db}. \tag{7.12}$$

One would hope this to converge to the limiting density of equation (7.7) when $n \to \infty$, but this is yet to be confirmed.

Consider a fictitious data set of $x = 276$ differences out of $n = 500$ sites. The MLE of b is $\hat{b} = 0.5$, and the normalizing constant is $C_x = 5.3482 \times 10^{-141}$. The density of equation (7.12) is shown in Fig. 7.8(c). As expected, this has longer tails than the density of Fig. 7.8(b) for infinite data. Note that for the uniform distribution of t, r has positive support for the whole positive half-line because of uncertainties in b, while r must be in the interval $(1/3, 1)$ if $n = \infty$.

A second approach is the Bayesian approach. This is discussed in the next section, for application to general cases on large trees. Here we use it for comparison. One may assign an improper prior for r, with $f(r) = 1$ for $0 < r < \infty$. The posterior is then

$$f(r|x) = \frac{\int_0^\infty \left(\tfrac{3}{4} - \tfrac{3}{4}e^{-8tr/3}\right)^x \left(\tfrac{1}{4} + \tfrac{3}{4}e^{-8tr/3}\right)^{n-x} g(t)\,dt}{\int_0^\infty \int_0^\infty \left(\tfrac{3}{4} - \tfrac{3}{4}e^{-8tr/3}\right)^x \left(\tfrac{1}{4} + \tfrac{3}{4}e^{-8tr/3}\right)^{n-x} g(t)\,dt\,dr}. \tag{7.13}$$

240 • 7 Molecular clock and species divergence times

This is shown in Fig. 7.8(d) for the examples of uniform and gamma priors on t. The limiting posterior density when $n \to \infty$ can be obtained by changing variables from (r, t) to (r, b), and then conditioning on b, as

$$f(r|x) = \frac{g(b/r)/r}{\int_0^\infty g(b/r)/r \, dr} \tag{7.14}$$

In the example of Fig. 7.8, equation (7.14) (plot not shown) is similar but not identical to equation (7.7).

A third approach may be to estimate r from the following likelihood, integrating over the uncertainties in t:

$$L(r|x) = f(x|r) = \int_0^\infty \phi(x|tr)g(t) \, dt$$

$$= \int_0^\infty \left(\tfrac{3}{4} - \tfrac{3}{4} e^{-8tr/3}\right)^x \left(\tfrac{1}{4} + \tfrac{3}{4} e^{-8tr/3}\right)^{n-x} g(t) \, dt. \tag{7.15}$$

This formulation assumes that the data consist of one sample point from the following composite sampling scheme. For any parameter value r,

(1) generate time t by sampling from $g(t)$;
(2) use branch length $b = tr$ to generate two sequences, of n sites long, under the JC69 model.

According to this scheme, a sequence alignment of n sites constitutes one data point. If we were to generate another data point, we would sample a different t and use $b = tr$ to generate another sequence alignment. This formulation clearly does not match the description of our problem. In fact, $g(t)$ represents our state of incomplete knowledge about t, and is not a frequency distribution of times from which one can take samples; we would generate data using a fixed branch length b, and demand inference on r given incomplete information about t. The likelihood of equation (7.15) is thus logically hard to justify, but it does account for uncertainties in t and may be practically useful. This is plotted in Fig. 7.8(e) when $g(t)$ is a uniform or gamma density. A 95% likelihood (confidence) interval may be constructed by lowering the log likelihood from the maximum by 1.92 (Table 7.1). Similarly the curvature of the log likelihood surface may be used to construct a confidence interval, relying on a normal approximation to the MLE. However, the reliability of such asymptotic approximations is uncertain. Note that the integrated likelihood (equation 7.15) and the Bayesian posterior density (equation 7.13) are proportional, although their interpretations are different.

7.3 Likelihood estimation of divergence times • 241

Table 7.1 Estimates of rate r by different methods

Method	Uniform prior on t		Gamma prior on t	
	Mean	(95% interval) and width	Mean	(95% interval) and width
Limiting ($n = \infty$) (Fig. 7.8b)	0.5493	(0.3390, 0.9524) 0.6134	0.5556	(0.2927, 1.0427) 0.75
Fiducial (Fig. 7.8c)	0.5563	(0.3231, 0.9854) 0.6623	0.5626	(0.2880, 1.0722) 0.7842
Posterior (Fig. 7.8d)	0.6145	(0.3317, 1.0233) 0.6916	0.6329	(0.3125, 1.2479) 0.9354
Likelihood (Fig. 7.8e)	0.3913	(0.3030, 1.0363) 0.7333	0.5025	(0.3027, 1.0379) 0.7352

The mean and the 2.5th and 97.5th percentiles of the limiting, fiducial, and posterior distributions are shown. For likelihood, the MLEs and the 95% likelihood intervals are shown.

A few remarks may be made in relation to the likelihood of equation (7.15). First, note that it is unworkable to consider $\phi(x|tr)$ as the likelihood and to maximize it to estimate both t and r under the constraint $t_L < t < t_U$, as in Sanderson's (1997) approach of constrained optimization. The integrated likelihood of equation (7.15) assumes that $t \sim U(t_L, t_U)$. This distributional assumption is stronger than merely saying that t lies in the interval (t_L, t_U) and appears necessary to incorporate fossil uncertainties for statistical estimation of divergence times. Second, the sampling scheme implied by the integrated likelihood of equation (7.15) suggests a heuristic nonparametric bootstrap procedure for constructing a confidence interval for estimated divergence times. One may resample sites to generate a bootstrap data set, and sample a fossil date t from $g(t)$ and use it as a fixed calibration point in the likelihood analysis of each bootstrap sample; that is, r is estimated from $\phi(x|tr)$ with t fixed. The MLEs of r from the bootstrap data sets will then be used to construct a confidence interval (for example, by using the 2.5th and 97.5th percentiles). This is similar to the multifactor bootstrap approach suggested by Kumar et al. (2005b). Note that it appears incorrect to resample sites only without sampling t and then to use equation (7.15) to get the MLE of r; such a procedure ignores uncertainties in the fossil date t. It also appears feasible to take a parametric bootstrap approach to construct a confidence interval. One may use the MLE of t from equation (7.15) and a random rate r from $g(r)$ to simulate every bootstrap data set, analyse each using the likelihood of (7.15), and collect the MLEs into a confidence interval. All those procedures suffer from conceptual difficulties associated with the likelihood of equation (7.15) and it remains to be seen whether they are practically useful in assessing uncertainties in likelihood estimates of divergence times.

The point and interval estimates of r from the three approaches discussed above are summarized in Table 7.1. The true rate is 0.5. Overall the approaches produced similar and reasonable results, but there are some interesting differences between them and between the uniform and gamma priors on t. Because the curves are not symmetrical, especially under the uniform prior for t, the means (of the fiducial and posterior distributions) and modes (of the likelihood) of the rate are quite different. As a result, the MLE under the uniform prior is 0.39 while all other point estimates are $\sim$ 0.5 or 0.6. All intervals include the true value (0.5). We first consider the uniform prior of t. In this case, the 95% fiducial, credibility, and confidence intervals in finite data ($n = 500$) are all wider than in the limiting case of infinite data ($n = \infty$), as one would expect. The Bayesian (credibility) interval is shifted to the right relative to the fiducial interval; this appears to be due to the improper prior for r used in Bayesian inference, which assigns density for unreasonably large values of r. The likelihood interval is wider than the fiducial and Bayesian intervals.

Under the gamma prior for t, the fiducial interval is wider than in the limiting case, as one would expect. The Bayesian interval is wider and shifted to the right relative to the fiducial interval, as in the case of the uniform prior on t. The likelihood interval is even narrower than the interval for infinite data, which is somewhat surprising. Note that the limiting density of Fig. 7.8(b) does not assume a prior on r, as in the Bayesian or integrated likelihood analysis; the limiting distribution of the Bayesian analysis is given in equation (7.14), not plotted in Fig. 7.8. The Bayesian posterior density (Fig. 7.8d) and the integrated likelihood (Fig. 7.8e) are proportional; the intervals produced by these two approaches should be more similar if the Bayesian HPD intervals are used instead of the equal-tail intervals.

7.3.5.6 Three species, infinite data under the clock

With three species, there are two divergence times: t_1 and t_2 (Fig. 7.1). Suppose the clock holds, so that there is one rate r. The model thus involves three parameters. Let $g(t_1)$ be the density for t_1, specified using fossil data. As a concrete example, suppose the true node ages are $t_1 = 1$ and $t_2 = 0.5$, and the true rate is $r = 0.1$. Two priors on t_1 are shown in Fig. 7.9(a): $t_1 \sim U(0.5, 1.5)$, and $t_1 \sim G(10, 10)$, both with mean 1. Our objective is to estimate t_2 and r, with t_2 to be of special interest. Under the JC69 model, the data are summarized as the counts n_0, n_1, n_2, n_3, and n_4 of site patterns xxx, xxy, xyx, yxx, and xyz, where x, y, z are any distinct nucleotides. Let the data be $x = (n_0, n_1, n_2, n_3, n_4)$.

First, consider an infinite amount of sequence data ($n \to \infty$), in which case the two branch lengths $b_1 = t_1 r$ and $b_2 = t_2 r$ are known without error. The limiting distributions of $r = b_1/t_1$ and $t_2 = t_1 b_2/b_1$ are easily derived through a change of variables, as $f(r|x) = g(b_1/r) \times b_1/r^2$ and $f(t_2|x) = g(t_2 b_1/b_2) \times b_1/b_2$. The densities for t_2 are shown in Fig. 7.9(b) when t_1 has uniform or gamma priors. The 95% equal-tail interval is $(0.2625, 0.7375)$ for the uniform prior and $(0.2398, 0.8542)$ for the gamma prior.

7.3 Likelihood estimation of divergence times • 243

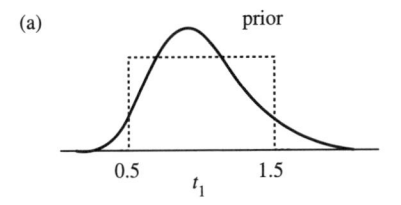

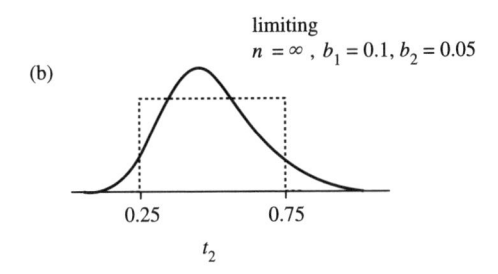

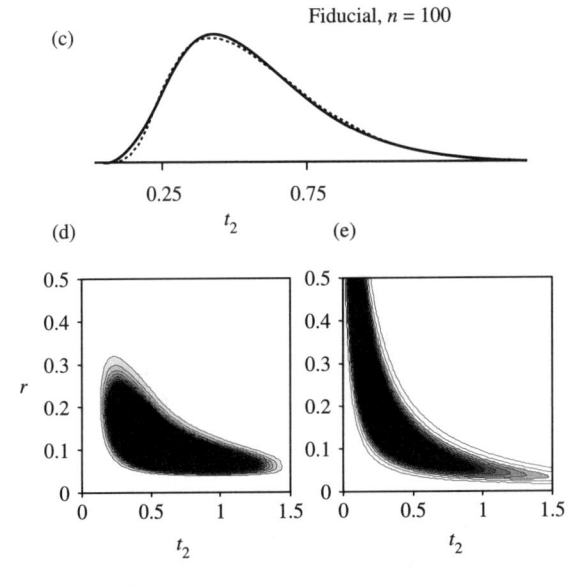

Fig. 7.9 Estimation of divergence time t_2 using fossil calibration t_1 in the tree of Fig. 7.1. The molecular clock is assumed, with rate r applied to all branches. The true parameter values are $t_1 = 1, t_2 = 0.5, r = 0.1$. (a) Two distributions for t_1 are shown. The first (dotted line) has $t_1 \sim U(0.5, 1.5)$, while the second (solid line) has $t_1 \sim G(10, 10)$, both with mean 1. (b) The limiting distributions of t_2 when the sequences are infinitely long so that the branch lengths, $b_1 = t_1 r = 0.1$ and $b_2 = t_2 r = 0.05$, are fixed. (c) The fiducial distributions of t_2 for a data set of 100 sites, calculated using equation (7.18). (d) The log likelihood contour plot for t_2 and r when t_1 has the uniform distribution of (a), calculated using equation (7.19). (e) The log likelihood contour plot for t_2 and r when t_1 has the gamma distribution of (a).

7.3.5.7 Three species, finite data under the clock

The likelihood given branch lengths $b_1 = t_1 r$ and $b_2 = t_2 r$ is

$$\phi(x|b_1, b_2) = p_0^{n_0} p_1^{n_1} p_2^{n_2+n_3} p_4^{n_4}, \tag{7.16}$$

where p_i ($i = 0, 1, \ldots, 4$) are the probabilities of the four site patterns, given as

$$p_0(t_0, t_1) = \tfrac{1}{16}(1 + 3e^{-8b_2/3} + 6e^{-8b_1/3} + 6e^{-4(2b_1+b_2)/3}),$$

$$p_1(t_0, t_1) = \tfrac{1}{16}(3 + 9e^{-8b_2/3} - 6e^{-8b_1/3} - 6e^{-4(2b_1+b_2)/3}),$$

$$p_2(t_0, t_1) = \tfrac{1}{16}(3 - 3e^{-8b_2/3} + 6e^{-8b_1/3} - 6e^{-4(2b_1+b_2)/3}), \tag{7.17}$$

$$p_3(t_0, t_1) = p_2,$$

$$p_4(t_0, t_1) = \tfrac{1}{16}(6 - 6e^{-8b_2/3} - 12e^{-8b_1/3} + 12e^{-4(2b_1+b_2)/3})$$

(Saitou 1988; Yang 1994c).

One may construct the fiducial distribution of branch lengths as $f(b_1, b_2|x) = C_x \phi(x|b_1, b_2)$, where $C_x = 1/\int_0^\infty \int_0^{b_1} \phi(x|b_1, b_2) \, db_2 \, db_1$ is a normalizing constant. Thus the density of b_1, b_2, t_1 is $f(b_1, b_2, t_1|x) = C_x \phi(x|b_1, b_2) g(t_1)$. Changing variables from (b_1, b_2, t_1) to (t_1, t_2, r), and integrating out t_1 and r, we obtain the fiducial distribution of t_2 as

$$f(t_2|x) = C_x \int_{t_2}^\infty \int_0^\infty \phi(x|t_1 r, t_2 r) g(t_1) \times r t_1 \, dr \, dt_1. \tag{7.18}$$

As an example, suppose the observed counts of site patterns at 100 sites are $n_0 = 78$, $n_1 = 12$, $n_2 + n_3 = 9$, $n_4 = 1$. The MLEs of branch lengths under the JC69 model are $\hat{b}_1 = 0.1$ and $\hat{b}_2 = 0.05$. The mean of t_2 calculated from the fiducial density of equation (7.18) is 0.5534. The 95% equal-tail interval is (0.2070, 1.0848) for the uniform prior and (0.1989, 1.1337) for the gamma prior. The interval under the gamma prior is wider, as expected.

Application of the Bayesian approach requires specifying priors on t_2 and r as well as on the prior on t_1. Given the prior, Bayesian computation is straightforward. The next section discusses the Bayesian approach, with MCMC algorithms used to achieve efficient computation.

Similarly to the two-species case, one may also define a likelihood function by integrating over the uncertainties in t_1, as

$$L(t_2, r) = \int_{t_2}^\infty \phi(x|t_1 r, t_2 r) g(t_1) \, dt_1. \tag{7.19}$$

This shares the conceptual difficulties of equation (7.15) for the two-species case, but incorporates the uncertainties in the fossil calibration. If prior information is available on rate r, one may similarly integrates over r to define a likelihood for t_2 only. The likelihood contours calculated using equation (7.19) are shown for the uniform and gamma priors in Figs. 7.9(d) and (e), respectively. In theory, one can use profile or relative likelihood to identify the MLE of t_2 and construct approximate likelihood intervals. This is not pursued here due to difficulties in numerical calculation of the integrals. From the contour plots, the MLEs appear close to the true values: $t_2 = 0.5$, $r = 0.1$, although the estimates clearly involve large sampling errors.

7.3.5.8 General case of arbitrary tree with variable rates

The ideas discussed above appear generally applicable. We may partition the nodes on the tree into those with and those without fossil calibration information. Let $\mathbf{t}_C$ be the ages of calibration nodes, with distribution $g(\mathbf{t}_C)$. Let $\mathbf{t}_{-C}$ be all other node ages, to be estimated. Suppose $h(\mathbf{r})$ represent the prior information on the rates, possibly specified using a rate-drift model. One can then construct the fiducial distribution $f(\mathbf{b}, \mathbf{t}_C, \mathbf{r}|x) = C_x \phi(x|\mathbf{b}) g(\mathbf{t}_C) h(\mathbf{r})$, where $C_x = \int \phi(x|\mathbf{b}) \, d\mathbf{b}$ is a normalizing constant. One can then derive the fiducial distribution of $\mathbf{t}_{-C}$ as functions of $\mathbf{b}$, $\mathbf{r}$, and $\mathbf{t}_C$, through variable transformations. If a molecular clock is assumed so that there is only one rate r, the rate can be estimated as a parameter together with $\mathbf{t}_{-C}$. The integrated likelihood approach may be applied as well, by integrating over prior distributions $g(\mathbf{t}_C)$ and $h(\mathbf{r})$ in the likelihood function. Both approaches involve high-dimensional integrals, which are computationally challenging.

7.4 Bayesian estimation of divergence times

7.4.1 General framework

A Bayesian MCMC algorithm was developed by Thorne and colleagues (Thorne et al. 1998; Kishino et al. 2001) to estimate species divergence times under a geometric Brownian motion model of rate drift over time. A slightly different model using a compound-Poisson process to model rate drift was described by Huelsenbeck et al. (2000a). Yang and Rannala (2006) and Rannala and Yang (2006) developed a similar algorithm to allow for arbitrary distributions to be used to describe uncertainties in fossil dates. Here we describe the general structure of the models and then comment on the details of the different implementations.

Let x be the sequence data, $\mathbf{t}$ the $s-1$ divergence times and $\mathbf{r}$ the rates. The rates can be either for the $(2s-2)$ branches, as in Thorne et al. (1998) and Yang and Rannala (2006), or for the $(2s-1)$ nodes, as in Kishino et al. (2001). In the former case, the rate of the midpoint of a branch is used as the approximate average rate for the whole branch. In the latter case, the average of the rates at the two ends of the branch is used to approximate the average rate for the branch. Let $f(\mathbf{t})$ and $f(\mathbf{r}|\mathbf{t})$ be the priors. Let θ be parameters in the substitution model and in the priors for $\mathbf{t}$ and $\mathbf{r}$, with prior $f(\theta)$.

The joint conditional (posterior) distribution is then given as

$$f(\mathbf{t},\mathbf{r},\theta|X) = \frac{f(X|\mathbf{t},\mathbf{r},\theta)f(\mathbf{r}|\mathbf{t},\theta)f(\mathbf{t}|\theta)f(\theta)}{f(X)}. \tag{7.20}$$

The marginal probability of the data, $f(X)$, is a high-dimensional integral over $\mathbf{t}$, $\mathbf{r}$, and θ. The MCMC algorithm generates samples from the joint posterior distribution. The marginal posterior of $\mathbf{t}$,

$$f(\mathbf{t}|X) = \int \int f(\mathbf{t},r,\theta|X)\,\mathrm{d}\mathbf{r}\,\mathrm{d}\theta, \tag{7.21}$$

can be constructed from the samples taken during the MCMC.

The following is a sketch of the MCMC algorithm implemented in the MCMC-TREE program in the PAML package (Yang 1997a). See Chapter 5 and especially Sections 5.6 and 5.7 for discussions of MCMC algorithms. In the next few subsections we discuss the individual terms involved in equation (7.20).

- Start with a random set of divergence times $\mathbf{t}$, substitution rates $\mathbf{r}$, and parameters θ.
- In each iteration do the following:
  o Propose changes to the divergence times $\mathbf{t}$.
  o Propose changes to the substitution rates for different loci.
  o Propose changes to substitution parameters θ.
  o Propose a change to all times and rates, multiplying all times by a random variable c close to one and dividing all rates by c.
- For every k iterations, sample the chain: save $\mathbf{t}$, $\mathbf{r}$, and θ to disk.
- At the end of the run, summarize the results.

7.4.2 Calculation of the likelihood

7.4.2.1 Likelihood given times and rates

The likelihood $f(X|\mathbf{t},\mathbf{r},\theta)$ can be calculated under any substitution model on the sequence alignment, as discussed in Chapter 4. This approach is straightforward but expensive, and is taken by Yang and Rannala (2006). To achieve computational efficiency, Thorne et al. (1998) and Kishino et al. (2001) used a normal approximation to the MLEs of branch lengths in the rooted tree of the ingroup species, estimated without the clock, with the variance–covariance matrix calculated using the local curvature of the log likelihood surface (see Subsection 1.4.1). They used outgroup species to break the branch around the root into two parts, such as branch lengths b_7 and b_8 in the tree of Fig. 7.10(a). If no outgroups are available, or if the outgroups are too far away from the ingroup species to be useful, an alternative may be to estimate the $2s - 3$ branch lengths in the unrooted tree for the s ingroup species only, without outgroups (Fig. 7.10b). In the likelihood calculation, the predicted branch length will be the sum of the two parts (that is, b_7 in Fig. 7.10b instead of b_7 and b_8 in Fig. 7.10a). The use of the normal approximation means that the sequence alignment

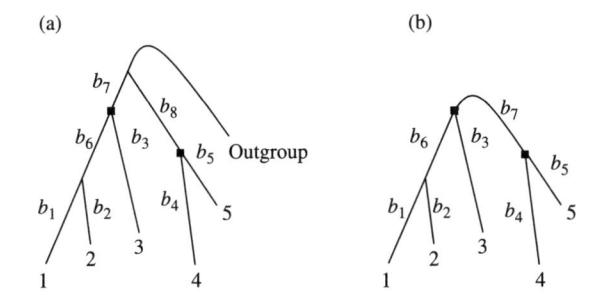

Fig. 7.10 Calculation of the likelihood function for data of s sequences (species) by the normal approximation to the branch lengths estimated without assuming the molecular clock. (a) Thorne et al. (1998) used outgroup species to locate the root on the ingroup tree, estimating $2s - 2$ branch lengths for the rooted tree of s ingroup species. Here the ingroup tree is that of Fig. 7.3. The likelihood is then approximated by the normal density of the MLEs of those $2s - 2$ branch lengths ($b_1, b_2, \ldots, b_8$ in the example). (b) An alternative may be to estimate $2s - 3$ branch lengths in the unrooted tree of s ingroup species, without the need for outgroups. The likelihood will then be approximated by the normal density of the MLEs of the $2s - 3$ branch lengths ($b_1, b_2, \ldots, b_7$ in the example).

is not needed in the MCMC algorithm, after the MLEs of branch lengths and their variance–covariance matrix have been calculated.

7.4.2.2 Multiple genes

In a combined analysis of sequences from multiple genes, differences in the evolutionary dynamics of different genes, such as different rates, different transition/transversion rate ratios, or different levels of rate variation among sites, should be accounted for in the likelihood calculation. In the implementation of Thorne and Kishino (2002), branch lengths on the unrooted tree and substitution parameters are estimated separately for different loci, so that the different parameter estimates naturally accommodate differences among genes. In the implementation of Yang and Rannala (2006), the likelihood is calculated using the sequence alignment under models that allow different genes or site partitions to have different parameters (Yang 1996b).

7.4.3 Prior on rates

Thorne et al. (1998) and Kishino et al. (2001) used a recursive procedure to specify the prior for rates, proceeding from the root of the tree towards the tips. The rate at the root is assumed to have a gamma prior. Then the rate at each node is specified by conditioning on the rate at the ancestral node. Specifically, given the log rate, $\log(r_A)$, of the ancestral node, the log rate of the current node, $\log(r)$, follows a normal distribution with mean $\log(r_A) - \sigma^2 t/2$ and variance $\sigma^2 t$, where t is the time duration separating the two nodes. Note that if the log rate $y \sim N(\mu, \sigma^2)$, then

$E(e^y) = e^{\mu+\sigma^2/2}$, so the correction term $-\sigma^2 t/2$ is used to remove the positive drift in rate, to have $E(r) = r_A$ (Kishino et al. 2001). In other words, given the rate r_A at the ancestral node, the rate r at the current node has a log-normal distribution with density

$$f(r|r_A) = \frac{1}{r\sqrt{2\pi t \sigma^2}} \exp\left\{-\frac{1}{2t\sigma^2}[\log(r/r_A) + t\sigma^2/2]^2\right\}, \quad 0 < r < \infty. \quad (7.22)$$

Parameter σ^2 controls how rapidly the rate drifts or how clock-like the tree is *a priori*. A large σ^2 means that the rates vary over time or among branches and the clock is seriously violated, while a small σ^2 means that the clock roughly holds. The prior for rates on the tree $f(\mathbf{r})$ is calculated by multiplying the prior densities across the nodes.

This model is called *geometric Brownian motion*, as the logarithm of the rate drifts over time according to a *Brownian motion* process (Fig. 7.5). Geometric Brownian motion is the simplest model for random drift of a positive variable, and is widely used in financial modelling and resource management, for example, to describe the fluctuations of the market value of a stock.

In the implementation of Rannala and Yang (2006), the same geometric Brownian motion model is used to specify the distribution of rates at the midpoints of the branches on the tree. In addition, those authors implemented an independent-rates model, in which the rate for any branch is a random draw from the same log normal distribution with density

$$f(r|\mu,\sigma^2) = \frac{1}{r\sqrt{2\pi\sigma^2}} \exp\left\{-\frac{1}{2\sigma^2}[\log(r/\mu) + \tfrac{1}{2}\sigma^2]^2\right\}, \quad 0 < r < \infty. \quad (7.23)$$

Here μ is the average rate for the locus, and σ^2 measures departure from the clock model. The prior density for all rates $f(\mathbf{r})$ is simply the product of the densities like equation (7.23) across all branches on the tree.

7.4.4 Uncertainties in fossils and prior on divergence times

Kishino et al. (2001) devised a recursive procedure for specifying the prior for divergence times, proceeding from the root towards the tips. A gamma density is used for the age of the root (t_1 in the example tree of Fig. 7.3) and a Dirichlet density is used to break the path from an ancestral node to the tip into time segments, corresponding to branches on that path. For example, along the path from the root to tip 1 (Fig. 7.3), the proportions of the three time segments, $(t_1 - t_2)/t_1$, $(t_2 - t_4)/t_1$, and t_4/t_1, follow a Dirichlet distribution with equal means. Next, the two proportions $(t_1 - t_3)/t_1$ and t_3/t_1 follow a Dirichlet distribution with equal means. The Dirichlet distribution is a generalization of the uniform or beta distribution to multiple dimensions. If $x \sim U(0, 1)$, then x and $1-x$ are two proportions with a Dirichlet distribution with equal means.

Uncertainties in the fossil calibrations are incorporated in the prior for times $f(\mathbf{t}|\theta)$. Thorne *et al.* (1998) allowed a lower bound (minimum age) and upper bound (maximum age) on node ages, which are implemented in the MCMC algorithm by rejecting proposals of divergence times that contradict such bounds. This strategy in effect specifies a uniform prior on the fossil calibration age: $t \sim U(t_L, t_U)$. Note that this distributional assumption is conceptually different from Sanderson's (1997) use of lower and upper bounds in constrained optimization in the likelihood framework: $t_L < t < t_U$.

One could change parameters in the Dirichlet distribution to generate trees of different shapes, such as trees with long ancient branches and short young branches. A drawback of the procedure is that the prior density $f(\mathbf{t})$ is intractable analytically, making it difficult to incorporate statistical distributions for fossil calibrations more complex than the uniform distribution. The bounds specified by the uniform prior may be called 'hard' bounds, which assign zero probability for any ages outside the interval. Such priors represent strong conviction on the part of the researcher and may not always be appropriate. In particular, fossils often provide good lower bounds and rarely good upper bounds. As a result, the researcher may be forced to use an unrealistically high upper bound to avoid precluding an unlikely (but not impossible) ancient age for the node. Such a 'conservative' approach may be problematic as the bounds imposed by the prior may influence posterior time estimation.

Yang and Rannala (2006) developed a strategy to incorporate arbitrary distributions for describing fossil calibration uncertainties. These are called 'soft' bounds and assign nonzero probabilities over the whole positive half-line ($t > 0$). A few examples are shown in Fig. 7.11. The basic model used is the birth–death process (Kendall 1948), generalized to account for species sampling (Rannala and Yang 1996; Yang and Rannala 1997). Here we describe the main features of this model, as it is also useful for generating random trees and branch lengths. Let λ be the per-lineage birth (speciation) rate, μ the per-lineage death (extinction) rate, and ρ the sampling fraction. Then conditional on the root age t_1, the $(s-2)$ node ages of the tree are order statistics from the following kernel density

$$g(t) = \frac{\lambda p_1(t)}{v_{t_1}}, \tag{7.24}$$

where

$$p_1(t) = \frac{1}{\rho} P(0,t)^2 e^{(\mu-\lambda)t} \tag{7.25}$$

is the probability that a lineage arising at time t in the past leaves exactly one descendant in the sample, and

$$v_{t_1} = 1 - \frac{1}{\rho} P(0,t_1) e^{(\mu-\lambda)t_1}, \tag{7.26}$$

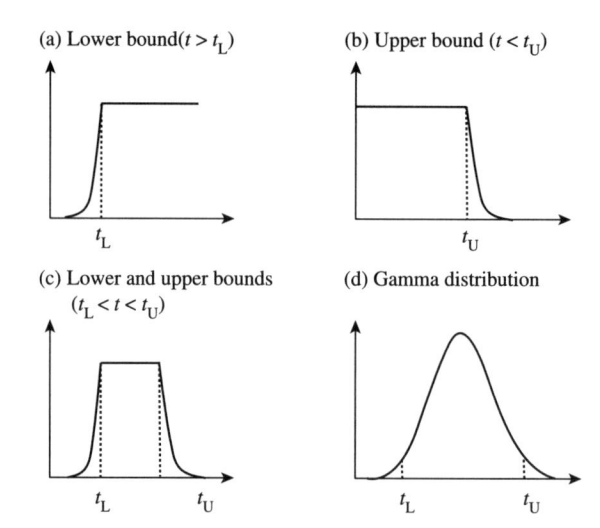

Fig. 7.11 Probability densities for node ages implemented by Yang and Rannala (2006) to describe uncertainties in fossils. The bounds are 'soft', and the node age has nonzero probability density over the whole positive line: that is, $g(t) > 0$ for $0 < t < \infty$. (a) A lower bound, specified by $t > t_L$, implemented with $P(t < t_L) = 2.5\%$. This is an improper prior density and should not be used alone. (b) An upper bound, specified by $t < t_U$, with $P(t > t_U) = 2.5\%$. (c) Both lower and upper bounds, specified as $t_L < t < t_U$, with $P(t < t_L) = P(t > t_U) = 2.5\%$. (d) A gamma distribution, with (t_L, t_U) to be the 95% equal-tail prior interval.

with $P(0, t)$ to be the probability that a lineage arising at time t in the past leaves one or more descendants in a present-day sample

$$P(0,t) = \frac{\rho(\lambda - \mu)}{\rho\lambda + [\lambda(1-\rho) - \mu]e^{(\mu-\lambda)t}}. \tag{7.27}$$

When $\lambda = \mu$, equation (7.24) becomes

$$g(t) = \frac{1 + \rho\lambda t_1}{t_1(1 + \rho\lambda t)^2}. \tag{7.28}$$

In other words, to generate a set of $s - 2$ random node ages under the model, one generates the root age t_1 and then $s - 2$ independent random variables from the kernel density (equation 7.24) conditional on t_1, and then orders them. Sampling from the kernel density can be achieved by using the cumulative distribution function given in Yang and Rannala (1997).

The prior density of times **t** can thus be calculated analytically under the model based on the theory of order statistics. It is too optimistic to expect this model to provide an accurate description of the biological process of speciation, extinction, and sampling of species by biologists. However, by changing parameters λ, μ, and ρ

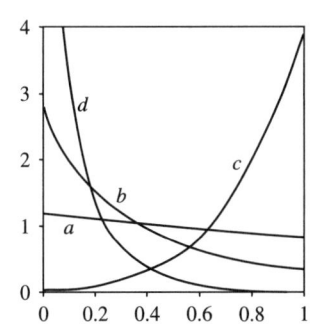

Fig. 7.12 Kernel densities of the birth–death process with species sampling, with per-lineage birth rate λ, death rate μ, and sampling fraction ρ (equation 7.24). The parameters used are (a): $\lambda = 2$, $\mu = 2$, $\rho = 0.1$; (b): $\lambda = 2$, $\mu = 2$, $\rho = 0.9$; (c): $\lambda = 10$, $\mu = 5$, $\rho = 0.001$; (d): $\lambda = 10$, $\mu = 5$, $\rho = 0.999$.

in the model, one can easily change the shape of the tree, and examine the sensitivity of posterior time estimates to prior assumptions about the tree shape. A few example densities are shown in Fig. 7.12, with the age of the root fixed at $t_1 = 1$. In (a), the density is nearly uniform between 0 and 1. This appears close to the prior specified by Thorne et al. (1998) using Dirichlet densities with equal means. In (b) and (d), the densities are highly skewed towards 0, so the tree will tend to have long internal and short external branches. This is the shape of trees generated by the standard coalescent prior or the Yule process without species sampling. In (c), the density is skewed towards 1, so that the tree will be star-like, with short internal branches and long terminal branches.

Fossil calibration information is incorporated in the prior for times, $f(\mathbf{t})$. Let $\mathbf{t}_C$ be ages of nodes for which fossil calibration information is available, and $\mathbf{t}_{-C}$ be the ages of the other nodes, with $\mathbf{t} = (\mathbf{t}_C, \mathbf{t}_{-C})$. The prior $f(\mathbf{t})$ is constructed as

$$f(\mathbf{t}) = f(\mathbf{t}_C, \mathbf{t}_{-C}) = f_{BD}(\mathbf{t}_{-C}|\mathbf{t}_C) f(\mathbf{t}_C), \tag{7.29}$$

where $f(\mathbf{t}_C)$ is the density of the ages of the fossil calibration nodes, specified by summarizing fossil information, and $f_{BD}(\mathbf{t}_{-C}|\mathbf{t}_C)$ is the conditional distribution of $\mathbf{t}_{-C}$ given $\mathbf{t}_C$, specified according to the birth–death process with species sampling.

It should be noted that as long as the uncertainties in fossil calibrations are incorporated in the prior on time $f(\mathbf{t})$, they will be automatically incorporated in the posterior density. There is no need to use bootstrap resampling to incorporate fossil uncertainties in Bayesian time estimation, as did Kumar et al. (2005b). Effort should instead be spent on developing objective priors that best summarize fossil records to represent our state of knowledge concerning the ages of calibration nodes. Studies of fossilization, fossil preservation, and of errors in fossil dating techniques may all contribute to this goal (Tavaré et al. 2002).

7.4.5 Application to primate and mammalian divergences

We apply the Bayesian methods to two data sets. The first is that of Steiper *et al.* (2004), for comparison with the likelihood analysis discussed in Subsection 7.3.4. The second is that of Springer *et al.* (2003), who used the method of Thorne *et al.* (1998) and Kishino *et al.* (2001) to date divergences among mammals. Here we use the method of Yang and Rannala (2006) and Rannala and Yang (2006) for comparison.

7.4.5.1 Analysis of the data of Steiper et al.

The MCMCTREE program in the PAML package is used to analyse the data of Steiper *et al.* under the JC69 and HKY85 + Γ_5 models. Three rate models are assumed: the global clock assuming one rate throughout the tree, the independent-rates model and the autocorrelated-rates model. The overall rate is assigned the gamma prior $G(2, 2)$ with mean 1 and variance $1/2$. Here one time unit is 100 MY so a rate of 1 means 10^{-8} substitutions per site per year. Parameter σ^2 in equations (7.22) and (7.23) is assigned the gamma prior $G(0.01, 1)$. Under HKY85 + Γ_5, the transition/transversion rate ratio κ is assigned the gamma prior $G(6, 2)$, while the shape parameter α for rate variation among sites is assigned the prior $G(1, 1)$. These two parameters are reliably estimated in the data, so that the prior has little significance.

The prior for divergence times is specified using the birth–death process with species sampling, with the birth and death rates $\lambda = \mu = 2$, and sampling fraction $\rho = 0.1$. The two fossil calibrations, 6–8 MY for t_{ape} and 5–7 MY for t_{mon}, are specified using gamma distributions that match those lower and upper bounds with the 2.5th and 97.5th percentiles of the density (Fig. 7.11). In other words, $t_{\text{ape}} \sim G(186.2, 2672.6)$, with mean 0.07 and the 95% prior interval $(0.06, 0.08)$, and $t_{\text{mon}} \sim G(136.2, 2286.9)$, with prior mean 0.06 and the 95% prior interval $(0.05, 0.07)$. In addition, the age of the root is constrained to be less than 60 MY: $t_1 < 0.6$. Using the notation of the MCMCTREE program, the rooted tree and the fossil calibrations are specified as "((human, chimpanzee) '> 0.06 = 0.0693 < 0.08', (baboon, macaque) '>0.05=0.0591<0.07') '<0.6' ", with calibration information specified as node labels.

The posterior means and 95% credibility intervals in the different analyses are listed in Table 7.2. The following observations can be made. First, posterior estimates of the root age t_1 under the three rate models are similar to each other and also to the ML estimates. In these data, the molecular clock does not appear to be violated, so that the different models are expected to produce similar results. Second, the Bayesian analysis has the advantage of providing CI intervals that take into account fossil uncertainties. Indeed the Bayesian CIs are all much wider than the likelihood CIs. For the two fossil calibration nodes (t_{ape} and t_{mon}), the posterior CIs are no narrower than the prior intervals. The intervals are particularly wide under the autocorrelated-rates model. Third, the two substitution models provided almost identical estimates, with the posterior means of t_1 at ~ 33 MY under JC69 and 34 MY under HKY85+Γ_5. Fourth, there seem to be some conflict between fossils and molecules: the posterior mean of t_{ape} is often smaller than the posterior mean of t_{mon} although in the prior the opposite is true. The posterior intervals of the two node ages overlap, so it is uncertain

Table 7.2 Likelihood and Bayesian estimates of divergence times from the data of Steiper et al. (2004)

	ML			Bayesian		
	Global clock	Local clock	Prior	Global clock (clock1)	Independent rates (clock2)	Autocorrelated rates (clock3)
JC69						
t_0 (root)	32.8 (31.5, 34.1)	32.3 (31.0, 33.5)	760 (86, 3074)	32.9 (29.2, 36.9)	33.1 (24.1, 44.5)	33.6 (25.4, 46.4)
t_{ape} (ape)	7	7	7 (6, 8)	**5.9 (5.2, 6.6)**	**6.3 (5.4, 7.5)**	**6.5 (5.5, 7.6)**
t_{mon} (monkey)	6	6	6 (5, 7)	**7.2 (6.4, 8.0)**	**6.7 (5.4, 7.8)**	**6.5 (5.3, 7.7)**
r	6.6 (6.3, 6.9)	r_{ape}: 5.4 (5.1, 5.7) r_{mon}: 8.0 (7.6, 8.4)	100 (12, 278)	6.6 (5.9, 7.4)		
HKY85 + Γ_5						
t_0 (root)	33.8 (32.4, 35.2)	33.3 (31.9, 34.7)	760 (86, 3074)	34.0 (30.1, 38.0)	34.1 (25.3, 44.1)	34.8 (26.2, 47.8)
t_{ape} (ape)	7	7	7 (6, 8)	**5.9 (5.2, 6.6)**	**6.2 (5.3, 7.5)**	**6.6 (5.6, 7.7)**
t_{mon} (monkey)	6	6	6 (5, 7)	**7.2 (6.3, 8.0)**	**6.8 (5.5, 7.9)**	**6.4 (5.3, 7.6)**
r	6.6 (6.4, 6.9)	r_{ape}: 5.4 (5.1, 5.8) r_{mon}: 8.0 (7.6, 8.5)	100 (12, 278)	6.6 (5.9, 7.4)		

Divergence times (in million years) are defined in Fig. 7.6. Rate is $\times 10^{-10}$ substitutions per site per year. For ML, the MLEs and the 95% confidence intervals are shown. For Bayesian inference, the posterior mean and 95% credibility intervals are shown. Note that the ML analysis ignores uncertainties in the fossil dates and so the CIs are too narrow. The likelihood and Bayesian results under the global clock are taken from Yang and Rannala (2006), while those under the local-clock models are obtained by using programs BASEML and MCMCTREE in the PAML package (Yang 1997a).

which of the two nodes is more ancient. In all analyses, the posterior age of the root is older than 25 MY, consistent with the analysis of Steiper *et al.* (2004).

7.4.5.2 Divergence times of mammals

Springer *et al.* (2003) estimated divergence times among 42 mammalian species using 19 nuclear and three mitochondrial genes, totalling 16 397 sites in the alignment. The species phylogeny is shown in Fig. 7.13. The authors used the Bayesian MCMC program MULTIDIVTIME of Thorne *et al.* (1998), which relaxes the molecular clock by using a geometric Brownian motion model of rate drift. The results of that analysis were reproduced in Table 7.3. Here we apply the Bayesian program MCMCTREE. The focus here is on the illustration and comparison of methods. For more detailed discussions of molecular estimates of mammalian divergence times, see Hasegawa *et al.* (2003) and Douzery *et al.* (2003).

Nine nodes are used as fossil calibrations, listed in Table 7.3; see Springer *et al.* (2003) for detailed discussions of the fossil data. These are specified using densities of the types shown in Fig. 7.11, assuming soft bounds. Besides the calibration information, a diffuse prior is specified for the age of the root, in the interval (25 MY, 185 MY). Parameters of the birth–death process with species sampling are $\lambda = \mu = 2$, and $\rho = 0.1$. The kernel density corresponding to these parameter values (equation 7.24) is shown in Fig. 7.12(a). The density is nearly flat, and the prior on time appears to be similar to that assumed in MULTIDIVTIME. The overall substitution rate μ is assigned the gamma prior $G(2, 2)$ with mean 1 and variance 1/2. Parameter σ^2 is assigned the prior $G(1, 10)$, with mean 0.1, variance 0.01.

The HKY85 + Γ_5 substitution model is assumed, with κ assigned the gamma prior $G(6, 2)$, and α the gamma prior $G(1, 1)$. The nucleotide frequencies are estimated using the observed frequencies. The data set is very informative about those parameters and the prior has little effect on the posterior.

The MCMC is run for 200 000 iterations, after a burn-in of 10 000 iterations. For each analysis, the MCMC algorithm is run twice using different starting values to confirm convergence to the same posterior.

Table 7.3 shows the posterior means and 95% credibility intervals of divergence times. MCMCTREE/clock3 implements the geometric Brownian motion model, as in MULTIDIVTIME. Results obtained from these two analyses are in general similar. For example, MULTIDIVTIME estimated the age of the base of Placentalia (the root of the tree) to be at 107 MY with the 95% CI to be (98–117), while MCMCTREE/clock3 estimated the same node to be at 107 MY with the CI to be (94, 140). For some nodes, MCMCTREE/clock3 produced wider credibility intervals than MULTIDIVTIME. For example, the rat–mouse divergence is dated to 16 MY (13–21) by MULTIDIVTIME and to 23 MY (13–36) by clock3. This difference may reflect the use of hard bounds in MULTIDIVTIME, which tends to overestimate the confidence in time estimates when there are conflicts between fossil calibrations or between fossils and molecules.

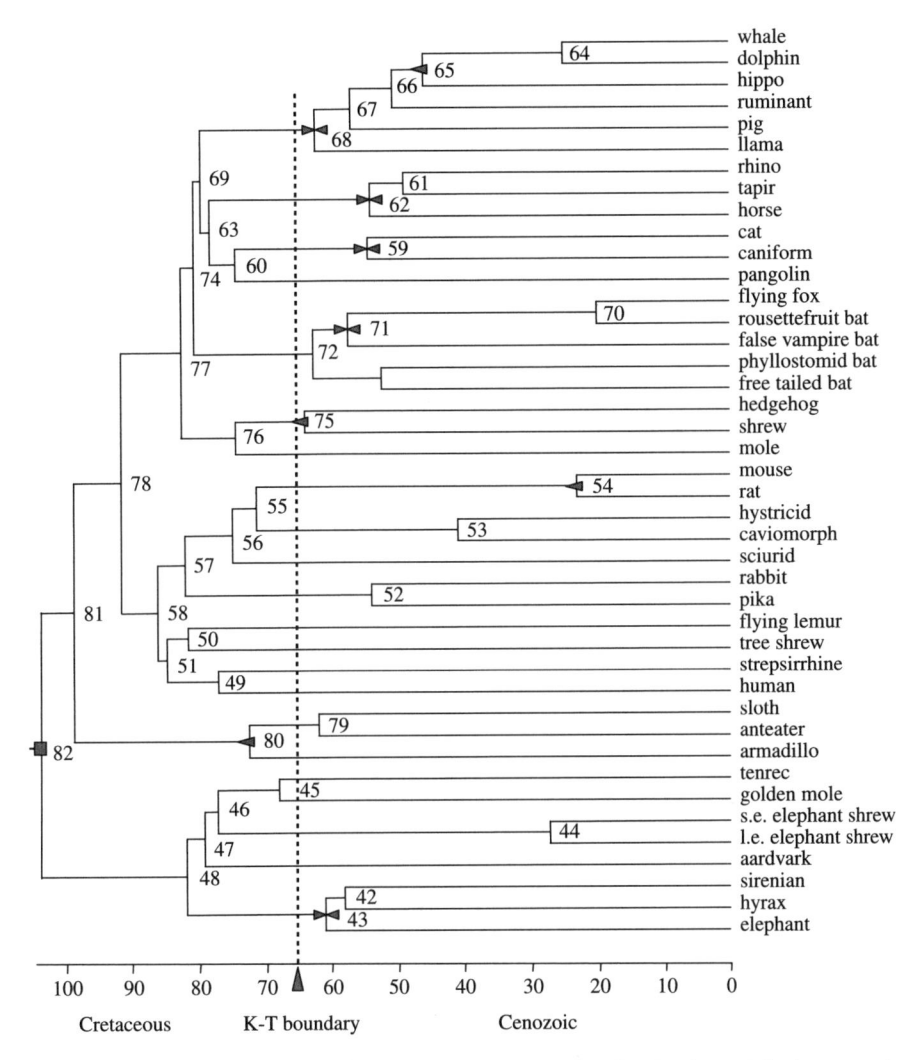

Fig. 7.13 The phylogenetic tree of 42 mammalian species for the data of Springer et al. (2003). Fossil calibrations are indicated on the tree, described in Table 7.3. The branches are drawn in proportion to the posterior means of divergence times estimated under the Bayesian local-clock model (clock 3 in Table 7.3).

MCMCTREE/clock2 assumes that the rates are independent variables from the same distribution. This model produced older time estimates for most nodes than under clock3. The lower bounds of the CIs are similar to those under clock3, but the upper bounds are much older, producing much wider CIs.

MCMCTREE/clock1 implements the global-clock model. Most estimates do not look very poor in comparison with estimates under other models, considering that

Table 7.3 Posterior mean and 95% CIs of divergence times (MY) for the mammalian data of Springer et al. (2004)

Node		Fossil	MULTIDIVTIME	MCMCTREE clock1	clock2	clock3
82	Base of Placentalia		107 (98, 117)	126 (122, 129)	121 (100, 146)	107 (95, 140)
81	Xenarthra–Boreoeutheria		102 (94, 111)	124 (120, 128)	116 (97, 140)	102 (94, 110)
80	Base of Xenarthra	>60	71 (63, 79)	72 (69, 76)	91 (63,113)	74 (64, 87)
78	Base of Boreoeutheria		94 (88, 101)	118 (115, 122)	106 (89, 127)	94 (87, 101)
77	Base of Laurasiatheria		85 (80, 90)	107 (103, 110)	95 (81, 111)	84 (79, 90)
76	Base of Eulipotyphla		76 (71, 81)	100 (97, 104)	85 (71, 102)	76 (71, 82)
75		>63		94 (90, 98)	76 (61, 93)	65 (59, 72)
73	Base of Chiroptera		65 (62, 68)	73 (70, 76)	63 (54, 75)	64 (60, 69)
71		43–60		62 (60, 65)	55 (47, 60)	59 (55, 61)
68	Base of Cetartiodactyla	55–65	64 (62, 65)	67 (64, 69)	71 (62, 83)	64 (59, 69)
65		>52		44 (42, 47)	51 (44, 59)	47 (42, 52)
62	Base of Perissodactyla	54–58	56 (54, 58)	54 (53, 54)	56 (54, 58)	55 (53, 57)
59	Base of Carnivora	50–63	55 (51, 60)	54 (51, 57)	56 (49, 62)	56 (50, 62)
58	Base of Euarchontoglires		87 (81, 94)	114 (110, 118)	98 (82, 118)	88 (81, 95)
56	Base of Rodentia		74 (68, 81)	107 (103, 110)	85 (69, 102)	77 (68, 83)
54	Mus–Rattus	>12	16 (13, 21)	34 (32, 36)	32 (19, 50)	23 (13, 36)
53	Hystricidae–Caviomorpha		38 (31, 46)	64 (62, 67)	51 (34, 64)	42 (24, 55)
52	Base of Lagomorpha		51 (43, 59)	76 (72, 80)	59 (45, 80)	55 (39, 65)
49	Base of Primates		77 (71, 84)	85 (81, 90)	84 (63, 103)	79 (68, 86)
45	Base of Afrosoricida		66 (60, 73)	87 (84, 91)	73 (54, 95)	70 (57, 116)
43	Base of Paenungulata	54–65	62 (57, 65)	65 (63, 66)	60 (54, 65)	62 (57, 65)

Node numbering refers to the tree of Fig. 7.13. 'Fossil' refers to fossil calibrations, also shown in the tree. The bounds are hard in MULTIDIVTIME and soft in MCMCTREE. MCMCTREE/clock1 refers to the global clock, clock2 the independent-rates model, and clock3 the geometric Brownian motion model.

the clock is seriously violated. However, the rat–mouse divergence is dated to be at 32–36 MY, much older than estimates under other models. This old estimate is apparently due to high substitution rates in rodents. More seriously, the posterior credibility intervals under the global clock tend to be much narrower than those under clock2 and clock3, and the method is overconfident.

7.5 Perspectives

The greatest obstacle to reliable estimation of divergence times is perhaps the confounding effect of time and rate; sequences provide information about the distance but not about time and rate separately. A fixed amount of evolution is compatible with infinitely many assumptions about the rates, as they can fit the data equally well by an appropriate adjustment of the times (Thorne and Kishino 2005). Relaxation of the clock thus makes time estimation a tricky business. This problem may be alleviated to some extent by analysing multiple gene loci simultaneously and by using multiple calibration points. If different genes have different patterns of evolutionary rate change but share the same divergence times, analysing multiple gene loci simultaneously may allow one locus to 'borrow information' from other loci, making inference about times possible. A long branch at one locus may be due to either a long time duration or a high rate of evolution, but a high rate becomes a more likely explanation if the same branch is short at other loci. Similarly, simultaneous application of multiple fossil calibrations may be critical in characterizing local evolutionary rates on the phylogeny.

The critical importance of reliable and precise fossil calibrations is most evident when one considers the limiting case of an infinite amount of sequence data. In the Bayesian framework, an 'infinite-sites theory' has been developed to specify the limiting distributions of times and rates when the length of sequence approaches infinity (Yang and Rannala 2006; Rannala and Yang 2006). An important and depressing result from this theoretical analysis is that in typical data sets with more than a few thousand sites in the sequence, uncertainties in time estimates mainly reflect uncertainties in fossil calibrations rather than uncertainties in branch lengths due to finite sequence data. The observation that the posterior and prior intervals are often of equal or comparable widths, made above in the analysis of the primate and mammalian data sets, is quite typical.

Recent methodological developments in both the likelihood and Bayesian frameworks make it possible to estimate divergence times without relying on a global molecular clock. Current implementations of those methods allow integrated analysis of heterogeneous data sets from multiple loci and simultaneous incorporation of multiple fossil calibrations. The Bayesian method can naturally incorporate uncertainties in node ages provided by fossils, but developing objective approaches to describing fossil uncertainties using statistical distributions may be important. Statistically justifiable and computationally feasible likelihood methods for dealing with fossil uncertainties have yet to be developed. Even with the most powerful methodology, one has to bear in mind that estimation of divergence times without the clock is an extremely difficult exercise.

Below are a few recommendations on the use of the Bayesian programs MULTIDIVTIME and MCMCTREE. First, there is no hope of estimating absolute divergence times without fossil calibration. Indeed one should have at least one lower bound and at least one upper bound on the tree. The lower and upper bounds may be on different nodes but should be more informative if they are on the same node. A single-moded distribution like the gamma plays a similar role to a pair of lower and upper bounds. Second, MULTIDIVTIME requires the specification of a gamma prior on the age of the root. Although it is not required by MCMCTREE, it has been noted that placing an upper bound on the root age is very useful. Third, both programs make a number of prior assumptions, often without the option of allowing the user to change them. For the present, it is not well understood which factors are most important for posterior time estimation. Analysis of both real and simulated data sets may help us to understand the strengths and weakness of the programs and more generally what is achievable and what is not in this difficult estimation problem.

8

Neutral and adaptive protein evolution

8.1 Introduction

Adaptive evolution of genes and genomes is ultimately responsible for adaptation in morphology, behavior, and physiology, and for species divergence and evolutionary innovations. Molecular adaptation is thus an exciting topic in molecular evolutionary studies. While natural selection appears to be ubiquitous in shaping morphological and behavioural evolution, its role in evolution of genes and genomes is more controversial. Indeed the neutral theory of molecular evolution claims that much of the observed variation within and between species is not due to natural selection but rather to random fixation of mutations with little fitness significance (Kimura 1968; King and Jukes 1969). The last four decades have seen the development of a number of tests of neutrality. In this chapter, we will introduce the basic concepts of negative and positive selection and the major theories of molecular evolution, and review tests of neutrality developed in population genetics (Section 8.2). The rest and bulk of the chapter follows up on Chapter 3, in which we described Markov models of codon substitution and their use to estimate d_S and d_N, the synonymous and nonsynonymous distances between two protein-coding sequences. Here we discuss the use of codon models in phylogenetic analysis to detect positive selection driving the fixation of advantageous replacement mutations.

For understanding the roles of natural selection, protein-coding sequences offer a great advantage over introns or noncoding sequences in that they allow us to distinguish synonymous and nonsynonymous substitutions (Miyata and Yasunaga 1980; Li et al. 1985; Nei and Gojobori 1986). With the synonymous rate used as a benchmark, one can infer whether fixation of nonsynonymous mutations is aided or hindered by natural selection. The nonsynonymous/synonymous rate ratio, $\omega = d_N/d_S$, measures selective pressure at the protein level. If selection has no effect on fitness, nonsynonymous mutations will be fixed at the same rate as synonymous mutations, so that $d_N = d_S$ and $\omega = 1$. If nonsynonymous mutations are deleterious, purifying selection will reduce their fixation rate, so that $d_N < d_S$ and $\omega < 1$. If nonsynonymous mutations are favoured by Darwinian selection, they will be fixed at a higher rate than synonymous mutations, resulting in $d_N > d_S$ and $\omega > 1$. A significantly higher nonsynonymous rate than the synonymous rate is thus evidence for adaptive protein evolution.

Early studies using this criterion took the approach of pairwise sequence comparison, averaging d_S and d_N over all codons in the gene sequence and over the whole time period separating the two sequences. However, one may expect most sites in a functional protein to be constrained during most of the evolutionary time. Positive selection, if it occurs, should affect only a few sites and occur in an episodic fashion (Gillespie 1991). Thus the pairwise averaging approach rarely detects positive selection (e.g. Sharp 1997). Recent efforts have focused on detecting positive selection affecting particular lineages on the phylogeny or individual sites in the protein. These methods are discussed in Sections 8.3 and 8.4. Section 8.5 discusses methods aimed at detecting positive selection affecting only a few sites along particular lineages. In Section 8.6, we discuss assumptions and limitations of methods based on the ω ratio, in comparison with the tests of neutrality. Section 8.7 reviews examples of genes detected to be undergoing adaptive evolution.

8.2 The neutral theory and tests of neutrality

8.2.1 The neutral and nearly neutral theory

Here we introduce the basic concepts of positive and negative selection as well as the major theories of molecular evolution. We also describe briefly a few commonly used tests of neutrality developed in population genetics. This discussion will not do justice to the fundamental role that population genetics plays in studies of molecular evolution. The reader is referred to textbooks of population genetics (e.g. Hartl and Clark 1997; Gillespie 1991; Gillespie 1998; Li 1997). Ohta and Gillespie (1996) provide a historical perspective of the developments of the neutral theory.

In population genetics, the relative fitness of a new mutant allele a relative to the predominant wild-type allele A is measured by the selective coefficient (Malthusian parameter) s. Let the relative fitness of genotypes AA, Aa, and aa be 1, $1 + s$, and $1 + 2s$, respectively. Then $s < 0, = 0, > 0$ correspond to negative (purifying) selection, neutral evolution, and positive selection, respectively. The frequency of the new mutant allele goes up or down over generations, affected by natural selection as well as random genetic drift. Whether random drift or selection dominates the fate of the mutation depends on Ns, where N is the effective population size. If $|Ns| \gg 1$, natural selection dominates the fate of the allele while if $|Ns|$ is close to 0, random drift will be very important and the mutation is effectively neutral or nearly neutral.

Much of the theoretical work studying the dynamics of allele frequency changes was completed by the early 1930s by Fisher (1930a), Haldane (1932), and Wright (1931). The consensus generally accepted until the 1960s was that natural selection was the driving force of evolution. Natural populations were believed to be in nearly optimal state with little genetic variation. Most new mutations were deleterious and quickly removed from the population. Rarely an advantageous mutation occurred and spread over the entire population. However, in 1966, high levels of genetic variation were detected in allozymes using electrophoresis in *Drosophila* (Lewontin and Hubby 1966; Harris 1966). The *neutral theory*, or *the neutral-mutation random-drift hypothesis*, was proposed by Kimura (1968) and King and Jukes (1969) mainly to accommodate this surprising finding.

According to the neutral theory, the genetic variation we observe today — both the polymorphism within a species and the divergence between species — is not due to fixation of advantageous mutations driven by natural selection but to random fixation of mutations with effectively no fitness effect, that is, neutral mutations. Below are some of the claims and predictions of the theory (Kimura 1983):

- Most mutations are deleterious and are removed by purifying selection.
- The nucleotide substitution rate is equal to the neutral mutation rate, that is, the total mutation rate times the proportion of neutral mutations. If the neutral mutation rate is constant among species (either in calendar time or in generation time), the substitution rate will be constant. This prediction provides an explanation for the molecular-clock hypothesis.
- Functionally more important genes or gene regions evolve more slowly. In a gene with a more important role or under stronger functional constraint there will be a smaller proportion of neutral mutations so that the nucleotide substitution rate will be lower. The negative correlation between functional significance and substitution rate is now a general observation in molecular evolution. For example, replacement substitution rates are almost always lower than silent substitution rates; third codon positions evolve more quickly than first and second codon positions; and amino acids with similar chemical properties tend to replace each other more often than dissimilar amino acids. If natural selection drives the evolutionary process at the molecular level, we would expect functionally important genes to have higher evolutionary rates than functionally unimportant genes.
- Within-species polymorphism and between-species divergence are two phases of the same process of neutral evolution.
- Evolution of morphological traits (including physiology, behaviour, etc.) is indeed driven by natural selection. The neutral theory concerns evolution at the molecular level.

The controversy surrounding the neutral theory has generated a rich body of population genetics theory and analytical tools. The neutral theory makes simple and testable predictions. This fact as well as the fast accumulation of DNA sequence data in the last few decades has prompted development of a number of tests of the theory. Some of them are discussed below. See Kreitman and Akashi (1995), Kreitman (2000), Nielsen (2001b), Fay and Wu (2001, 2003), Ohta (2002), and Hein et al. (2005) for reviews.

The strict neutral model assumes only two kinds of mutations: strictly neutral mutations with $s = 0$ and highly deleterious mutations that are wiped out by purifying selection as soon as they occur ($Ns \ll -1$). The strict neutral model is often rejected when tested against real data. Tomoko Ohta (1973) proposed the *slightly deleterious mutation hypothesis*, which allows for mutations with small negative selective coefficients so that their fate is influenced by both random drift and selection. The fixation probability of such mutations is positive but smaller than for neutral mutations. This model later evolved into the *nearly neutral hypothesis* (Ohta and Tachida 1990; Ohta

1992), which allows for both slightly deleterious and slightly advantageous mutations. In contrast to the strict neutral model, in which the dynamics depends on the neutral mutation rate alone and not on other factors such as population size and selective pressure, the dynamics of slightly deleterious or nearly neutral mutations depend on all these parameters. As a result, the modified theories are very difficult to test or refute. This is also the case for various selection models (Gillespie 1991). See Fig. 8.1 for an illustration of these different theories.

8.2.2 Tajima's D statistic

The amount of genetic variation at a neutral locus maintained in a random mating population is determined by $\theta = 4N\mu$, where N is the (effective) population size and μ is the mutation rate per generation. Defined on a per-site basis, θ is also the expected site heterozygosity between two sequences drawn at random from the population. For example, in human noncoding DNA, $\hat{\theta} \sim 0.0005$, which means that between two random human sequences, about 0.05% of the sites are different (Yu et al. 2001; Rannala and Yang 2003). Population data typically involve little variation so that the *infinite-sites model* is commonly used, which assumes that every mutation occurs at a different site in the DNA sequence and there is no need for correction for multiple hits. Note that both a large population size and a high mutation rate cause greater genetic variation to be maintained in the population.

Two simple approaches can be used to estimate θ using a random sample of DNA sequences from the population. First the number of segregating (polymorphic) sites S in a sample of n sequences is known to have the expectation $E(S) = L\theta a_n$, where L is the number of sites in the sequence and $a_n = \sum_{i=1}^{n-1} 1/i$ (Watterson 1975). Thus θ can be estimated by $\hat{\theta}_S = S/(La)$. Second, the average proportion of nucleotide differences over all pairwise comparisons of the n sequences has the expectation θ and can thus be used as an estimate; let this be $\hat{\theta}_\pi$ (Tajima 1983). Both estimates of θ are unbiased under the neutral mutation model, assuming no selection, no recombination, no population subdivision or population size change, and equilibrium between mutation and drift. However, when the model assumptions are violated, different factors have different effects on $\hat{\theta}_S$ and $\hat{\theta}_\pi$. For example, slightly deleterious mutants are maintained in a population at low frequencies and can greatly inflate S and $\hat{\theta}_S$, while having little effect on $\hat{\theta}_\pi$. The direction and magnitude of the difference between the two estimators of θ can provide insights into factors and mechanisms that caused departure from the strict neutral model. Thus Tajima (1989) constructed the following test statistic

$$D = \frac{\hat{\theta}_\pi - \hat{\theta}_S}{\text{SE}(\hat{\theta}_\pi - \hat{\theta}_S)}, \tag{8.1}$$

where SE is the standard error. Under the null neutral model, D has mean 0 and variance 1. Tajima suggested the use of the standard normal and beta distributions to determine whether D is significantly different from 0.

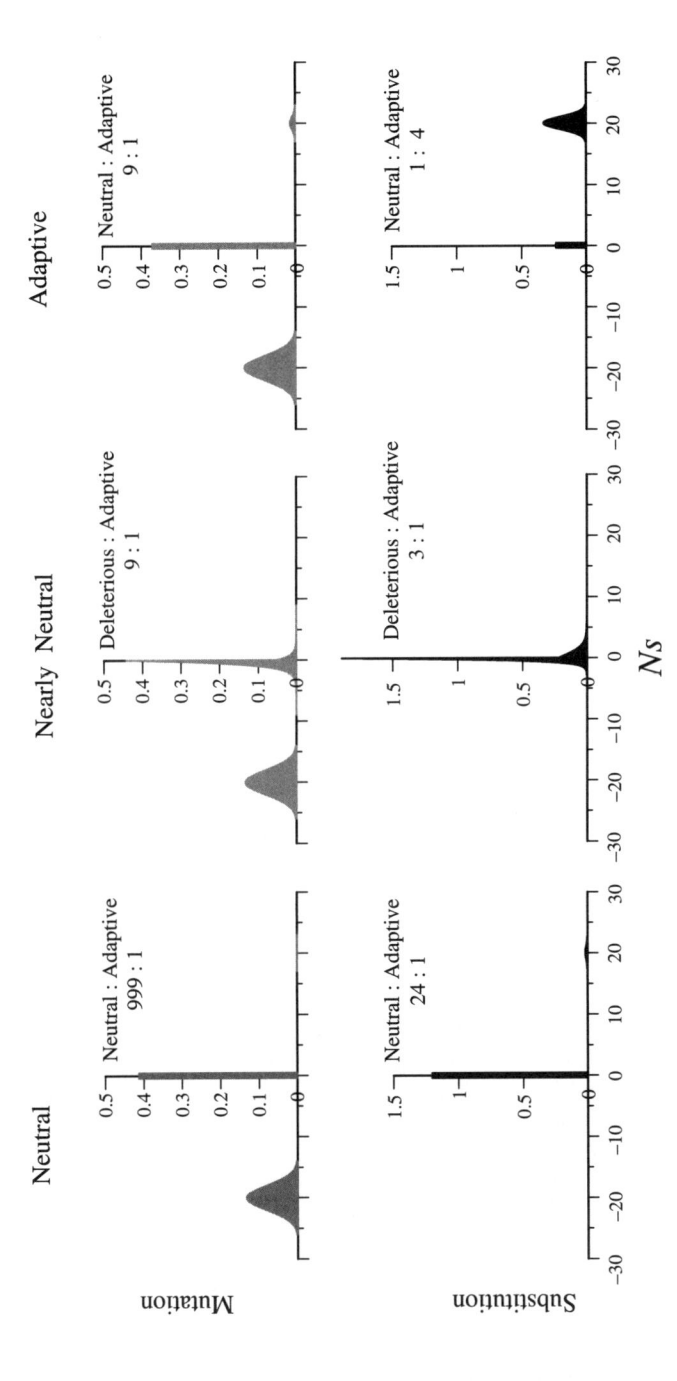

Fig. 8.1 Distributions of fitness effects (Ns) among mutations and substitutions under neutral, nearly neutral, and adaptive models of molecular evolution. One possible distribution is shown for each model, with the area under the curves summing to 1. The bars at $Ns = 0$ for neutral mutations are broadened for visualization. The sampling formulae under the Poisson random field model of Sawyer and Hartl (1992) were used to transform the density of fitness effects of mutations to that of fixations (substitutions). Redrawn after Akashi (1999b); data courtesy of Hiroshi Akashi.

Statistical significance of Tajima's D test may be compatible with several different explanations, and it can be difficult to distinguish among them. As discussed above, a negative D is indicative of purifying selection or presence of slightly deleterious mutations segregating in the population. However, a negative D can also be caused by population expansion. In an expanding population, many new mutations may be segregating and will be observed in the data as *singletons*, sites at which only one sequence has a different nucleotide while all other sequences are identical. Singletons inflate the number of segregating sites and cause D to be negative. Similarly, a positive D is compatible with balancing selection, which maintains mutations at intermediate frequencies. However, a shrinking population can also generate a positive D.

8.2.3 Fu and Li's D and Fay and Wu's H statistics

In a sample of n sequences, the frequency of the mutant nucleotide at a polymorphic site can be $r = 1, 2, \ldots$, or $n - 1$. The observed distribution of such frequencies of mutations in a sample is called the *site-frequency spectrum*. Often a closely related outgroup species is used to infer the ancestral and derived nucleotide state. For example, if the observed nucleotides in a sample of $n = 5$ are AACCC and if the outgroup has A, which is assumed to be the ancestral type, then $r = 3$. Fu (1994) referred to r as the *size* of the mutation. If the ancestral state is unknown, it will be impossible to distinguish mutations of sizes r and $n - r$, so that those mutations are grouped into the same class. The site-frequency spectrum is then said to be *folded* (Akashi 1999a). Needless to say, folded configurations can be much less informative than unfolded ones. Thus use of the outgroup to infer ancestral states should improve the power of the test, but it has the drawback that the test may be affected by errors in ancestral reconstruction (Baudry and Depaulis 2003; Akashi *et al.* 2006). Different types of natural selection will cause the site-frequency spectrum to deviate from the neutral expectation in distinct ways. Thus the site-spectrum distributions or their summary statistics can be used to construct statistical tests of neutrality. For example, Tajima's D is such a test; it contrasts singletons (which mostly contribute to S) with mutations of intermediate frequencies (which mostly contribute to π). Two more popular tests explore similar ideas.

Fu and Li (1993) distinguished *internal* and *external* mutations, which occur along internal or external branches of the genealogical tree, respectively. Let the numbers of such mutations be η_I and η_E. Note that η_E is the number of singletons. Fu and Li constructed the following statistic

$$D = \frac{\eta_I - (a_n - 1)\eta_E}{\text{SE}(\eta_I - (a_n - 1)\eta_E)}, \tag{8.2}$$

where $a_n = \sum_{i=1}^{n-1} 1/i$ and SE is the standard error. Similar to Tajima's D, this statistic is also constructed as the difference of two estimates of θ under the neutral model, divided by the standard error of the difference. Fu and Li (1993) argue that deleterious mutations segregating in the population tend to be recent and reside on external branches of the tree, and contribute to η_E, while mutations in the internal branches are most likely neutral and contribute to η_I (Williamson and Orive 2002). Besides D in equation (8.2), Fu and Li constructed several other tests of this kind. The power of such tests depends on how different the two estimates of θ used in the test are when there is selection. Braverman *et al.* (1995), Simonsen *et al.* (1995), Fu (1997), Akashi (1999a, b), and McVean and Charlesworth (2000) conducted simulations to examine the power of the tests.

Fay and Wu (2000) explored a similar idea and constructed an estimate of θ as

$$\hat{\theta}_H = \sum_{i=1}^{n} \frac{2 S_i i^2}{n(n-1)}, \qquad (8.3)$$

where S_i is the number of mutations of size i. They then defined a statistic $H = \hat{\theta}_\pi - \hat{\theta}_H$, which has expectation 0 under the strict neutral model. The null distribution is generated by computer simulation. Note that in this notation, $\hat{\theta}_\pi = \sum_{i=1}^{n} [2 S_i i (n-i)/(n(n-1))]$, so that mutations of intermediate frequencies (with i close to $n/2$) make the greatest contribution to $\hat{\theta}_\pi$, while mutations of high frequencies (with i close to n) make the greatest contribution to $\hat{\theta}_H$. Thus Fay and Wu's H statistic compares mutations at intermediate and high frequencies. Under selective neutrality, the site-frequency spectrum is L-shaped, with low-frequency mutants to be common and high-frequency mutants to be rare. When a neutral mutation is tightly linked to a locus under positive selection, the mutant may rise to high frequencies due to selection driving the advantageous allele at the selected locus to fixation. The neutral mutation is said to be under *genetic hitchhiking* (Maynard Smith and Haigh 1974; Braverman *et al.* 1995). Fay and Wu (2000) point out that an excess of high-frequency mutants, indicated by a significantly negative H, is a distinct feature of hitchhiking. The test requires the sequence from an outgroup species to be used to infer the ancestral and derived states at segregating sites.

8.2.4 McDonald–Kreitman test and estimation of selective strength

The neutral theory claims that both the diversity (polymorphism) within a species and the divergence between species are two phases of the same evolutionary process; that is, both are due to random drift of selectively neutral mutations. Thus if both synonymous and nonsynonymous mutations are neutral, the proportions of synonymous and nonsynonymous polymorphisms within a species should be the same as the proportions of synonymous and nonsynonymous differences between species. The McDonald–Kreitman (1991) test examines this prediction.

Variable sites in protein-coding genes from closely related species are classified into four categories in a 2×2 contingency table, depending on whether the site has a polymorphism or a fixed difference, and whether the difference is synonymous or nonsynonymous (Table 8.1). Suppose we sample five sequences from species 1 and four sequences from species 2. A site with data AAAAA in species 1 and GGGG in species 2 is called a fixed difference. A site with AGAGA in species 1 and AAAA in

Table 8.1 Numbers of silent and replacement divergences and polymorphisms in the *Adh* locus in *Drosophila* (from McDonald and Kreitman 1991)

Type of change	Fixed	Polymorphic
Replacement (nonsynonymous)	7	2
Silent (synonymous)	17	42

species 2 is called a polymorphic site. Note that under the infinite-sites model there is no need to correct for hidden changes. The neutral null hypothesis is equivalent to independence between the row and column in the contingency table and can be tested using the χ^2 distribution or Fisher's exact test if the counts are small. McDonald and Kreitman (1991) sequenced the alcohol dehydrogenase (*Adh*) gene from three species in the *Drosophila melanogaster* subgroup and obtained the counts of Table 8.1. The *P* value is < 0.006, suggesting significant departure from the neutral expectation. There are far more replacement differences between species than within species. McDonald and Kreitman interpreted the pattern as evidence for positive selection driving species differences.

To see the reasoning behind this interpretation, consider the effects of selection on nonsynonymous mutations that occurred after the divergence of the species, assuming that synonymous mutations are neutral. Advantageous replacement mutations are expected to go to fixation quickly and become fixed differences between species. Thus an excess of replacement fixed differences, as observed at the *Adh* locus, is indicative of positive selection. As pointed out by McDonald and Kreitman, an alternative explanation for the pattern is a smaller population size in the past than in the present, combined with presence of slightly deleterious replacement mutations. Such mutations might have gone to fixation due to random drift in the past and became fixed differences between species, while they are removed by purifying selection in the large present-day populations. McDonald and Kreitman argued that such a scenario is unlikely for the *Adh* gene they analysed.

In mammalian mitochondrial genes, an excess of replacement polymorphism is observed (e.g. Rand *et al.* 1994; Nachman *et al.* 1996). This is indicative of slightly deleterious replacement mutations under purifying selection. Deleterious mutations are removed by purifying selection and will not be seen in between-species comparisons but might still be segregating within species.

The infinite-sites model assumed in the McDonald and Kreitman test can break down if the different species are not very closely related. A likelihood ratio test for testing the equality of the within-species ω_W and the between-species ω_B was implemented by Hasegawa *et al.* (1998), which uses the codon model to correct for multiple hits. However, the test uses a genealogical tree, and it is unclear how robust the test is to errors in the estimated tree topology.

An idea very similar to that behind the McDonald and Kreitman test is used by Akashi (1999b) to test for natural selection on silent sites driving synonymous codon usage. Rather than the synonymous and nonsynonymous categories, Akashi classified variable silent sites according to whether the changes are to *preferred* (commonly used) or *unpreferred* (rarely used) codons. If both types of mutations are neutral, their proportions should be the same within species and between species. Akashi detected significant departure from this neutral expectation, providing evidence that evolution at silent sites is driven by natural selection, possibly to enhance the efficiency and accuracy of translation.

The idea underlying the McDonald and Kreitman test has been extended to estimate parameters measuring the strength of natural selection, using the so-called *Poisson random field* theory (Sawyer and Hartl 1992; Hartl *et al.* 1994; Akashi 1999a). The model assumes free recombination among sites within the same gene. Then the counts in the 2×2 contingency table (see Table 8.1) are independent Poisson random variables with means to be functions of parameters in the model, which include the population sizes of the current species, the divergence time between the species, and the selection coefficient of new replacement mutations. The test of selection using this model becomes more powerful if multiple loci are analysed simultaneously, as the species divergence time and population sizes are shared among loci. Power is also improved by using the full site-frequency spectrum at polymorphic sites instead of the counts in the 2×2 table. Likelihood and Bayesian methods can be used to test for adaptive amino acid changes and to estimate the strength of selection (Bustamante *et al.* 2002, 2003). Bustamante *et al.* (2002) demonstrated evidence for beneficial substitutions in *Drosophila* and deleterious substitutions in the mustard weed *Arabidopsis*. They attributed the difference to partial self-mating in *Arabidopsis*, which makes it difficult for the species to weed out deleterious mutations.

The Poisson random-field model is currently the only tractable framework that explicitly incorporates selection and is applicable to molecular sequence data; in contrast, the neutrality tests only detect violation of the null neutral model but do not consider selection explicitly in the model. While the theory provides a powerful framework for estimating mutation and selection parameters in various population genetics settings, the assumption of free recombination within locus is highly unrealistic, and can greatly influence the test, especially if the full site-frequency spectrum is analysed (Bustamante *et al.* 2001). Approximate likelihood methods incorporating recombination have recently been implemented, which showed promising performance (Zhu and Bustamante 2005).

8.2.5 Hudson–Kreitman–Aquade test

The Hudson–Kreitman–Aquade test or HKA test (Hudson *et al.* 1987) examines the neutral prediction that polymorphism within species and divergence between species are two facets of the same process. The test uses sequence data at multiple unlinked loci (typically non-coding) from at least two closely related species, and tests whether the polymorphisms and divergences at those loci are compatible. The rationale is that at a locus with high mutation rate, both polymorphism and divergence should be high, while at a locus with low mutation rate, both polymorphism and divergence should be low.

Suppose there are L loci. Let the numbers of segregating sites in the two species A and B be S_i^A and S_i^B at locus i, and the number of differences between the two species at the locus be D_i. S_i^A and S_i^B measure within-species polymorphism while D_i measures between-species divergence. Hudson et al. (1987) assumed that S_i^A, S_i^B, and D_i are independent normal variates and used the neutral theory to derive their expectations and variances to construct a goodness-of-fit test statistic

$$X^2 = \sum_{i=1}^{L}(S_i^A - E(S_i^A))^2/V(S_i^A) + \sum_{i=1}^{L}(S_i^B - E(S_i^B))^2/V(S_i^B)$$
$$+ \sum_{i=1}^{L}(D_i - E(D_i))^2/V(D_i). \tag{8.4}$$

The null neutral model involves $L+2$ parameters: θ_i for each locus defined for species A, the ratio of the two population sizes, and the species divergence time T. There are $3L$ observations: S_i^A, S_i^B, and D_i for all loci i. Thus the test statistic is compared with a χ^2 distribution with $2L - 2 = 3L - (L+2)$ degrees of freedom to test whether the data fit the neutral expectation.

8.3 Lineages undergoing adaptive evolution

8.3.1 Heuristic methods

In this section we discuss phylogenetic methods for detecting positive selection along prespecified lineages on the phylogenetic tree, indicated by $d_N > d_S$. The classic example is the analysis of lysozyme evolution in primates, by Messier and Stewart (1997). These authors inferred the gene sequences in extinct ancestral species and used them to calculate d_N and d_S for every branch on the tree. (See Sections 3.4 and 4.4 for ancestral sequence reconstruction under the parsimony and likelihood criteria.) Positive selection along a branch is identified by testing whether d_N is significantly greater than d_S for that branch, using a normal approximation to the statistic $d_N - d_S$. The authors were able to detect positive selection along the branch ancestral to the leaf-eating colobine monkeys, supporting the hypothesis that acquisition of a new function in the colobine monkeys (i.e. digestion of bacteria in the foreguts of these animals) caused accelerated amino acid substitutions in the enzyme. (Another branch, ancestral to the hominoids, was also found to have a very high d_N/d_S ratio, but no biological explanation has yet been offered for the result, which appears to be a chance effect due to multiple testing.) By focusing on a single branch rather than averaging over the whole time period separating the sequences, the approach of Messier and Stewart has improved power in detecting episodic adaptive evolution.

Zhang et al. (1997) were concerned about the reliability of the normal approximation in small data sets, as the lysozyme gene has only 130 codons. They suggested instead the use of Fisher's exact test applied to counts of synonymous and nonsynonymous sites and synonymous and nonsynonymous differences along each branch. A drawback of this approach is that it fails to correct for multiple hits as the inferred differences are not substitutions. Both the approaches of Messier and Stewart (1997) and Zhang et al. (1997) use reconstructed ancestral sequences without accommodating their errors (see Section 4.4).

Another simple approach, suggested by Zhang et al. (1998), is to calculate d_N and d_S in all pairwise comparisons of current species, and to fit branch lengths using least squares for synonymous and nonsynonymous rates separately. The synonymous and nonsynonymous branch lengths, called b_S and b_N, can then be compared to test whether $b_N > b_S$ for the branch of interest, using a normal approximation to the statistic $b_N - b_S$. This approach has the benefit of avoiding the use of reconstructed ancestral sequences. It may be important to use a realistic model to estimate d_N and d_S in pairwise comparisons.

8.3.2 Likelihood method

The simple approaches discussed above have an intuitive appeal and are useful for exploratory analysis. They can be made more rigorous by taking a likelihood approach under models of codon substitution (Yang 1998a), analysing all sequences jointly on a phylogenetic tree. The likelihood calculation averages over all possible states at ancestral nodes, weighting them according to their probabilities of occurrence. Thus random and systematic errors in ancestral sequence reconstruction are avoided if the assumed substitution model is adequate. It is also straightforward in a likelihood model to accommodate different transition and transversion rates and unequal codon usage; as a result, likelihood analysis can be conducted under more realistic models than heuristic analysis based on pairwise sequence comparisons.

When a model of codon substitution (see Section 2.4) is applied to a tree, one can let the ω ratio differ among lineages. Yang (1998a) implemented several models that allow for different levels of heterogeneity in the ω ratio among lineages. The simplest model ('one-ratio') assumes the same ω ratio for all branches. The most general model ('free-ratio') assumes an independent ω ratio for each branch in the phylogeny. This model involves too many parameters except on small trees with few lineages. Intermediate models with two or three ω ratios are implemented as well. These models can be compared using the likelihood ratio test to examine interesting hypotheses. For example, the one-ratio and free-ratio models can be compared to test whether the ω ratios are different among lineages. We can fit a two-ratio model in which the branches of interest (called foreground branches) has a different ω ratio from all other branches (called the background branches). This two-ratio model can be compared with the one-ratio model to test whether the two ratios are the same. Similarly one can test the null hypothesis that the ratio for the lineages of interest is 1. This test directly examines the possibility of positive selection along specific lineages.

270 • 8 Neutral and adaptive protein evolution

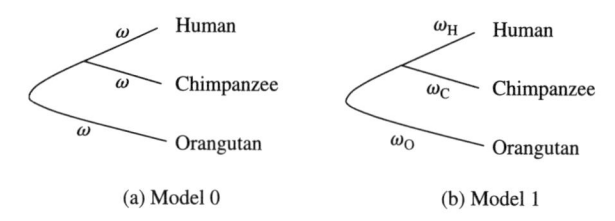

Fig. 8.2 The unrooted phylogenetic tree of human (H), chimpanzee (C), and orangutan (O), showing two evolutionary models. (a) In model 0, the same ω ratio is assumed for all branches in the tree. (b) In model 1, three different ω ratios (ω_H, ω_C, ω_O) are assumed for the three branches. Twice the log likelihood difference between the two models can be compared with the χ^2 distribution with d.f. = 2. After Zhang (2003).

Likelihood calculation under such *branch models* is very similar to that under the standard codon model (one-ratio), discussed in Section 4.2; the only difference is that the transition probabilities for different branches need to be calculated from different rate matrices (Qs) generated using different ωs. As an illustration, consider the two models of Fig. 8.2, used in Zhang's (2003) analysis of the *ASPM* gene, a major determinant of human brain size. In model 0, the same ω ratio is assumed for all branches in the tree, while in model 1, three different ω ratios (ω_H, ω_C, ω_O) are assumed for the three branches. Let x_H, x_C, x_O be codons observed at a particular site in the three sequences, and t_H, t_C, t_O be the three branch lengths in the tree. The probability of observing this site is

$$f(x_H, x_C, x_O) = \sum_i \pi_i p_{ix_H}(t_H; \omega) p_{ix_C}(t_C; \omega) p_{ix_O}(t_O; \omega) \tag{8.5}$$

under model 0 and

$$f(x_H, x_C, x_O) = \sum_i \pi_i p_{ix_H}(t_H; \omega_H) p_{ix_C}(t_C; \omega_C) p_{ix_O}(t_O; \omega_O) \tag{8.6}$$

under model 1. Here we use $p_{ij}(t; \omega)$ to emphasize that the transition probability from codon i to j over time t is calculated using the rate ratio ω. The likelihood, i.e. the probability of observing the whole sequence alignment, is the product of the probabilities across all sites in the sequence.

The estimates obtained by Zhang (2003) under model 1 for the *ASPM* gene are $\hat{\omega}_H = 1.03$, $\hat{\omega}_C = 0.66$, $\hat{\omega}_O = 0.43$. Zhang provided evidence that the high ω_H estimate is due to positive selection driving enlargement of the human brain, and not due to relaxed selective constraint along the human lineage. Another application of the branch-based test is discussed in Section 8.5, where the evolution of the angiosperm phytochrome gene family is analysed.

Two remarks are in order concerning the branch-based tests. First, the lineages of interest should be identified *a priori*. It is inappropriate, for example, to analyse the data to identify lineages of high ω ratios and then to apply the branch test to examine whether these ω ratios are significantly higher than 1. Because of the problem of multiple testing, the null distribution will not be correct since the hypothesis is derived from the data. In this regard, it may be noted that Kosakovsky Pond and Frost (2005a) described a genetic algorithm to assign ω ratios to lineages on the tree, effectively 'evolving' the model to be tested out of the genetic algorithm by maximizing the model's fit to data. This approach does not require *a priori* specification of the lineages to be tested, but does not appear to be a valid statistical test as it fails to account for multiple testing.

Second, variation in the ω ratio among lineages is a violation of the strictly neutral model, but is itself not sufficient evidence for adaptive evolution. Similarly, if the ω ratio on the foreground branch is higher than that on the background branches, but not higher than 1, the result cannot be taken as convincing evidence for positive selection. Besides positive selection, two other compatible explanations are relaxation of purifying selection due to loss or diminishment of the protein function and reduced efficacy of purifying selection removing deleterious mutations due to reduction in population size (Ohta 1973). The criterion that the ω ratio, averaged over all sites in the sequence, should be greater than 1 is very stringent; as a result, the branch test often has little power to detect positive selection.

8.4 Amino acid sites undergoing adaptive evolution

8.4.1 Three strategies

The assumption that all amino acid sites in a protein are under the same selective pressure, with the same underlying nonsynonymous/synonymous rate ratio (ω), is grossly unrealistic. Most proteins have highly conserved amino acid positions at which the underlying ω ratio is close to zero. The requirement that the ω ratio, averaged over all sites in the protein, is > 1 is thus a very stringent criterion for detecting adaptive evolution. It seems most sensible to allow the ω ratio to vary among sites. Positive selection is then indicated by presence of sites at which $\omega > 1$ rather than the ω ratio averaged over all sites being > 1.

Three different strategies appear possible to detect positive selection affecting individual amino acid sites. The first is to focus on amino acid sites that are likely to be under positive selection, as indicated by external information such as the crystal structure of the protein. This is workable only if such information is available. The classic example is the analysis of the human major histocompatibility complex (MHC) loci by Hughes and Nei (1988) and Hughes *et al.* (1990). The d_N/d_S ratio for the entire gene, although higher than in most other protein-coding genes, is < 1, providing no evidence for positive selection. However, studies of the tertiary structure of the molecule (Bjorkman *et al.* 1987a, b) identified 57 amino acid residues that make up the antigen recognition site (ARS), a groove in the structure involved in binding foreign antigen. Hughes and Nei (1988) thus focused on these 57 codons only, and found that the d_N/d_S ratio for them was significantly > 1. Hughes and Nei (1988) used an approach of pairwise sequence comparison. A likelihood approach for joint analysis of multiple sequences on a phylogeny was implemented by Yang and Swanson (2002), in which codons in different partitions are assigned different ω ratios, estimated from the data. In the case of the MHC, two independent ω ratios can be assigned and estimated for codons in the ARS region and those outside, and a likelihood ratio test can be used to test whether ω_{ARS} is significantly greater than 1. Such models are referred to as *fixed-sites* models by Yang and Swanson (2002). They are similar to the models of different substitution rates for different codon positions or different gene loci, discussed in Section 4.3.2. The likelihood calculation under the models is similar to that under the model of one ω ratio for all sites; the only difference is that the correct ω ratio is used to calculate the transition probabilities for data in each partition. Such models can also be used to test for different selective constraints, indicated by the ω ratio, among multiple genes. Muse and Gaut (1997) termed such tests *relative-ratio tests*.

The second strategy is to estimate one ω ratio for every site. There is then no need for *a priori* partitioning of sites. Fitch *et al.* (1997) and Suzuki and Gojobori (1999) reconstructed ancestral sequences on the phylogeny by parsimony, and used the reconstructed sequences to count synonymous and nonsynonymous changes at each site along branches of the tree. Fitch *et al.* (1997) analysed the human influenza virus type A haemagglutinin (HA) genes and considered a site to be under positive selection if the calculated ω ratio for the site is higher than the average across the whole sequence. Suzuki and Gojobori (1999) used a more stringent criterion and considered the site to be under positive selection only if the estimated ω ratio for the site is significantly greater than 1. The test is conducted by applying the method of Nei and Gojobori (1986) to analyse counts of sites and of differences at each site. Suzuki and Gojobori analyzed three large data sets: the HLA gene, the V3 region of the HIV-1 *env* gene, and the human influenza virus type A haemagglutinin (HA) gene, detecting sites under positive selection in each data set. Computer simulations (Suzuki and Gojobori 1999; Wong *et al.* 2004) demonstrate that a large number of sequences are needed for the test to have any power. The simple approach has intuitive appeal and is useful in exploratory analysis of large data sets.

The use of reconstructed ancestral sequences in the approaches of Fitch *et al.* (1997) and Suzuki and Gojobori (1999) may be a source of concern since these sequences are not real observed data. In particular, positively selected sites are often the most variable sites in the alignment, at which ancestral reconstruction is the least reliable. This problem can be avoided by taking a likelihood approach, averaging over all possible ancestral states. Indeed Suzuki (2004), Massingham and Goldman (2005), and Kosakovsky Pond and Frost (2005b) implemented methods to estimate one ω parameter for each site using ML. Typically, other parameters in the model such as branch lengths are estimated for all sites and fixed when the ω ratio is estimated for every site. Use of a model of codon substitution enables the method to incorporate the transition/transversion rate difference and unequal codon frequencies. At each site, the null hypothesis $\omega = 1$ is tested using either the χ_1^2 distribution or a null distribution generated by computer simulation. Massingham and Goldman (2005) called this the *site-wise likelihood ratio* (SLR) test. While the model allows the ω ratio to vary freely among sites, the number of parameters increases without bound with the increase of the sequence length. The model thus does not have the well-known asymptotic properties of MLEs (Stein 1956, 1962; Kalbfleisch 1985, pp. 92–95). The standard procedure for pulling ourselves out of this trap of infinitely many parameters is to assign a prior on ω, and to derive the conditional (posterior) distribution of ω given the data. This is the Bayesian (Lindley 1962) or empirical Bayes approach (Maritz and Lwin 1989; Carlin and Louis 2000), the third strategy to be discussed below. Despite those theoretical criticisms, computer simulations of Massingham and Goldman (2005) suggested that the SLR test achieved good false-positive rates as well as reasonably high power, even though the ω ratios for sites were not estimated reliably.

Note that all methods discussed in this category (Fitch *et al.* 1997; Suzuki and Gojobori 1999; Suzuki 2004; Massingham and Goldman 2005; Kosakovsky Pond and Frost 2005b) are designed to test for positive selection on a single site. To test whether the sequence contains any sites under positive selection (that is, whether the protein is under positive selection), a correction for multiple testing should be applied. For example, Wong *et al.* (2004) applied the modified Bonferroni procedure of Simes (1986) to the test of Suzuki and Gojobori (1999) to detect positive selection on the protein.

The third strategy, as indicated above, is to use a statistical distribution (prior) to describe the random variation of ω over sites (Nielsen and Yang 1998; Yang *et al.* 2000). The model assumes that different sites have different ω ratios but we do not know which sites have high ωs and which low ωs. The null hypothesis of no positive selection can be tested using a likelihood ratio test comparing two statistical distributions, one of which assumes no sites with $\omega > 1$ while the other assumes the presence of such sites. When the likelihood ratio test suggests the presence of sites with $\omega > 1$, an empirical Bayes approach is used to calculate the conditional (posterior) probability distribution of ω for each site given the data at the site. Such models are referred to as the *random-sites* models by Yang and Swanson (2002). They are discussed in the next two subsections.

8.4.2 Likelihood ratio test of positive selection under random-sites models

We now discuss likelihood calculation under the random-sites models. The ω ratio at any site is a random variable from a distribution $f(\omega)$. Thus inference concerning ω is based on the conditional (posterior) distribution $f(\omega|X)$ given data X. In simple statistical problems, the prior $f(\omega)$ can be estimated from the data without assuming any distributional form, leading to so-called nonparametric empirical Bayes (Robbins 1955, 1983; Maritz and Lwin 1989, pp. 71–78; Carlin and Louis 2000, pp. 57–88). However, the approach appears intractable in the present case. Instead Nielsen and Yang (1998) and Yang *et al.* (2000) implement parametric models $f(\omega)$ and estimate parameters involved in those densities. The synonymous rate is assumed to be homogeneous among sites, and only the nonsynonymous rates are variable. Branch length t is defined as the expected number of nucleotide substitutions per codon, averaged over sites.

The model has the same structure as the gamma and discrete-gamma models of variable rates among sites (Yang 1993, 1994a) (see Subsection 4.3.1). The likelihood is a function of parameters in the ω distribution (and not of the ωs themselves), as well as other parameters in the model such as the branch lengths and the transition/transversion rate ratio κ. The probability of observing data at a site, say data $\mathbf{x}_h$ at site h, is an average over the ω distribution:

$$f(\mathbf{x}_h) = \int_0^\infty f(\omega) f(\mathbf{x}_h | \omega) \, d\omega \cong \sum_{k=1}^K p_k f(\mathbf{x}_h | \omega_k). \tag{8.7}$$

If $f(\omega)$ is a discrete distribution, the integral becomes a summation. The integral for a continuous $f(\omega)$ is intractable analytically. Thus we apply a discrete approximation, with $K = 10$ equal-probability categories used to approximate the continuous density, so that p_k and ω_k in equation (8.7) are all calculated as functions of the parameters in the continuous density $f(\omega)$.

Positive selection is tested using a likelihood ratio test comparing a null model that does not allow $\omega > 1$ with an alternative model that does. Two pairs of models are found to be particularly effective in computer simulations (Anisimova et al. 2001, 2002; Wong et al. 2004). They are summarized in Table 8.2. The first pair involves the null model M1a (neutral), which assumes two site classes in proportions p_0 and $p_1 = 1 - p_0$ with $0 < \omega_0 < 1$ and $\omega_1 = 1$, and the alternative model M2a (selection), which adds a proportion p_2 of sites with $\omega_2 > 1$ estimated from the data. As M2a has two more parameters than M1a, the χ_2^2 distribution may be used for the test. However, the regularity conditions for the asymptotic χ^2 approximation are not met, and the correct null distribution is unknown. First, M1a is equivalent to M2a by fixing $p_2 = 0$, which is at the boundary of the parameter space. Second, when $p_2 = 0$, ω_2 is not identifiable. Thus the difference between M1a and M2a is not as large as two free parameters and the use of χ_2^2 is conservative.

Table 8.2 Parameters in models of variable ω ratios among sites

Model	p	Parameters
M0 (one ratio)	1	ω
M1a (neutral)	2	p_0 ($p_1 = 1 - p_0$), $\omega_0 < 1, \omega_1 = 1$
M2a (selection)	4	p_0, p_1 ($p_2 = 1 - p_0 - p_1$), $\omega_0 < 1, \omega_1 = 1, \omega_2 > 1$
M3 (discrete)	5	p_0, p_1 ($p_2 = 1 - p_0 - p_1$) $\omega_0, \omega_1, \omega_2$
M7 (beta)	2	p, q
M8 (beta&ω)	4	p_0 ($p_1 = 1 - p_0$), $p, q, \omega_s > 1$

p is the number of parameters in the ω distribution.

The second pair of models consists of the null model M7 (beta), which assumes a beta distribution for ω, and the alternative model M8 (beta&ω), which adds an extra class of sites under positive selection with $\omega_s > 1$. The beta distribution restricts ω to the interval (0, 1), but can take a variety of shapes depending on its two parameters p and q (Fig. 8.3). It is thus a flexible null model. M8 has two more parameters than M7, so that χ_2^2 may be used to conduct the LRT. As in the comparison between M1a and M2a, the use of χ_2^2 is expected to make the test conservative.

Another model, called M3 (discrete), may be mentioned here as well. This assumes a general discrete model, with the frequencies and the ω ratios (p_k and ω_k in equation 8.7) for K site classes estimated as free parameters. All models discussed here may be considered special cases of this general mixture model. As is typical in such mixture models, often only a few classes can be fitted to real data sets. Model

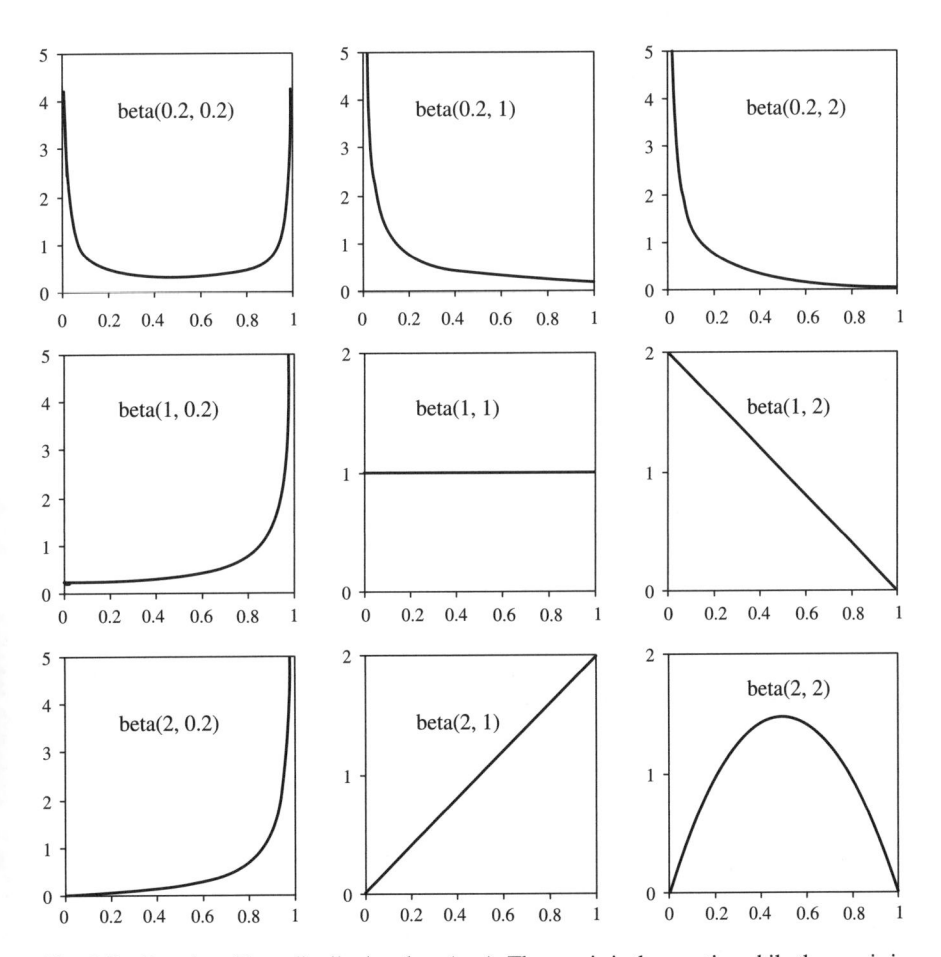

Fig. 8.3 Density of beta distribution: beta(p, q). The x-axis is the ω ratio, while the y-axis is proportional to the number of sites with that ω ratio.

M3 may be compared with model M0 (one-ratio) to construct an LRT. However, this is more a test of variability in selective pressure among sites and should not be considered a reliable test of positive selection.

8.4.3 Identification of sites under positive selection

When the LRTs suggest the presence of sites under positive selection, a natural question to ask is where these sites are. The empirical Bayes (EB) approach can be used to calculate the posterior probability that each site is from a particular site class, and sites with high posterior probabilities (say, with $P > 95\%$) coming from the class with $\omega > 1$ are most likely under positive selection. This approach makes it possible to detect positive selection and identify sites under positive selection even if the average ω ratio over all sites is much less than 1

$$f(\omega_k|\mathbf{x}_h) = \frac{p_k f(\mathbf{x}_h|\omega_k)}{f(\mathbf{x}_h)} = \frac{p_k f(\mathbf{x}_h|\omega_k)}{\sum_j p_j f(\mathbf{x}_h|\omega_j)}. \quad (8.8)$$

This is the same methodology as used in estimating substitution rates under the gamma or discrete-gamma models of rates for sites (see equation 4.13). Equation (8.8) requires the knowledge of model parameters, such as the proportions and ω ratios for the site classes as well as the branch lengths on the tree. Nielsen and Yang (1998) and Yang et al. (2000) replaced those parameters by their MLEs. This approach, known as naïve empirical Bayes (NEB), ignores sampling errors in parameters. It may be unreliable in small data sets, which do not contain enough information to estimate the parameters reliably. This deficiency has been explored in several computer simulation studies (e.g. Anisimova et al. 2002; Wong et al. 2004; Massingham and Goldman 2005; Scheffler and Seoighe 2005). A more reliable approach is implemented by Yang et al. (2005), known as the Bayes empirical Bayes (BEB; Deely and Lindley 1981). BEB accommodates uncertainties in the MLEs of parameters in the ω distribution by integrating numerically over a prior for the parameters. Other parameters such as branch lengths are fixed at their MLEs, as these are expected to have much less effect on inference concerning ω. An hierarchical (full) Bayesian approach is implemented by Huelsenbeck and Dyer (2004), using MCMC to average over tree topologies, branch lengths, as well as other substitution parameters in the model. This approach involves more intensive computation but may produce more reliable inference in small noninformative data sets, where the MLEs of branch lengths may involve large sampling errors (Scheffler and Seoighe 2005).

8.4.4 Positive selection in the human major histocompatability (MHC) locus

Here we use the random-sites models to analyse a data set of class I major histocompatability locus (MHC or HLA) alleles from humans, compiled by Yang and Swanson (2002). The data set consists of 192 alleles from the A, B, and C loci, with 270 codons in each sequence, after the removal of alignment gaps. The apparent selective force acting upon the class I MHC is to recognize and bind a large number of foreign peptides. The ARS is the cleft that binds foreign antigens (Bjorkman et al. 1987a, b). The identification of the ARS enabled previous researchers to partition the data into ARS and non-ARS sites and to demonstrate positive selection in the ARS (Hughes and Nei 1988; Hughes et al. 1990). Without partitioning the data, positive selection was not detected in pairwise comparisons averaging rates over the entire sequence.

The random-sites models do not make use of the structural information. The tree topology is estimated using the neighbour-joining method (Saitou and Nei 1987) based on MLEs of pairwise distances (Goldman and Yang 1994). To save computation, the branch lengths are estimated under model M0 (one-ratio) and then fixed when other models are fitted to the data. Equilibrium codon frequencies are calculated using the base frequencies at the three codon positions (the F3 × 4 model), with the frequencies observed in the data used as estimates. The MLEs of parameters and the log likelihood values are given in Table 8.3 for several random-sites models. For example, the MLEs under M2a suggest that about 8.4% of sites are under positive selection with $\omega = 5.4$. The likelihood ratio test statistic comparing models M1a and M2a is $2\Delta\ell = 518.68$, much greater than critical values from the χ_2^2. The test using models M7 and M8 leads to the same conclusion. Parameter estimates of Table 8.3 can be used to calculate posterior probabilities that each site is from the different site classes, using the NEB procedure (equation 8.8). The results obtained under M2a are presented in Fig. 8.4, while amino acid residues with posterior probability $P > 95\%$ of coming from the site

Table 8.3 Log likelihood values and parameter estimates under models of variable ω ratios among sites for 192 MHC alleles

Model	p	ℓ	Estimates of parameters	Positively selected sites
M0 (one-ratio)	1	−8225.15	0.612	None
M1a (neutral)	2	−7490.99	$\hat{p}_0 = 0.830, (\hat{p}_1 = 0.170),$ $\hat{\omega}_0 = 0.041, (\omega_1 = 1)$	Not allowed
M2a (selection)	4	−7231.15	$\hat{p}_0 = 0.776,$ $\hat{p}_1 = 0.140 (\hat{p}_2 = 0.084)$ $\hat{\omega}_0 = 0.058 (\omega_1 = 1),$ $\hat{\omega}_2 = 5.389$	9F 24A 45M 62G 63E 67V 70H 71S 77D 80T 81L 82R 94T **95V 97R** 99Y 113Y **114H 116Y** 151H 152V 156L 163T 167W
M7 (beta)	2	−7498.97	$\hat{p} = 0.103, \hat{q} = 0.354$	Not allowed
M8 (beta&ω)	4	−7238.01	$\hat{p}_0 = 0.915 (\hat{p}_1 = 0.085),$ $\hat{p} = 0.167, \hat{q} = 0.717,$ $\hat{\omega}_s = 5.079$	9F 24A 45M 63E 67V 69A 70H 71S 77D 80T 81L 82R 94T **95V 97R** 99Y 113Y **114H 116Y** 151H 152V 156L 163T 167W

p is the number of parameters in the ω distribution. Branch lengths are fixed at their MLEs under M0 (one-ratio). Estimates of the transition/transversion rate ratio κ range from 1.5 to 1.8 among models. Positively selected sites are inferred at the cutoff posterior probability $P \geq 95\%$ with those reaching 99% shown in bold. Amino acid residues are from the reference sequence in the PDB structure file 1AKJ. (adapted from Yang and Swanson 2002; Yang et al. 2005)

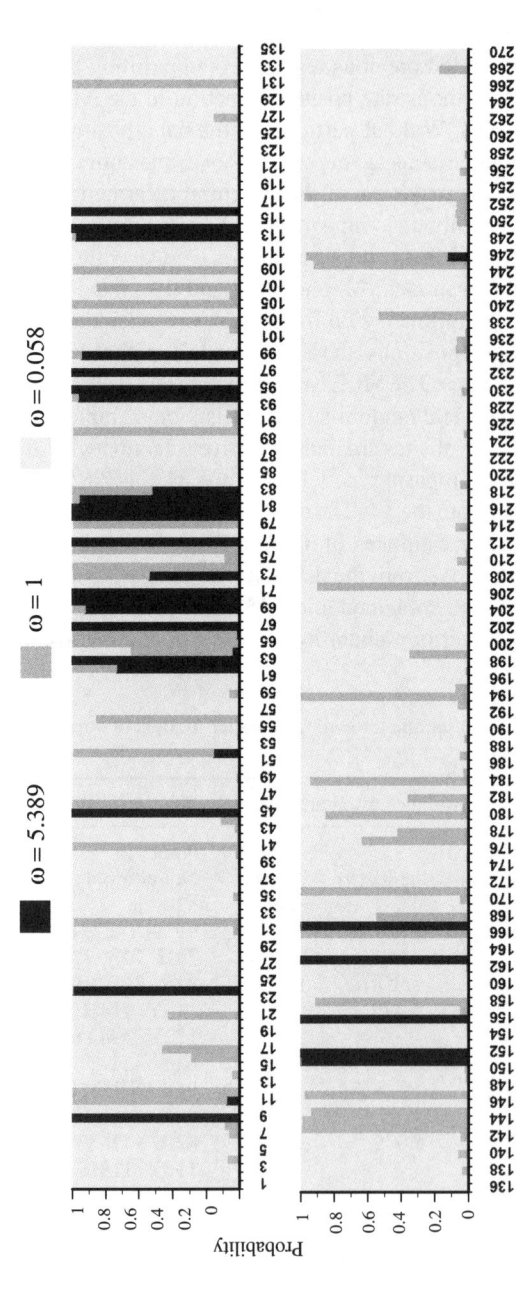

Fig. 8.4 Posterior probabilities that each site is from the three site classes under the M2a (selection) model, calculated using the naïve empirical Bayes (NEB) procedure. The MLEs of parameters under the model suggest three site classes with $\omega_0 = 0.058$, $\omega_1 = 1$, $\omega_2 = 5.389$ in proportions $p_0 = 0.776$, $p_1 = 0.140$, $p_2 = 0.084$ (Table 8.3). These proportions are the prior probabilities that any site belongs to the three classes. The data at a site (codon configurations in different sequences) alter the prior probabilities dramatically, so that the posterior probabilities can be very different from the prior probabilities. For example, at site 4, the posterior probabilities are 0.925, 0.075, and 0.000, and thus the site is likely to be under strong purifying selection. At site 9, the posterior probabilities are 0.000, 0.000, and 1.000, and thus the site is almost certainly under positive selection. Sites with posterior probability $P > 0.95$ for site class 2 are listed in Table 8.3 and also shown on the protein structure in Fig. 8.5. The BEB procedure produced virtually identical posterior probabilities.

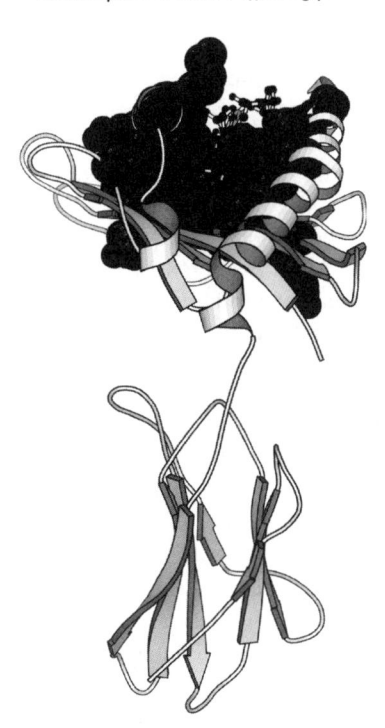

Fig. 8.5 The structure of the class I MHC allele H-2Db (Protein Data Bank file 1AKJ, chain A), with a bound antigen shown in stick and ball format. Amino acid residues identified to be under positive selection under the random-sites model M2a are shown in spacefill, and all fall in the ARS domain. See also Table 8.3 and Fig. 8.4. From Yang and Swanson (2002).

class of positive selection are shown on the three-dimensional structure in Fig. 8.5. Most of these sites are on the list of 57 amino acid residues making up the ARS (Bjorkman *et al.* 1987a, b). Three of them are not on the list but are nevertheless in the same region of the structure. It may be noted that the inferred sites are scattered along the primary sequence (Fig. 8.4) but are clustered on the structure (Fig. 8.5). This data set is large, so that the parameters are reliably estimated. The NEB and BEB procedures produced almost identical posterior probabilities and lists of positively selected sites (see Yang *et al.* 2005). Furthermore, models M2a and M8 produced highly similar results (Table 8.3).

8.5 Adaptive evolution affecting particular sites and lineages

8.5.1 Branch-site test of positive selection

For many genes, both the branch- and site-based tests of positive selection are expected to be conservative. In the branch test, positive selection is detected along the branch only if the ω ratio averaged over all sites is significantly greater than 1. If most sites in a protein are under purifying selection with ω close to 0, the ω ratio averaged over all sites may not exceed 1 even if a few sites are evolving fast, driven by positive selective pressure. Similarly, the site test detects positive selection only if the ω ratio averaged over all branches on the tree is greater than 1. This assumption may be reasonable for genes involved in a genetic arms race such as host–pathogen antagonism, in which the genes may be under constant pressure of diversifying selection. This might explain the fact that the site test has been more successful than the branch test in identifying genes under positive selection (see Section 8.7 for examples). For most other genes, one might expect positive selection to affect only a few amino acid residues along particular lineages. The *branch-site* models (Yang and Nielsen 2002) attempt to detect signals of such local episodic natural selection.

In the models of Yang and Nielsen (2002), branches in the tree are divided *a priori* into foreground and background categories, and a likelihood ratio test is constructed to compare an alternative model that allows for some sites under positive selection on the foreground branches with a null model that does not. However, in a simulation study, Zhang (2004) found that the tests were sensitive to the underlying models, and, when the model assumptions were violated, could lead to high false positives. A slight modification was introduced and found to lead to much better performance (Yang *et al.* 2005; Zhang *et al.* 2005). Here we describe the modified test.

Table 8.4 summarizes the model, called the branch-site model A. Along the background lineages, there are two classes of sites: conserved sites with $0 < \omega_0 < 1$ and neutral sites with $\omega_1 = 1$. Along the foreground lineages, a proportion $(1 - p_0 - p_1)$ of sites become under positive selection with $\omega_2 \geq 1$. Likelihood calculation under this model can be easily adapted from that for the site models. As we do not know *a priori* which site class each site is from, the probability of data at a site is an average over the four site classes. Let $I_h = 0, 1, 2a, 2b$ be the site class that site h is from. We have

$$f(\mathbf{x}_h) = \sum_{I_h} p_k f(\mathbf{x}_h | I_h). \tag{8.9}$$

The conditional probability $f(\mathbf{x}_h|I_h)$ of observing data $\mathbf{x}_h$ at site h, given that the site comes from site class I_h, is easy to calculate, because the site evolves under the one-ratio model if $I_h = 0$ or 1, and under the branch model if $I_h = 2a$ or 2b.

Table 8.4 The ω ratios assumed in branch-site model A

Site class	Proportion	Background ω	Foreground ω
0	p_0	$0 < \omega_0 < 1$	$0 < \omega_0 < 1$
1	p_1	$\omega_1 = 1$	$\omega_1 = 1$
2a	$(1 - p_0 - p_1)p_0/(p_0 + p_1)$	$0 < \omega_0 < 1$	$\omega_2 > 1$
2b	$(1 - p_0 - p_1)p_1/(p_0 + p_1)$	$\omega_1 = 1$	$\omega_2 > 1$

The model involves four parameters: $p_0, p_1, \omega_0, \omega_2$.

In the branch-site test of positive selection, model A is the alternative hypothesis, while the null hypothesis is the same model A but with $\omega_2 = 1$ fixed (Table 8.4). The null model has one fewer parameter but since $\omega_2 = 1$ is fixed at the boundary of the parameter space of the alternative model, the null distribution should be a 50:50 mixture of point mass 0 and χ_1^2 (Self and Liang 1987). The critical values are 2.71 and 5.41 at the 5% and 1% levels, respectively. One may also use χ_1^2 (with critical values 3.84 at 5% and 5.99 at 1%) to guide against violations of model assumptions. In computer simulations, the branch-site test was found to have acceptable false-positive rates, and also had more power than the branch test (Zhang et al. 2005). However, the model allows only two kinds of branches, the foreground and background, which may be unrealistic for many data sets.

The NEB and BEB procedures were implemented for the branch-site model as well, allowing identification of amino acid sites that are under positive selection along the foreground lineages, as in the site-based analysis. However, the posterior probabilities often do not reach high values such as 95%, indicating a lack of power in such analysis.

Similar to the branch test, the branch-site test requires the foreground branches to be specified *a priori*. This may be easy if a well-formulated biological hypothesis exists, for example if we want to test adaptive evolution driving functional divergences after gene duplication. The test may be difficult to apply if no *a priori* hypothesis is available. To apply the test to several or all branches on the tree, one has to correct for multiple testing (Anisimova and Yang 2006).

8.5.2 Other similar models

Several other models of codon substitution have been implemented that allow the selective pressure indicated by the ω ratio to vary both among lineages and among sites. Forsberg and Christiansen (2003) and Bielawski and Yang (2004) implemented the *clade models*. Branches on the phylogeny are *a priori* divided into two clades, and a likelihood ratio test is used to test for divergences in selective pressure between the two clades indicated by different ω ratios. Clade model C, implemented by Bielawski and Yang (2004), is summarized in Table 8.5. This assumes three site classes. Class 0 includes conserved sites with $0 < \omega_0 < 1$, while class 1 includes neutral sites with $\omega_1 = 1$; both apply to all lineages. Class 2 includes sites that are under different selective pressures in the two clades, with ω_2 for clade 1 and ω_3 for clade 2. There may not be any sites under positive selection with $\omega > 1$. The model involves five parameters in the ω distribution: $p_0, p_1, \omega_0, \omega_2$, and ω_3. A likelihood ratio test can be constructed by comparing model C with the site model M1a (neutral), which assumes two site classes with two free parameters: p_0 and ω_0. The χ_3^2 distribution may be used for the test. The clade models are similar to the models of functional divergence of Gu (2001) and Knudsen and Miyamoto (2001), in which the amino acid substitution rate is used as an indicator of functional constraint.

A switching model of codon substitution is implemented by Guindon et al. (2004), which allows the ω ratio at any site to switch among three different values

Table 8.5 The ω ratios assumed in clade model C

Site class	Proportion	Clade 1	Clade 2
0	p_0	$0 < \omega_0 < 1$	$0 < \omega_0 < 1$
1	p_1	$\omega_1 = 1$	$\omega_1 = 1$
2	$p_2 = 1 - p_0 - p_1$	ω_2	ω_3

The model involves five parameters: $p_0, p_1, \omega_0, \omega_2, \omega_3$.

$\omega_1 < \omega_2 < \omega_3$. Besides the Markov-process model of codon substitution, a hidden Markov chain runs over time and describes the switches of any site between different selective regimes (i.e. the three ω values). The model is similar in structure to the covarion model for variable substitution rates, in which every site switches between high and low rates (Tuffley and Steel 1998; Galtier 2001; Huelsenbeck 2002). The switching model is an extension of the site models (Nielsen and Yang 1998; Yang et al. 2000), which had a fixed ω for every site. A likelihood ratio test can thus be used to compare them. Guindon et al. (2004) found that the switching model fitted a data set of the HIV-1 *env* genes much better than the site models. An empirical Bayes procedure can be used to identify lineages and sites with high ω ratios.

8.5.3 Adaptive evolution in angiosperm phytochromes

Here we apply the branch and branch-site tests to detect positive selection in the evolution of the phytochrome gene *phy* subfamily. Phytochromes are the best characterized plant photosensors. They are chromoproteins that regulate the expression of a large number of light-responsive genes and many events in plant development, including seed germination, flowering, fruit ripening, stem elongation, and chloroplast development (Alba et al. 2000). All angiosperms characterized to date contain a small number of PHY apoproteins encoded by a small *phy* gene gamily. The data analysed here are from Alba et al. (2000). We use the alignment of the 16 sequences from the A and C/F subfamilies. The phylogenetic tree of Alba et al. (2000) is shown in Fig. 8.6. We test whether the gene was under positive selection along the branch separating the A and the C/F subfamilies, which represents a gene duplication.

The one-ratio model (M0) gives an estimate $\hat{\omega} = 0.089$ ((a) in Table 8.6). This is an average over all branches and sites, and the small value reflects the dominating role of purifying selection in the evolution of the *phy* gene family. In the branch model ((b) in Table 8.6), two ω ratios are assigned for the background (ω_0) and foreground (ω_1) branches. Although $\hat{\omega}_1 > \hat{\omega}_0$, $\hat{\omega}_1$ is not greater than one. Furthermore, the null hypothesis $\omega_0 = \omega_1$ is not rejected by the data; the likelihood ratio test statistic for comparing the one-ratio and two-ratio models is $2\Delta\ell = 2 \times 0.64 = 1.28$, with $p = 0.26$ at d.f. $= 1$. Thus the branch test provides no evidence for positive selection.

We then conducted the branch-site test of positive selection. The test statistic is calculated to be $2\Delta\ell = 2 \times 9.94 = 19.88$, with $p = 8.2 \times 10^{-6}$ if we use χ_1^2 or $p = 8.2 \times 10^{-6}/2 = 4.1 \times 10^{-6}$ if we use the 50:50 mixture of 0 and χ_1^2.

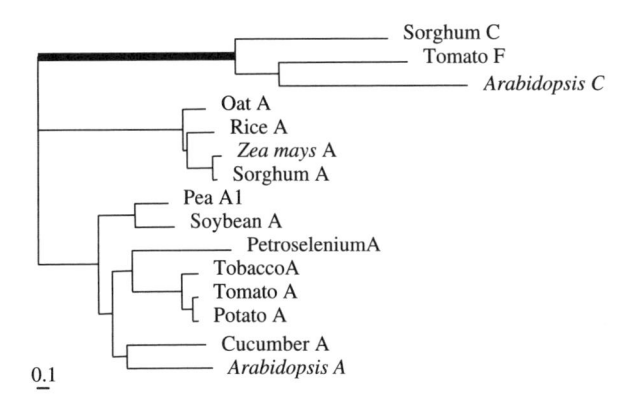

Fig. 8.6 Phylogeny for the phytochrome *phy* gene family in angiosperms. The branch separating the A and the C/F subfamilies is postulated to be under positive selection after gene duplication. This is the foreground branch in the branch and branch-site tests, while all other branches are the background branches (Table 8.6).

Table 8.6 Log likelihood values and parameter estimates under various models for the phytochrome gene *phy* AC&F subfamilies

Model	p	ℓ	Estimates of parameters
(a) One-ratio (M0)	1	−29 984.12	$\hat{\omega} = 0.089$
(b) Branch model (2-ratios)	2	−29 983.48	$\hat{\omega}_0 = 0.090$, $\hat{\omega}_1 = 0.016$
(c) Branch-site model A, with $\omega_2 = 1$ fixed	3	−29 704.74	$\hat{p}_0 = 0.774$, $\hat{p}_1 = 0.073$ ($\hat{p}_2 + \hat{p}_3 = 0.153$), $\hat{\omega}_0 = 0.078$, $\omega_1 = \omega_2 = 1$
(d) Branch-site model A, with $\omega_2 > 1$	4	−29 694.80	$\hat{p}_0 = 0.813$, $\hat{p}_1 = 0.075$ ($\hat{p}_2 + \hat{p}_3 = 0.111$), $\hat{\omega}_0 = 0.080$, $\hat{\omega}_2 = 131.1$

p is the number of free parameters in the ω distribution. Estimates of κ are around 2.1. Estimates of branch lengths are not shown (see Fig. 8.6). Codon frequencies are calculated using the base frequencies at the three codon positions under the F3 × 4 model.

Thus the test provides strong evidence for positive selection. The parameter estimates under branch-site model A suggests about 11% of sites to be under positive selection along the foreground lineage with very high ω_2. Application of the BEB procedure under the model identified 27 amino acid sites potentially under positive selection along the foreground lineage at the cutoff posterior probability 95%. These are 55R, 102T, 105S, 117P, 130T, 147S, 171T, 216E, 227F, 252I, 304I, 305D, 321L, 440L, 517A, 552T, 560Y, 650T, 655S, 700A, 736K, 787N, 802V, 940T, 986M, 988Q, and 1087D (the amino acids refer to the *Zea mays* sequence). These are a subset of sites predicted to have changed along the branch using ancestral sequence reconstruction.

8.6 Assumptions, limitations, and comparisons

In this section I discuss some of the limitations of the codon-substitution methods used for detecting positive selection. I also attempt to compare them with the tests of neutrality developed in population genetics, reviewed in Section 8.2.

8.6.1 Limitations of current methods

Codon-substitution models developed for detecting signals of natural selection in protein-coding genes involve a number of assumptions, the effects of which are not all well understood. First, like the many models of nucleotide substitution (see Chapter 2), the codon models describe substitutions and do not model mutation and selection explicitly as in a population genetics model. This formulation is both a drawback and an advantage. The drawback is that codon models do not appear to be useful for detecting natural selection at silent sites (Akashi 1995). If the silent substitution rate is high, the model cannot tell whether it is due to a high mutation rate or a relaxed selective constraint, or even positive selection. The advantage is that the model, when used to detect selection acting on the protein, is not so sensitive to whether mutational bias or natural selection drives evolution at silent sites. In this regard, a number of authors have suggested that the use of the ω ratio to detect selection in protein-coding genes requires the assumption of neutral evolution at silent sites (e.g. Kreitman and Akashi 1995; Fay *et al.* 2001; Bierne and Eyre-Walker 2003). I believe this assumption is unnecessary. The ω ratio measures, to quote Benner (2001), 'the difference between how proteins have divergently evolved in their past, and how they would have evolved had they been formless, functionless molecules'. Whether mutation or selection drives evolution at the silent sites should not affect this interpretation of the ω ratio or its use to detect positive selection at the protein level.

Second, even when considered as models of *substitution* and not of mutation and selection, the codon models may be highly unrealistic. For example, the same ω ratio is assumed for substitutions between different amino acids, whereas it is known that the relative substitution rates between amino acids are strongly influenced by their chemical properties (e.g. Zuckerkandl and Pauling 1962; Dayhoff *et al.* 1978). Incorporating chemical properties into codon models did significantly improve the model's fit to data (Goldman and Yang 1994; Yang *et al.* 1998), but the improvement was not extraordinary. This appears partly due to our poor understanding of how chemical properties affect amino acid substitution rates. It is also unclear how to define positive selection in a model incorporating chemical properties. Some authors distinguish between radical and conservative amino acid replacements (i.e., between amino acids with very different and very similar chemical properties, respectively), and suggest that a higher radical than conservative rate is evidence for positive selection (Hughes *et al.* 1990; Rand *et al.* 2000; Zhang 2000). However, this criterion appears less convincing than the ω ratio and is also much more sensitive to model assumptions, such as unequal transition and transversion rates and unequal amino acid compositions (Dagan *et al.* 2002).

Third, tests based on the ω ratio are most likely to be conservative. The branch models average the ω ratio over all sites while the site models average it over all lineages, so that both may be expected to be conservative. The models appear effective in detecting recurrent diversifying selection driving fixation of nonsynonymous mutations, but may lack power in detecting one-off directional selection that drives a new advantageous mutation quickly to fixation. In this regard, it is remarkable that the tests have successfully detected adaptive evolution in a number of genes and organisms where positive selection was not suspected before. See the next section for a sample of such cases. By allowing the selective pressure to vary both among sites and among lineages, the branch-site and switching models (Yang and Nielsen 2002; Guindon et al. 2004; Yang et al. 2005; Zhang et al. 2005) appear to have more power. However the utility of these methods is yet to be tested extensively in real data analysis. Intuitively it is sensible to focus on a short time period and a few amino acid sites so that the signal of adaptive evolution will not be overwhelmed by the almost ubiquitous purifying selection. However, a short time period and a few amino acid sites may not offer enough opportunities for evolutionary changes to generate a signal detectable by statistical tests.

Fourth, another assumption that may be unrealistic is the constant rate of silent substitutions among sites. This assumption may be violated in some genes. A concern is that by ignoring possible silent rate variation, a site model may misidentify a site as being under positive selection not because the replacement rate is elevated but because the silent rate is reduced. Models relaxing this assumption have been implemented by Kosakovsky Pond and Muse (2005), who suggested that it may be worthwhile to account for silent rate variation among sites. However, this result is based on comparison with the site model M3 (discrete), which was found not to perform well in simulations. Site models M2 (selection) and M8 (beta&ω) are more robust; in particular, the use of a site class with $\omega_1 = 1$ in M2 is effective in reducing false positives in detecting positively selected sites.

Finally, current codon models assume that the same phylogenetic tree applies to all sites in the sequence. If intragenic recombination is frequent, as in some viral data sets, the likelihood ratio tests may be misled into falsely claiming positive selection (Anisimova et al. 2003; Shriner et al. 2003). Unfortunately, most recombination detection methods (e.g. Hudson 2001; McVean et al. 2002) assume strict neutrality and may mistake selection as evidence of recombination. Methods that can deal with both selection and recombination have yet to be developed (Wilson and McVean 2006).

For practical data analysis, and in particular in large-scale genome-wide analysis, the most common problem in applying codon-based models of positive selection appears to be the wrong levels of sequence divergence. Codon-based analysis contrasts the rates of synonymous and nonsynonymous substitutions and relies on information on both types of changes. If the compared species are too close or the sequences are too similar, the sequence data will contain too little variation and thus too little information. The methods appear to have no power in population data and are expected to be useful for species data and fast-evolving viral sequences only (Anisimova et al.

2001, 2002). On the other hand, if the sequences are too divergent, silent changes may have reached saturation and the data will contain too much noise. In computer simulations, the methods are quite tolerant of high divergences. However, in real data high divergences are often associated with a host of other problems, such as difficulties in alignment and different codon usage patterns in different sequences.

8.6.2 Comparison between tests of neutrality and tests based on d_N and d_S

It is interesting to compare phylogenetic tests of positive selection based on the ω ratio and population genetics tests of neutrality, to understand what kind of selection each class of methods are likely to detect and how powerful the tests are (Nielsen 2001b; Fay and Wu 2003). First, the phylogenetic tests rely on an excess of amino acid replacements to detect positive selection and thus require the sequences to be quite divergent. They have virtually no power when applied to population data (Anisimova *et al.* 2002). In contrast, tests of neutrality are designed for DNA samples taken in a population or from closely related species. Most tests also assume the infinitely many sites model, which may break down when sequences from different species are compared. Second, a test based on the ω ratio may provide more convincing evidence for positive selection than a test of neutrality. The strict neutral model underlying a neutrality test is typically a composite model, with a number of assumptions such as neutral evolution, constant population size, no population subdivision, no linkage to selected loci, and so on. Rejection of such a composite model may be caused by violations of any of the assumptions, besides selection at the locus. It is often difficult to distinguish between the multiple factors that can cause violoation of the strict neutral model. For example, patterns of sequence variation at a neutral locus undergoing genetic hitchhiking are similar to those in an expanding population (Simonsen *et al.* 1995; Fay and Wu 2000). Similarly, patterns of sequence variation at a locus linked to a locus under balancing selection can be similar to sequence variations sampled from subdivided populations (Hudson 1990).

8.7 Adaptively evolving genes

Table 8.7 provides a sample of genes inferred to be undergoing adaptive evolution based on comparisons of synonymous and nonsynonymous rates. The list, by no means exhaustive, should give a flavour of the kind of genes potentially under positive selection. Hughes (1999) discussed several case studies in great detail. Yokoyama (2002) reviewed studies on visual pigment evolution in vertebrates. Vallender and Lahn (2004) provide a comprehensive summary of genes under positive selection, although with an anthropocentric bias. See also the Adaptive Evolution Database, compiled by Roth *et al.* (2005).

Most genes detected to be under positive selection based on the ω ratio fall into the following three categories. The first includes host genes involved in defence or immunity against viral, bacterial, fungal, or parasitic attacks, as well as viral or pathogen genes involved in evading host defence. The former includes, for example, the major

Table 8.7 Selected examples of proteins inferred to be under positive selection through synonymous and nonsynonymous rate comparison

Proteins	Organisms	References
Proteins involved in defensive systems or immunity		
Antiviral enzyme APOBEC3G	Primates	Sawyer et al. (2004), Zhang and Webb (2004)
Class I chitinase	Arabidopsis	Bishop et al. (2000)
Defensin-like peptides	Termites	Bulmer and Crozier (2004)
Glycophorin A	Mammals	Baum et al. (2002)
Immunoglobulin V_H	Mammals	Tanaka and Nei (1989), Su et al. (2002)
MHC	Mammals, fish	Hughes and Nei (1988), Yang and Swanson (2002), Consuegra et al. (2005)
α_1-Proteinase inhibitors	Rodents	Goodwin et al. (1996)
Plant disease-resistant proteins	Arabidopsis	Mondragon-Palomino et al. (2002)
Plant endo-β-1,3-glucanases (EGases)	Soybean	Bishop et al. (2005)
Polygalacturonase inhibitors	Legume and dicots	Stotz et al. (2000)
RH blood group and RH50	Primates and rodents	Kitano et al. (1998)
Retroviral inhibitor TRIM5α	Rhesus monkey	Sawyer et al. (2005)
Transferrin	Salmonid fishes	Ford et al. (1999)
Proteins involved in evading defensive systems or immunity		
Antigen HDAg	Hepatitis D virus	Anisimova and Yang (2004)
Capsid protein	Foot and mouth disease virus	Haydon et al. (2001)
Capsid VP2	Canine parvovirus	Shackelton et al. (2005)
env, gag, pol, vif, vpr	HIV-1	Bonhoeffer et al. (1995), Mindell (1996), Zanotto et al. (1999), Yamaguchi-Kabata and Gojobori (2000), Yang et al. (2003), Choisy et al. (2004)
Envelope glycoprotein	Dengue virus	Twiddy et al. (2002)
Haemagglutinin	Human influenza A virus	Fitch et al. (1997), Bush et al. (1999), Suzuki and Gojobori (1999), Pechirra et al. (2005)
Lymphocyte protein CD45	Primates	Filip and Mundy (2004)
Outer membrane protein wsp	Wolbachia bacteria	Jiggins et al. (2002)
Proline-rich antigen (PRA)	Human pathogen Coccidioides	Johannesson et al. (2004)

Table 8.7 (*Contd.*)

Proteins	Organisms	References
Polygalacturonase	Fungal pathogens	Stotz *et al.* (2000)
Trichothecene mycotoxin	Fungi	Ward *et al.* (2002)
Toxins		
Conotoxin	*Conus* gastropods	Duda and Palumbi (1999)
Sodium channel toxin α and β	Scorpians	Zhu *et al.* (2004)
Proteins involved in reproduction		
Acp26Aa	*Drosophila*	Tsaur and Wu (1997)
Androgen-binding protein Abp	Rodents	Karn and Nachman (1999), Emes *et al.* (2004)
Bindin	*Echinometra*	Metz and Palumbi (1996)
Dntf-2r, male-specific expression	Drosophila	Betran and Long (2003)
Floral regulatory protein	Hawaiian silversword	Barrier *et al.* (2001)
IEMT, floral scent enzyme	Plant Onagraceae	Barkman (2003)
Ods homeobox	*Drosophila*	Ting *et al.* (1998)
Pem homeobox	Rodents	Sutton and Wilkinson (1997)
Pheromone-binding proteins	Moths *Choristoneura*	Willett (2000)
Courtship pheromone (PRF)	Plethodontid salamanders	Palmer *et al.* (2005)
Protamine P1	Primates	Rooney and Zhang (1999)
Sperm lysin	Abalone *Haliotis*	Lee *et al.* (1995), Vacquier *et al.* (1997)
Sry, sex determination protein	Primates	Pamilo and O'neill (1997)
Testes-specific α4 proteasome subunit	Drosophila	Torgerson and Singh (2004)
V1RL, putative pheromone receptor	Primates	Mundy and Cook (2003)
V1r, pheromone receptor	Teleost fishes	Pfister and Rodriguez (2005)
Y-specific genes USP9Y and UTY	Primates	Gerrard and Filatov (2005)
Proteins involved in digestion		
κ-casein	Bovids	Ward *et al.* (1997)
Lysozyme	Primates	Messier and Stewart (1997), Yang (1998a)
Pancreatic ribonuclease	Leaf-eating monkeys	Zhang *et al.* 1998; Zhang *et al.* (2002)
Alanine : glyoxylate aminotransferase	Primates and carnivores	Holbrook *et al.* (2000), Birdsey *et al.* (2004)

Duplicated proteins

DAZ	Primates	Bielawski and Yang (2001)
Morpheus gene family	Hominoids	Johnson et al. (2001)
Chorionic gonadotropin	Primates	Maston and Ruvolo (2002)
Neural transmembrane protein Pcdh	Primates	Wu (2005)
Xanthine dehydrogenase	Vertebrates	Rodriguez-Trelles et al. (2003)
CCT chaperonin subunits	Eukaryotes	Fares and Wolfe (2003)

Miscellaneous

ASPM, for brain size	Humans	Zhang (2003)
Breast cancer gene BRCA1	Humans and chimps	Huttley et al. (2000)
FOXP2, for speech and language	Humans	Enard et al. (2002)
Heat-shock protein GroEL	Bacteria	Fares et al. (2002)
Haemoglobin β-chain	Antarctic teleost fish	Bargelloni et al. (1998)
Hemoglobin	Tubeworms	Bailly et al. (2003)
Interleukin genes	Mammals	Shields et al. (1996), Zhang and Nei (2000)
Iron-binding protein transferrin	Salmonids	Ford (2001)
Insulin	Rodents	Opazo et al. (2005)
Microcephalin for brain size	Primates	Evans et al. (2004), Wang and Su (2004)
Myc-like anthocyanin regulatory protein	Dogwoods	Fan et al. (2004)
Opsin	Cichlid fish	Terai et al. (2002), Spady et al. (2005)
Proteorhodopsin	Marine bacteria	Bielawski et al. (2004)
Sodium channel gene (Nav1.4a)	Electric fishes	Zakon et al. (2006)
srz family	Worms *Caenorhabditis*	Thomas (2005)
Testis-expressed homeobox	Primates	Wang and Zhang (2004)

histocompatibility complex (Hughes and Nei 1988, see also Subsection 8.4.4), lymphocyte protein CD45 (Filip and Mundy 2004), plant R-genes involved in pathogen recognition (Lehmann 2002), and the retroviral inhibitor *TRIM5α* in primates (Sawyer *et al.* 2005). The latter includes viral surface or capsid proteins (Haydon *et al.* 2001; Shackelton *et al.* 2005), *Plasmodium* membrane antigens (Polley and Conway 2001), and polygalacturonases produced by plant enemies, such as bacteria, fungi, oomycetes, nematodes, and insects (Götesson *et al.* 2002). One may expect the pathogen gene to be under selective pressure to evolve into new forms unrecognizable by the host defence system while the host has to adapt and recognize the pathogen. Thus an evolutionary arms race ensues, driving new replacement mutations to fixation in both the host and the pathogen (Dawkins and Krebs 1979). Toxins in snake or scorpion venoms are used to subdue prey and often evolve at fast rates under similar selective pressures (Duda and Palumbi 2000; Zhu *et al.* 2004).

The second main category includes proteins or pheromones involved in sexual reproduction. In most species, the male produces numerous sperm using apparently very little resource. The egg is in contrast a big investment. The conflict of interests between the two sexes espouses a genetic battle: it is best for the sperm to recognize and fertilize the egg as soon as possible, while it is best for the egg to delay sperm recognition to avoid fertilization by multiple sperm (polyspermy), which causes loss of the egg. A number of studies have detected rapid evolution of proteins involved in sperm–egg recognition (Palumbi 1994) or in other aspects of male or female reproduction (Tsaur and Wu 1997; Hellberg and Vacquier 2000; Wyckoff *et al.* 2000; Swanson *et al.* 2001a, b). It is also possible that natural selection on some of these genes may have accelerated or contributed to the origination of new species. See the excellent reviews by Swanson and Vacquier (2002a, b) and Clark *et al.* (2006).

The third category overlaps with the previous two and includes proteins that acquired new functions after gene duplications. Gene duplication is one of the primary driving forces in the evolution of genes, genomes, and genetic systems, and is believed to play a leading role in evolution of novel gene functions (Ohno 1970; Kimura 1983). The fate of duplicated genes depends on whether they bring a selective advantage to the organism. Most duplicated genes are deleted or degrade into pseudogenes, disabled by deleterious mutations. Sometimes the new copies acquire novel functions under the driving force of adaptive evolution, due to different functional requirements from those of the parental genes (Walsh 1995; Lynch and Conery 2000; Zhang *et al.* 2001; Prince and Pickett 2002). Many genes were detected to experience accelerated protein evolution following gene duplication, including the DAZ gene family in primates (Bielawski and Yang 2001), chorionic gonadotropin in primates (Maston and Ruvolo 2002), the pancreatic ribonuclease genes in leaf-eating monkeys (Zhang *et al.* 1998, 2002), and xanthine dehydrogenase genes (Rodriguez-Trelles *et al.* 2003). Population genetics tests also suggest the prominent role of positive selection in the evolutionary dynamics of the very early histories of duplicate nuclear genes (Moore and Purugganan 2003, 2005).

Many other proteins have been detected to be under positive selection as well, although they are not as numerous as those involved in evolutionary arms race such as the host–pathogen antagonism or reproduction (Table 8.7). This pattern appears partly to be due to the limitations of the detection methods based on the ω ratio, which may miss one-off adaptive evolution in which an advantageous mutation arose and spread in the population quickly, followed by purifying selection. It may be expected that improved methods for detecting episodic and local adaptation that affects a few sites along particular lineages (Guindon et al. 2004; Yang et al. 2005; Zhang et al. 2005) might lead to detection of more cases of positive selection.

Needless to say, a statistical test cannot prove that the gene is undergoing adaptive evolution. A convincing case may be built through experimental verification and functional assays, establishing a direct link between the observed amino acid changes with changes in protein folding, and with phenotypic differences such as different efficiencies in catalysing chemical reactions. In this regard, the statistical methods discussed in this chapter are useful for generating biological hypotheses for verification in the laboratory, and can significantly narrow down possibilities to be tested in the laboratory. A few recent studies demonstrate the power of such an approach combining comparative analysis with carefully designed experiment. For example, Ivarsson et al. (2002) inferred positively selected amino acid residues in glutathione transferase, multifunctional enzymes that provide cellular defence against toxic electrophiles. They then used site-directed mutagenesis to confirm that those mutations were capable of driving functional diversification in substrate specificities. The evolutionary comparison provided a novel approach to designing new proteins, reducing the need for extensive mutagenesis or structural knowledge. Bielawski et al. (2004) detected amino acid sites under positive selection in proteorhodopsin, a retinal-binding membrane protein in marine bacteria that functions as a light-driven proton pump. Site-directed mutagenesis and functional assay demonstrated that those sites were responsible for fine-tuning the light absorption sensitivity of the protein to different light intensities in the ocean. Sawyer et al. (2005) inferred amino acid residues under positive selection in TRIM5α, a protein in the cellular antiviral defence system in primates that can restrict retroviruses such as HIV-1 and SIV in a species-specific manner. In particular a 13-amino acid 'patch' had a concentration of positively selected sites, implicating it as an antiviral interface. By creating chimeric *TRIM5α* genes, Sawyer et al. demonstrated that this patch is responsible for most of the species-specific antiretroviral activity. Lastly, Bishop (2005) found that amino acids identified to be under positive selection in phylogenetic analysis in plant polygalacturonase inhibitor proteins (PGIPs) are also sites with natural mutations conferring novel defensive capabilities.

Relevant to this discussion of which genes are undergoing adaptive evolution, an interesting question is whether molecular adaptation is driven mainly by changes in regulatory genes or in structural genes. Developmental biologists tend to emphasize the importance of regulatory genes, sometimes nullifying the significance of structural genes to evolution. This view derives support from the observation that dramatic shifts in organismal structure can arise from mutations at key regulatory loci. For example, mutations in the homeotic (*hox*) gene cluster are responsible for diversification of body plans in major animal phyla (Akam 1995; Carroll 1995). Nevertheless, examples like those discussed here demonstrate the adaptive significance of amino acid changes in structural genes. Whether regulatory or structural genes are of greater importance to molecular adaptation, it is a common observation that a few changes in the key loci may bring about large changes in morphology and behaviour, while many molecular substitutions appear to have little impact (e.g. Stewart *et al.* 1987). As a result, rates of morphological evolution are in general not well correlated with rates of molecular evolution. This pattern is in particular conspicuous in species groups that have undergone recent adaptive radiations, such as the cichlid fish species in the lakes of East Africa (Meyer *et al.* 1990) or the plant species in the Hawaiian silversword alliance (Barrier *et al.* 2001). As the divergence time among the species is very short, there is very little genetic differentiation at most loci, while the dramatic adaptation in morphology and behaviour must be due to a few key changes at the molecular level. Identifying these key changes should provide exciting opportunities for understanding speciation and adaptation (Kocher 2004).

9

Simulating molecular evolution

9.1 Introduction

Computer simulation, also known as *stochastic simulation* or *Monte Carlo simulation*, is a virtual experiment in which we mimic the biological process on a computer to study its properties. It is particularly useful for studying complex systems or problems which are analytically intractable. Use of random numbers is a major feature of the method. Some authors use the term Monte Carlo simulation when the answer is deterministic, as in calculation of an integral by Monte Carlo integration, and stochastic simulation or computer simulation when the answer involves random variations, as in studies of biases and variances of an estimation method. We ignore such a distinction here.

Simulation is a useful approach for validating a theory or program implementation when the method of analysis is complex. When the model is intractable analytically, simulation provides a powerful way of studying it. Indeed, as it is easy to simulate even under a complex model, there is no need to make overly simplistic assumptions. Simulation is commonly used to compare different analytical methods, especially their robustness when the underlying assumptions are violated, in which case theoretical results are often unavailable. Simulation also plays a useful role in education. By simulating under a model and observing its behaviour, one gains intuitions about the system. Lastly, Monte Carlo simulation forms the basis of many modern computation-intensive statistical methods, such as *bootstrapping*, *importance sampling*, and *Markov-chain Monte Carlo* (see Chapter 5).

A consequence of the ease of simulating and the availability of cheap and fast computers is that simulation is widely used and misused. Many simulation studies are poorly designed and analysed, with results largely uninterpretable. One should bear in mind that simulation is experimentation, and the same care should be exercised in the design and analysis of a simulation experiment as an ordinary experiment. While it is perhaps the simplest thing one can do with a model, simulation does offer the uninitiated ample opportunities for mistakes. A major limitation of simulation is that one can examine only a small portion of the parameter space, and the behaviour of the system may be different in other unexplored regions of the parameter space. One should thus resist the temptation of over-generalization. Analytical results are in general superior as they apply to all parameter values.

This chapter provides a cursory introduction to simulation techniques. The reader is invited to implement ideas discussed here. Any programming language, such as

C/C++, Fortran, BASIC, Java, Perl, or Python, will serve the purpose well. The reader may refer to many of the textbooks on simulation, such as Ripley (1987) and Ross (1997).

9.2 Random number generator

Random variables from the uniform distribution $U(0, 1)$, called *random numbers*, are fundamentally important in computer simulation. For example, to simulate an event (E) that occurs with probability 0.23, we draw a random number $u \sim U(0, 1)$. If $u < 0.23$, we decide that the event occurs; otherwise it does not (Fig. 9.1). It is obvious that the probability that $u < 0.23$ is 0.23. The $U(0, 1)$ random numbers form the basis for generating random variables from other distributions.

A mathematical algorithm is used to generate a sequence of numbers between 0 and 1 that appear random. The algorithm is deterministic and will produce the same fixed sequence of numbers every time it is run. The generated numbers are thus not random at all, and are more appropriately called *pseudo-random numbers* but often simply random numbers. The mathematical algorithm is called a *random-number generator*. A typical simulation study may need millions or billions of random numbers, so it may be important to have a reliable and efficient random-number generator.

A commonly used algorithm is the *multiplication-congruent method*, given by the following formulae

$$A_i = cA_{i-1} \bmod M, \qquad (9.1)$$

$$u_i = A_i/M. \qquad (9.2)$$

Here, A_i, c, and M are all positive integers: c is called the multiplier and M the modulus, with ($a \bmod M$) to be the remainder when a is divided by M. cA_{i-1} and A_i have the same remainder when divided by M and are said to be congruent. A_0 is the initial value, known as the *seed*. $A_1, A_2, \ldots$ will be a sequence of integers between 0 and $M - 1$, and $u_1, u_2, \ldots$ will be a sequence of (pseudo-) random numbers.

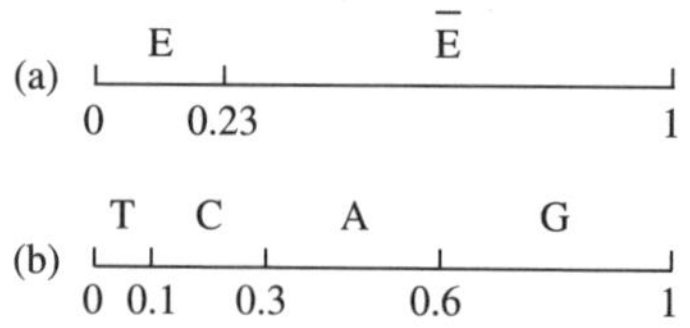

Fig. 9.1 Sampling from a discrete distribution. (a) To simulate an event that occurs with probability 0.23, draw a random number $u \sim U(0, 1)$. If it falls into the interval (0, 0.23), the event occurs (E). Otherwise the event does not occur ($\bar{E}$). (b) Similarly to sample a nucleotide with probabilities 0.1, 0.2, 0.3, and 0.4, for T, C, A, and G, respectively, draw a random number $u \sim U(0, 1)$ and choose the corresponding nucleotide depending on whether u falls into the four intervals of lengths 0.1, 0.2, 0.3, and 0.4.

As an example, imagine a computer using the familiar decimal system, with the register (storage) having only four digits. The computer can thus represent any integer in the range (0000, 9999). Let $M = 10^4$, $c = 13$, and $A_0 = 123$. Equation (9.1) then generates the A_i sequence 1599, 6677, 1271, 6333, 8959, ..., corresponding to the u_i sequence 0.1599, 0.6677, 0.1271, 0.6333, 0.8959, ... Note that if the product cA_{i-1} is greater than 9999, we simply ignore the high digits to get the remainder A_i; most computer systems allow such 'overflow' without any warning. The sequence u_i does not show an obvious trend and appears random. However, as this computer can represent only 10 000 integers and many of them will never occur in the sequence, very soon a certain number A_i will reoccur and then the sequence will repeat itself. A real computer, however, has more digits, and by intelligent choices of M and c (and, to a lesser extent, A_0 as well), the period of the sequence can be made very long.

Real computers use the binary system. Thus the natural choice of M is 2^d, often with d to be the number of bits in an integer (say, 31, 32, or 64). Note that such a choice of M eliminates the need for the division in equation (9.1) for the modulus calculation and also makes the division in equation (9.2) a trivial calculation.

How do we decide whether the random numbers generated are random enough. The requirement is that the numbers generated by the algorithm should be indistinguishable from random draws from the uniform distribution $U(0, 1)$. This can be examined using statistical tests, examining different departures from the expectation. For example, the random numbers should have the correct mean (1/2) and variance (1/12), and should not be autocorrelated. In general, it is not advisable for us biologists to devise random-number generators. Instead one should use algorithms that have been well tested (Knuth 1997; Ripley 1987). Ripley (1987) discussed various tests and commented that random-number generators supplied by computer manufacturers or compiler writers were not trustable. It is unclear whether the situation has improved in the last two decades. For the exercises of this chapter, it is sufficient to use a generator provided by the programming language.

9.3 Generation of continuous random variables

Random variables used in a computer simulation are all generated using $U(0, 1)$ random numbers. A very useful method is the *transformation method*, based on the fact that a function of a random variable is itself a random variable (see Appendix A). Thus we generate a random number $u \sim U(0, 1)$ and apply an appropriate variable transform, $x = g(u)$, to generate a random variable x with the desired distribution.

An important transformation method is the *inversion* method. Suppose random variable x has a cumulative density function (CDF) $F(x)$. Consider $u = F(x)$ as a function of x so that u is itself a random variable. Its distribution is known to be uniform: $u = F(x) \sim U(0, 1)$. Thus, if the inverse transform $x = F^{-1}(u)$ is available analytically, one can use it to generate x using uniform random number u.

1. *Uniform distribution.* To generate a random variable from the uniform distribution $x \sim U(a, b)$, one generates $u \sim U(0, 1)$ and apply the transform $x = a + u(b - a)$.
2. *Exponential distribution.* The exponential distribution with mean θ has density function $f(x) = \theta^{-1} e^{-x/\theta}$ and CDF $F(x) = \int_0^x e^{-y/\theta} dy = 1 - e^{-x/\theta}$. Let $u = F(x)$. This function is easy to invert, to give $x = F^{-1}(u) = -\theta \log(1 - u)$. Thus we generate $u \sim U(0, 1)$ and then $x = -\theta \log(1 - u)$ will have the exponential distribution with mean θ. Since $1 - u$ is also $U(0, 1)$, it suffices to use $x = -\theta \log(u)$.
3. *Normal distribution.* Very often the CDF is not easy to invert. This is the case with the CDF of the normal distribution, which can only be calculated numerically. However, Box and Muller (1958) described a clever method for generating standard normal variables using variable transform. Let u_1 and u_2 be two independent random numbers from $U(0, 1)$. Then

$$x_1 = \sqrt{-2\log(u_1)} \sin(2\pi u_2),$$
$$x_2 = \sqrt{-2\log(u_1)} \cos(2\pi u_2) \tag{9.3}$$

are two independent standard normal variables. (The reader may wish to confirm this using Theorem 1 in Appendix A.) A drawback of the algorithm is that it involves calculation of expensive functions log, sin and cos. Also two random variables are generated even if only one is needed. From a standard normal variable, one can easily sample from the normal distribution $N(\mu, \sigma^2)$; if $z \sim N(0, 1)$, then $x = \mu + z\sigma$ is the desired variable.

Random variables from other distributions such as the gamma and beta may be generated from special algorithms. These will not be discussed here; see Ripley (1987) and Ross (1997).

9.4 Generation of discrete random variables

9.4.1 Discrete uniform distribution

A random variable that takes n possible values (corresponding to n possible experimental outcomes, say), each with equal probability, is called a *discrete uniform distribution*. Suppose the values are $1, 2, \ldots, n$. To sample from this distribution, generate $u \sim U(0, 1)$ and then set $x = 1 + [nu]$, where $[a]$ means the integer part of a. Note that $[nu]$ takes values $0, 1, \ldots, n-1$, each with probability $1/n$. As an example, the JC69 and K80 models predict equal proportions of the four nucleotides. Suppose we let 1, 2, 3, and 4 represent T, C, A, and G, respectively. To sample a nucleotide, we generate $u \sim U(0, 1)$ and then set $x = 1 + [4u]$. Another use of discrete uniform distribution is nonparametric bootstrap, in which one samples sites in the sequence at random.

9.4.2 Binomial distribution

Suppose the probability of 'success' in a trial is p. Then the number of successes in n independent trials has a binomial distribution: $x \sim Bi(n, p)$. To sample from the binomial distribution, one may simulate n trials and count the number of successes. For each trial, generate $u \sim U(0, 1)$; if $u < p$, we count a 'success' and otherwise we count a failure.

An alternative approach is to calculate the probability of x successes

$$p_x = \binom{n}{x} p^x (1-p)^{n-x}, \tag{9.4}$$

for $x = 0, 1, \ldots, n$. One can then sample from the discrete distribution with $n + 1$ categories: $(p_0, p_1, \ldots, p_n)$. This involves the overhead of calculating p_xs but may be more efficient if many samples are to be taken from the same binomial distribution; see below about sampling from a general discrete distribution.

9.4.3 General discrete distribution

Suppose we want to sample a nucleotide at random with probabilities 0.1, 0.2, 0.3, 0.4 for T, C, A, G, respectively. We can break the line segment (0, 1) into four intervals: (0, 0.1), [0.1, 0.3), [0.3, 0.6), and [0.6, 1), corresponding to the probabilities (Fig. 9.1b). We then draw a random number $u \sim U(0, 1)$, and choose the nucleotide depending on which interval u falls into. Specifically, we compare u with 0.1, and choose T if $u < 0.1$. Otherwise we compare u with 0.3 and choose C if $u < 0.3$. Otherwise we compare u with 0.6 and choose A if $u < 0.6$. Otherwise we choose G. In general the comparison is repeated until u is smaller than the upper bound of the interval. It is obvious that the four nucleotides are sampled with the correct probabilities. Note that 0.1, 0.3, 0.6, and 1.0 are the cumulative probabilities, and that this algorithm is the discrete version of the inversion method:

Category	1 (T)	2 (C)	3 (A)	4 (G)
Probability	0.1	0.2	0.3	0.4
Cumulative probability (CDF)	0.1	0.3	0.6	1.0

We can sample from any discrete distribution in this way. However, if there are many categories and if many samples are needed this algorithm is very inefficient. One may rearrange the high-probability categories before the low-probability categories to reduce the number of comparisons. In the above example, it takes 1, 2, 3, or 3 comparisons to sample T, C, A, or G, respectively, with $0.1 \times 1 + 0.2 \times 2 + 0.3 \times 3 + 0.4 \times 3 = 2.6$ comparisons to sample a nucleotide on average. If we order the nucleotides as G, A, C, T, it takes on average 1.9 comparisons. If there are many categories, one may use a cruder classification to find the rough location of the sampled category before doing the finer comparisons.

In general this inversion method is inefficient. A clever algorithm for sampling from a general discrete distribution with a finite number of categories is the *alias method*. It requires only one comparison to generate a random variable, no matter what the number of categories n is. This is explained below in Subsection 9.4.6.

9.4.4 Multinomial distribution

In a binomial distribution, every trial has two outcomes: 'success' or 'failure'. If every trial has k possible outcomes, the distribution is called the *multinomial distribution*, represented as $MN(n, p_1, p_2, \ldots, p_k)$, where n is the number of trials. The multinomial variables are the counts of the k different outcomes $n_1, n_2, \ldots, n_k$, with $n_1 + n_2 + \ldots + n_k = n$. One can generate multinomial variables by sampling n times from the discrete distribution $(p_1, p_2, \ldots, p_k)$, and counting the number of times that each of the k outcomes is observed. If both n and k are large, it may be necessary to use an efficient algorithm to sample from the discrete distribution, such as the alias method described below.

Under most substitution models discussed in this book, the sequence data follow a multinomial distribution. The sequence length is the number of trials n, and the possible site patterns correspond to the categories. For nucleotide data, there are $k = 4^s$ categories for s species or sequences. One can thus generate a sequence data set by sampling from the multinomial distribution. See Subsection 9.5.1 below for details.

9.4.5 The composition method for mixture distributions

Suppose a random variable has a mixture distribution

$$f = \sum_{i=1}^{r} p_i f_i, \qquad (9.5)$$

where f_i represent a discrete or continuous distribution while $p_1, p_2, \ldots, p_r$ are the mixing proportions, which sum to 1. Then f is said to be a *mixture* or *compound* distribution. To generate a random variable with distribution f, first sample an integer (let it be I) from the discrete distribution $(p_1, p_2, \ldots, p_r)$, and then sample from f_I. This method is known as the *composition method*.

For example, the so-called 'Γ+I' model of rates for sites discussed in Subsection 4.3.1 assumes that a proportion p_0 of sites in the sequence are invariable with rate zero while all other sites have rates drawn from the gamma distribution. To generate the rate for a site, first sample from the distribution $(p_0, 1 - p_0)$ to decide whether the rate is zero or from the gamma distribution. Generate a random number $u \sim U(0, 1)$. If $u < p_0$, let the rate be 0; otherwise sample the rate from the gamma distribution. Mayrose *et al.* (2005) discussed a mixture of several gamma distributions with different parameters; rates can be sampled in a similar way from that model. Also the codon substitution model M8 (beta&ω) discussed in Section 8.4 is a mixture distribution, which assumes that a proportion p_0 of sites have the ω ratio drawn from the beta distribution while all other sites (with proportion $1 - p_0$) have the constant $\omega_s > 1$.

*9.4.6 The alias method for sampling from a discrete distribution

The alias method (Walker 1974; Kronmal and Peterson 1979) is a clever method for simulating random variables from an arbitrary discrete distribution with a finite number of cells (categories). It is an example of the *composition method* discussed above. It is based on the fact that any discrete distribution with n cells can be expressed as an equiprobable mixture of n two-point distributions. In other words, we can always find n distributions $q^{(m)}, m = 1, 2, \ldots, n$, such that $q_i^{(m)}$ is nonzero for at most two values of i, and that

$$p_i = \frac{1}{n} \sum_{m=1}^{n} q_i^{(m)}, \text{ for all } i. \tag{9.6}$$

The statement is proved below by construction when we describe how to construct those two-point distributions $q^{(m)}$. Table 9.1 shows an example with $n = 10$ cells. The target distribution p_i is expressed as a mixture, each with weight 1/10, of 10 distributions: $q^{(m)}, m = 1, 2, \ldots, 10$. The component distribution $q^{(i)}$ has nonzero probabilities on two cells: cell i (with probability F_i) and another cell L_i (with probability $1 - F_i$). F_i is called the cutoff and L_i the alias. It is a feature of the distribution $q^{(i)}$ considered here that one of the two points is cell i. Thus the distribution $q^{(i)}$ is fully specified by F_i and L_i. For example, $F_1 = 0.9$ and $L_1 = 3$ specify that the distribution $q^{(1)}$ has probability 0.9 on cell $i = 1$ and probability $1 - 0.9 = 0.1$ on cell $i = 3$.

We will describe how to set up the component distributions $q^{(m)}$ or the F and L vectors in a moment. Now suppose they are already given. It is then easy to sample from p_i: we generate a random integer k from $(1, 2, \ldots, n)$ and sample from $q^{(k)}$. The algorithm is shown in Box 9.1 (comments are in parentheses).

Box 9.1

Alias algorithm (To generate a random variable i from the specified discrete distribution $p_i, i = 1, 2, \ldots, n$, using the cutoff and alias vectors F and L.)

1. (Simulate a random integer k over $1, 2, \ldots, n$, and a random number $r \sim U(0, 1)$.)
   Generate random number $u \sim U(0, 1)$. Set $k \leftarrow [nu] + 1$ and $r \leftarrow nu + 1 - k$
2. (Sample from $q^{(k)}$.) If $r \leq F_k$, set $i \leftarrow k$; otherwise, set $i \leftarrow L_k$.

Step 1 uses a little trick to avoid the need for two random numbers: $nu+1$ is a random variable from $U(1, n + 1)$, so that its integer part can be used for k and the decimal

Table 9.1 Alias method for simulating a discrete distribution p_i with $n = 10$ cells

i	1	2	3	4	5	6	7	8	9	10	Sum
p_i	0.17	0.02	0.15	0.01	0.04	0.25	0.05	0.03	0.20	0.08	1
np_i	1.7	0.2	1.5	0.1	0.4	2.5	0.5	0.3	2	0.8	10
$q^{(2)}$	0.8	0.2									1
$q^{(1)}$	0.9		0.1								1
$q^{(4)}$			0.9	0.1							1
$q^{(3)}$			0.5			0.5					1
$q^{(5)}$					0.4	0.6					1
$q^{(7)}$						0.5	0.5				1
$q^{(6)}$						0.9			0.1		1
$q^{(8)}$								0.3	0.7		1
$q^{(10)}$									0.2	0.8	1
$q^{(9)}$									1		1
F_i	0.9	0.2	0.5	0.1	0.4	0.9	0.5	0.3	1	0.8	
L_i	3	1	6	3	6	9	6	9		9	

The target distribution $(p_1, p_2, \ldots, p_{10})$ is expressed as an equiprobable mixture of 10 two-point distributions $q^{(1)}, q^{(2)}, \ldots, q^{(10)}$. The component two-point distribution $q^{(i)}$, shown on the rows, has nonzero probabilities on two cells: cell i (with probability F_i) and another cell L_i (with probability $1 - F_i$). Cells with zero probabilities are left blank. For example, $q^{(1)}$ assigns probability 0.9 and 0.1 to cells 1 and 3 respectively, so that $F_1 = 0.9$ and $L_1 = 3$, shown on the bottom two rows. All information about the 10 component distributions is contained in the F and L vectors. F_i and L_i are called the cutoff probability and alias. An algorithm for generating F_i and L_i is described in the text, which produces the component distributions $q^{(2)}, q^{(1)}, q^{(4)}, \ldots, q^{(9)}$, in the order shown in the table.

part for r, with k and r being independent. In our example with $n = 10$, suppose we get $u = 0.421563$. Then $k = 4 + 1 = 5$ and $r = 0.21563$. Step 2 then uses r to sample from the component distribution $q^{(k)}$; since $r < F_5$, we get $i = L_5 = 6$. The alias algorithm thus uses one random number and does one comparison to generate a random variable from the discrete distribution p_i.

Now we describe how to set up the component distributions $q^{(m)}, m = 1, 2, \ldots, 10$ or the F and L vectors. The solution is not unique, and any will suffice. Note that the sum of the probabilities for cell i across the n component distributions is np_i (equation 9.6). For example, for the first cell $i = 1$, this sum is $10p_1 = 1.7$. The total sum across all cells is n. Our task is to divide up this sum of probabilities among n two-point distributions. Let F_i be the probability (or, more precisely, sum of probabilities) remaining in cell i. At the start, $F_i = np_i$. As the end of the algorithm, F_i will hold the cutoff probability for cell i, as defined earlier.

We identify a cell j for which $F_j < 1$ and a cell k for which $F_k \geq 1$. (Because $\sum_{i=1}^{n} F_i = n$, such j and k exist except when $F_i = 1$ for all i, in which case the job is done.) We then generate the distribution $q^{(j)}$, by using up all probability (F_j) on cell j, with the remaining probability needed, $1 - F_j$, contributed by cell k. In our example, $F_2 = 0.2 < 1$ and $F_1 = 1.7 > 1$, so let $j = 2$ and $k = 1$. The distribution $q^{(2)}$ assigns probability 0.2 on cell $j = 2$ and probability $1 - 0.2 = 0.8$ on $k = 1$. Thus $F_2 = 0.2$ and $L_2 = 1$ are finalized, and cell $j = 2$ will not be considered further in the algorithm. The remaining probability on cell $k = 1$ becomes $F_1 = 1.7 - 0.8 = 0.9$. At the end of this first round, the total remaining probability mass is $n - 1$, which will be divided up into $n - 1$ two-point distributions.

In the second round, we identify a cell j for which $F_j < 1$ and a cell k for which $F_k \geq 1$. Since now $F_1 = 0.9 < 1$ and $F_3 = 1.5 > 1$, we let $j = 1$ and $k = 3$. Generate $q^{(1)}$ and let it use up all probability on cell j (that is, $F_1 = 0.9$), with the rest of the probability (0.1) contributed by cell $k = 3$. Thus $F_1 = 0.9$ and $L_1 = 3$ for cell $j = 1$ are finalized. $F_3 = 1.5 - 0.1 = 1.4$ is reset.

The process is repeated. Each round generates a two-point distribution $q^{(j)}$ and finalizes F_j and L_j for cell j, reducing the total remaining probability by 1. The process ends after a maximum of n steps. Box 9.2 summarizes the algorithm. It uses an indicator I_j to record the status of cell j: $I_j = -1, 1,$ or 0, if $F_j < 1, F_j \geq 1$, or if cell j is not considered any further.

Box 9.2

Algorithm: generate the cutoff and alias vectors F and L for the alias method (Kronmal and Peterson 1979)
   (Summary: this creates two vectors F_i and L_i $i = 1, 2, \ldots, n$)

1. (Initialize.) Set $F_i \leftarrow np_i$, $i = 1, 2, \ldots, n$.
2. (initialize the indicator table I_i, $i = 1, 2, \ldots, n$.) Let $I_i = -1$ if $F_i < 1$ or $I_i = 1$ if $F_i \geq 1$.
3. (Main loop.) Repeat the following steps until none of I_i is -1. (Pick up a cell j with $I_j = -1$ and a cell k with $I_k = 1$. Generate distribution $q^{(j)}$, finalizing F_j and L_j for cell j.)
   3a. Scan the I vector to find a j such that $I_j = -1$ and a cell k such that $I_k = 1$.
   3b. Set $L_j \leftarrow k$. Set $F_k \leftarrow F_k - (1 - F_j)$. ($1 - F_j$ is the probability on cell k used up by distribution $q^{(j)}$.)
   3c. (Update I_j and I_k.) Set $I_j \leftarrow 0$. If $F_k < 1$, set $I_k \leftarrow -1$.

The alias method is efficient for generating many random variables from the same discrete distribution, as long as the F and L vectors have been set up. It requires only one comparison to generate one variable, irrespective of n. This contrasts with the inversion method, which requires more comparisons for larger n. Both the storage (for F, L, I) and the computation required in setting up the F and L vectors are proportional to n. It may not be worthwhile to use this algorithm to generate only a few random variables from the discrete distribution, but it is very useful for sampling from the multinomial distribution.

9.5 Simulating molecular evolution

9.5.1 Simulating sequences on a fixed tree

Here we consider generation of a nucleotide sequence alignment when the tree topology and branch lengths are given. The basic model assumes the same substitution process at all sites and along all branches, but we will also consider more complex models in which the evolutionary process may vary across sites or branches. We consider nucleotide models; amino acid or codon sequences can be generated using the same principles. Several approaches can be used and produce equivalent results.

9.5.1.1 Method 1. Sampling sites from the multinomial distribution of site patterns

If the substitution model assumes independent evolution at different sites in the sequence and all sites evolve according to the same model, data at different sites will have independent and identical distributions; they are said to be *i.i.d.*. The sequence data set will then follow a multinomial distribution, with every site to be a sample point, and every site pattern to be a category (cell) of the multinomial. For a tree of s species, there are 4^s, 20^s, or 64^s possible site patterns for nucleotide, amino acid, or codon sequences, respectively. Calculation of the probability for every site pattern is explained in Section 4.2, which describes the calculation of the likelihood function under the model. A sequence alignment can thus be generated by sampling from this multinomial distribution. The result will be the numbers of sites having the site patterns, many of which may be zero if the sequence is short. If a pattern, say TTTC (for four species), is observed 50 times, one simply writes out 50 sites with the same data TTTC, either with or without randomizing the sites. Most phylogeny programs, especially those for likelihood and Bayesian calculations, collapse sites into patterns to save computation, since the probabilities of observing sites with the same data are identical. As this simulation method generates counts of site patterns, those counts should in theory be directly usable by phylogeny programs.

The multinomial sampling approach is not feasible for large trees because the number of categories becomes too large. However, it is very efficient for small trees with only four or five species, especially when combined with an efficient algorithm for sampling from the multinomial, such as the alias method.

9.5.1.2 Method 2. Evolving sequences along the tree

This approach 'evolves' sequences along the given tree, and is the algorithm used in programs such as Seq-Gen (Rambaut and Grassly 1997) and EVOLVER (Yang 1997a). First, we generate a sequence for the root of the tree, by sampling nucleotides according to their equilibrium distribution under the model: $\pi_T, \pi_C, \pi_A, \pi_G$. Every nucleotide is sampled independently from the discrete distribution $(\pi_T, \pi_C, \pi_A, \pi_G)$. If the base frequencies are all equal, one can use the algorithm for discrete uniform distributions, which is more efficient. The sequence for the root is then allowed to evolve to produce sequences at the daughter nodes of the root. The procedure is repeated for every branch on the tree, generating the sequence at a node only after the sequence at its mother node has been generated. Sequences at the tips of the tree constitute the data, while sequences for ancestral nodes are discarded.

To simulate the evolution of a sequence along a branch of length t, calculate the transition probability matrix $P(t) = \{p_{ij}(t)\}$ (see Sections 1.2 and 1.5), and then simulate nucleotide substitutions at every site independently. For example, if a site is occupied by C in the source sequence, the nucleotide in the target sequence will be a random draw from the discrete distribution of T, C, A, G, with probabilities $p_{CT}(t), p_{CC}(t), p_{CA}(t), p_{CG}(t)$, respectively. This process is repeated to generate all sites in the target sequence. As the transition probabilities apply to all sites in the sequence for the branch, one has to calculate them only once for all sites for that branch.

9.5.1.3 Method 3. simulating the waiting times of a Markov chain

This is a variation to method 2. One generates a sequence for the root, and then simulates the evolution of any site along any branch as follows. Suppose the branch length is t, and the rate matrix of the Markov chain is $Q = \{q_{ij}\}$. Let $q_i = -q_{ii} = -\sum_{j \neq i} q_{ij}$ be the substitution rate of nucleotide i. Suppose the site is currently occupied by nucleotide i. Then the waiting time until the next substitution event has an exponential distribution with mean $1/q_i$. We draw a random waiting time s from the exponential distribution. If $s > t$, no substitution occurs before the end of the branch so that the target sequence has nucleotide i at the site as well. Otherwise a substitution occurs and we decide which nucleotide the site changes into. Given that the site with nucleotide i has experienced a change, the probability that it changes into nucleotide j is q_{ij}/q_i, and we can sample j from this discrete distribution. The remaining time for the branch becomes $t - s$. We then draw a random waiting time from the exponential distribution with mean $1/q_j$. The process is repeated until the time for the branch is exhausted.

This simulation procedure is based on the following characterization of a continuous-time Markov chain (Fig. 9.2). The waiting time until the next transition (change) is exponential with mean $1/q_i$. If we ignore the waiting times between transitions, the sequence of states visited by the process constitutes a discrete-time Markov

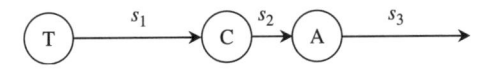

Fig. 9.2 Characterization of the Markov process as exponential waiting times and a jump chain. The waiting times until the next substitution s_1, s_2, and s_3 are independent random variables from the exponential distributions with means $1/q_T$, $1/q_C$, and $1/q_A$, respectively, where q_T, q_C, and q_A are the substitution rates of nucleotides T, C, and A, respectively.

chain, which is called the *jump chain*. The transition matrix of the jump chain is given as

$$M = \begin{bmatrix} 0 & \dfrac{q_{TC}}{q_T} & \dfrac{q_{TA}}{q_T} & \dfrac{q_{TG}}{q_T} \\ \dfrac{q_{CT}}{q_C} & 0 & \dfrac{q_{CA}}{q_C} & \dfrac{q_{CG}}{q_C} \\ \dfrac{q_{AT}}{q_A} & \dfrac{q_{AC}}{q_A} & 0 & \dfrac{q_{AG}}{q_A} \\ \dfrac{q_{GT}}{q_G} & \dfrac{q_{GC}}{q_G} & \dfrac{q_{GA}}{q_G} & 0 \end{bmatrix}. \tag{9.7}$$

Note that every row sums to 1.

The algorithm of simulating exponential waiting times and the jump chain may be applied to the whole sequence instead of one site. The total rate of substitution is the sum of the rates across sites, and the waiting time until a substitution occurs at any site in the whole sequence has an exponential distribution with the mean equal to the reciprocal of the total rate. If a substitution occurs, it is assigned to sites with probabilities proportional to the rates at the sites.

An advantage of this simulation procedure is that it does not require calculation of the transition-probability matrix $P(t)$ over branch length t, as both the waiting times and the transition matrix for the jump chain are fully specified by the instantaneous rates. As a result, this procedure can be adapted to simulate more complex sequence changes such as insertions and deletions. One simply calculates the total rate of all events (including substitutions, insertions, and deletions) for the whole sequence, and simulates the exponential waiting time until the next event. If an event occurs before the end of the branch, one assigns the event to a site and to an event type (a substitution, insertion, or deletion) with probabilities in proportion to the rates of those events at the sites.

9.5.1.4 Simulation under more complex models

The methods discussed above can be modified to simulate under more complex models, for example, to allow different substitution parameters among branches (such as different transition/transversion rate ratio κ, different base frequencies, or different ω ratios). The multinomial sampling approach (method 1) applies as long as the site pattern probabilities are calculated correctly under the model. The approaches of evolving sequences along branches on the tree, either by calculating the transition probabilities (method 2) or by simulating the waiting times and the jump chain (method 3) are also straightforward; one simply use the appropriate model and parameters for the branch when simulating the evolutionary process along that branch.

One may also simulate under models that allow heterogeneity among sites. We will consider as an example variable substitution rates but the approaches apply to other kinds of among-site heterogeneity. There are two kinds of models that incorporate rate variation among sites (see Subsections 4.3.1 and 4.3.2). The first is the so-called *fixed-sites model*, under which every site in the sequence belongs to a predetermined site partition. For example, one may simulate five genes evolving on the same tree but at different rates $r_1, r_2, \ldots, r_5$. The transition probability matrix for gene k is $p_{ij}(tr_k)$. Sites in different genes do not have the same distribution, but within each gene, the sites are *i.i.d.* Thus one can use either of the three methods discussed above to simulate the data for each gene separately and then merge them into one data set.

A second kind of heterogeneous-sites model are the so-called *random-sites models*. Examples include the gamma models of variable rates for sites (Yang 1993, 1994a), the codon models of variable ω ratios among sites (Nielsen and Yang 1998; Yang *et al.* 2000), and the covarion-like models (Galtier 2001; Huelsenbeck 2002; Guindon *et al.* 2004). The rates (or some other features of the substitution process) are assumed to be random variables drawn from a common statistical distribution, and we do not know a priori which sites have high rates and which have low rates. Data at different sites are *i.i.d.* The approach of multinomial sampling can be used directly, although the site pattern probabilities have to be calculated under the heterogeneous-sites model. One may also sample the rate for each site and apply the method of simulating evolution along branches. If a continuous distribution is used, one should in theory calculate the transition probability matrix $P(t)$ for every site and every branch. If a few site classes are assumed (as in the discrete-gamma model), one may sample rates for sites first, and then simulate data for the different rate classes separately, perhaps followed by a randomization of the sites. Note that under the random-sites model, the number of sites in any site class varies among simulated replicates, whereas in the fixed-sites model, the number is fixed.

9.5.2 Generating random trees

Several models can be used to generate random trees with branch lengths: the standard coalescent model, the Yule branching model, and the birth–death process model either with or without species sampling. All those models assign equal probabilities to all labelled histories. (A labelled history is a rooted tree with the internal nodes ranked according to their ages.) One may generate a random genealogical or phylogenetic tree by starting with the tips of the tree, and joining nodes at random until there is one lineage left.

The ages of the nodes are independent of the labelled history and can be attached to the tree afterwards. Under the coalescent model, the waiting time until the next coalescent event has an exponential distribution (see equation 5.41). Under the birth–death process model, the node ages on the labelled history are order statistics from a kernel density (see equation 7.24) and can be simulated easily (Yang and Rannala 1997). Trees generated in this way have branch lengths conforming to a molecular clock. One may modify the substitution rates to produce trees in which the clock is violated.

One may also generate random rooted or unrooted trees without assuming any biological model, by sampling at random from all possible trees. Branch lengths may also be sampled from arbitrary distributions such as the exponential or the gamma.

9.6 Exercises

9.1 Write a small simulation program to study the *birthday problem*. Suppose that there are 365 days in a year and that one's birthday falls on any day at random. Calculate the probability that at least two people out of a group of $k = 30$ people have the same birthday (that is, they were born on the same day and month but not necessarily in the same year). Use the following algorithm. (The answer is 0.706.)
1. Generate $k = 30$ birthdays, by taking 30 random draws from $1, 2, \ldots, 365$.
2. Check whether any two birthdays are the same.
3. Repeat the process 10^6 times and calculate the proportion of times in which at least two out of 30 people have the same birthday.

9.2 Monte Carlo integration (Subsection 5.3.1). Write a small program to calculate the integral $f(x)$ in the Bayesian estimation of sequence distance under the JC69 model, discussed in Subsection 5.1.2. The data are $x = 90$ differences out of $n = 948$ sites. Use the exponential prior with mean 0.2 for the sequence distance θ. Generate $N = 10^6$ or 10^8 random variables from the exponential prior: $\theta_1, \theta_2, \ldots, \theta_N$, and calculate

$$f(x) = \int_0^\infty f(\theta) f(x|\theta) d\theta \simeq \frac{1}{N} \sum_{i=1}^{N} f(x|\theta_i). \qquad (9.8)$$

Note that the likelihood $f(x|\theta_i)$ may be too small to represent in the computer, so scaling may be needed. One way is as follows. Compute the maximum log likelihood $\ell_m = \log\{f(x|\hat{\theta})\}$, where $\hat{\theta} = 0.1015$ is the MLE. Then multiply $f(x|\theta_i)$ in equation (9.8) by a big number $e^{-\ell_m}$ so that they are not all vanishingly small before summing them up; that is,

$$\sum_{i=1}^{N} f(x|\theta_i) = e^{\ell_m} \cdot \sum_{i=1}^{N} \exp\left(\log\{f(x|\theta_i)\} - \ell_m\right). \qquad (9.9)$$

9.3 Write a small simulation program to study the optimal sequence divergence when two sequences are compared to estimate the transition/transversion rate ratio κ under the K80 model. Assume $\kappa = 2$ and use a sequence length of 500 sites. Consider several sequence distances, say, $d = 0.01, 0.02, \ldots, 2$. For each d, simulate 1000 replicate data sets under the K80 model and analyse it under the same model to estimate d and κ using equation (1.11). Calculate the mean and variance of the estimate $\hat{\kappa}$ across replicate data sets. Each data set consists of a pair of sequences, which can be generated using any of the three approaches discussed in Subsection 9.5.1.

9.4 Long-branch attraction by parsimony. Use the JC69 model to simulate data sets on a tree of four species (Fig. 9.3a), with two different branch lengths $a = 0.1$ and $b = 0.5$. Simulate 1000 replicate data sets. For each data set, count the sites with the three site patterns *xxyy*, *xyxy*, *xyyx*, and determine the most parsimonious tree. To simulate a data set, reroot the tree at an interior node, as in, say, Fig. 9.3(b). Generate

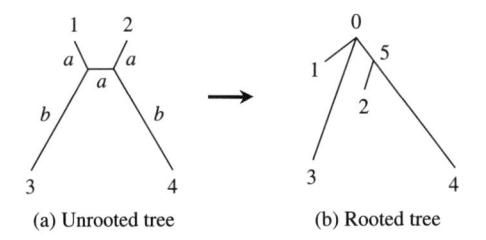

Fig. 9.3 (a) A tree of four species, with three short branches (of length *a*) and two long branches (with length *b*) for simulating data to demonstrate long-branch attraction. (b) The same tree rerooted at an ancestral node for simulation.

a sequence for the root (node 0) by random sampling of the four nucleotides, and then evolve the sequence along the five branches of the tree. You may also use the approach of multinomial sampling. Consider a few sequence lengths, such as 100, 1000, and 10 000 sites.

9.5 A useful test of a new and complex likelihood program is to generate a few data sets of very long sequences under the model and then analyse them under the same model, to check whether the MLEs are close to the true values used in the simulation. As MLEs are consistent, they should approach the true values when the sample size (sequence length) becomes larger and larger. Use the program written for Exercise 9.4 to generate a few data sets of 10^6, 10^7, or 10^8 sites and analyse them using a likelihood program (such as PHYLIP, PAUP, or PAML) under the same JC69 model, to see whether the MLEs of branch lengths are close to the true values. Beware that some programs may demand a lot of resources to process large data sets; save your important work before this exercise in case of a computer crash.

10

Perspectives

Molecular phylogenetics has experienced phenomenal growth in the past decade, driven by the explosive accumulation of genetic sequence data. This trend will most likely continue in the foreseeable future. Here I speculate on a few areas of research in theoretical molecular phylogenetics that appear important and are likely to be active in the next few years.

10.1 Theoretical issues in phylogeny reconstruction

In Chapter 6, I presented the argument that phylogeny reconstruction is a statistical problem of model selection rather than parameter estimation. Simulation results establish that the asymptotic efficiency of ML estimation of conventional parameters does not apply to tree reconstruction. My objective in that discussion was to highlight more the importance of distinguishing between parameter estimation and model selection than the possibility that under certain parameter combinations likelihood can be outperformed by some other methods such as parsimony. Both statistical and computational difficulties in tree reconstruction are precisely because the problem is not one of parameter estimation. In a parameter estimation problem, we have one likelihood function, and solve one multivariate optimization problem, similar to optimization of branch lengths on one tree topology. Computationally this is currently achievable on a personal computer for data sets of a few thousand species. The asymptotic theory of MLEs would apply, and the performance difficulties in likelihood tree reconstruction discussed in Subsection 6.2.3 would disappear.

Because tree reconstruction is a problem of model selection and the likelihood function differs among models or trees, one has to solve many optimization problems (in theory as many as the number of trees). Furthermore, model selection is a controversial and rapidly-changing area in statistics. A few attempts were made to extend the traditional likelihood ratio test to compare two or multiple nonnested models (e.g. Cox 1961, 1962), but the ideas have not led to any generally applicable methodology. Criteria such as the AIC (Akaike 1974) and BIC (Schwarz 1978) are not really useful for tree estimation because they simply rank trees according to their likelihood values and do not provide a measure of reliability of the ML tree. Similarly, clade support values provided by the bootstrap method currently do not have a clear interpretation. In Bayesian statistics, model selection is an equally difficult issue. The sensitivity of Bayesian model probabilities to vague priors on unknown parameters in the models poses serious difficulty, rendering objective Bayesian analysis virtually impossible.

In phylogenetic analysis of molecular data, posterior clade probabilities are often extremely high.

There exist some interesting differences between tree estimation and model selection. First, except for very small tree-reconstruction problems, there are typically a great many trees, while usually no more than a few candidate models are considered in model selection (White 1982; Vuong 1989). The trees are almost like different values of a parameter, except that changing the tree topology not only changes the value of the likelihood function but also the likelihood function itself (and the definitions of branch lengths). Second, in tree estimation, one of the possible trees is assumed to be true and the interest is to infer the true tree, while in model selection, the interest is often in inference concerning certain parameters of the model, and the different models are considered only because the inference may be sensitive to the assumed model. Third, there are intricate relationships among different trees as some trees are more similar than others. Summarizing shared features among a set of models does not appear to be something 'practised' by statisticians, but is the most commonly used procedure in bootstrap and Bayesian analyses. This practice does appear to cause difficulties, as discussed in Subsection 5.6.3.

Because of their importance, those conceptual problems will most likely be a topic of research in molecular phylogenetics for years to come. However, given that they are entwined with fundamental and philosophical controversies in statistics, it does not seem likely that they will easily be resolved.

10.2 Computational issues in analysis of large and heterogeneous data sets

Both likelihood and Bayesian methods involve intensive computation, and the problem will most certainly become more acute with the accumulation of sequence data. Even though sophistication of software algorithms and improvement of computing power will never catch up with the data sets that biologists will attempt to analyse, any improvements in computational algorithms may make a practical difference. These are areas for which progress are very likely. There is no conceptual difficulty for likelihood and Bayesian methods to analyse multiple heterogeneous data sets simultaneously, as the likelihood function combines information from heterogeneous data sets naturally (Yang 1996b; Pupko et al. 2002b; Nylander et al. 2004) (see Subsection 4.3.2). In the likelihood method, one can assign different parameters for different loci, but this may lead to the use of too many parameters if many genes are analysed. The Bayesian approach is more tolerant to the problem of multiple parameters as it assigns priors on parameters and integrates out the uncertainties in them in the MCMC algorithm.

10.3 Genome rearrangement data

Inference of genome rearrangement events, such as duplications, inversions, and translocations, is important in comparisons of multiple genomes. They may provide unique-event characters for phylogeny reconstruction (Belda *et al.* 2005), but perhaps more importantly, one can use the information to understand the process of genome evolution. Current analysis relies on parsimony-style arguments and aims to infer the path that transforms one genome into another by a minimum number of rearrangement events (e.g. Murphy *et al.* 2005). Even such a minimum-path approach poses serious computational difficulties. To an evolutionary biologist, it would be interesting to estimate the rates at which rearrangement events occur relative to nucleotide substitutions. Such statistical inference requires development of probabilistic models of genome evolution. The data generated from the various genome projects appear to contain sufficient information for such inference. This is an active research area and the next few years may well see exciting developments (York *et al.* 2002; Larget *et al.* 2005).

10.4 Comparative genomics

Phylogenetic methods have become increasingly important in interpreting genomic data. The genes and genomes we observe today are the product of complex processes of evolution, influenced by mutation, random drift, and natural selection. The comparative approach aims to detect signals of selection as an indicator of functional significance. Up to now, most genome-wide comparisons have relied on negative purifying selection to infer functionally conserved regions in the genome, with the rationale that sequence conservation across distant species implies functional significance. Similarity match and sequence alignment using programs such as BLAST (Altschul *et al.* 1990) and CLUSTAL (Higgins and Sharp 1988; Chenna *et al.* 2003) are effective in recognizing distant but related sequences. Protein-coding genes, RNA genes, and regulatory elements have different levels of sequence conservation and can be identified by comparison of genomes at different evolutionary distances (e.g. Thomas *et al.* 2003). However, high variability, if demonstrated to be driven by natural selection, also implies functional significance. As more genomes from closely related species are sequenced, the use of positive Darwinian selection to infer functional significance will become ever more important. For example, Clark *et al.* (2003) and Nielsen *et al.* (2005) detected a collection of genes under positive selection along the human lineage, potentially responsible for the differences between humans and the apes. Positive selection is particularly exciting to evolutionary biologists as it is ultimately responsible for species divergences and evolutionary innovations. Such studies may also provide a quantitative assessment of the relative roles of mutation and selection in driving the evolutionary process of genes and genomes.

Appendices

Appendix A: Functions of random variables

This section includes two theorems concerning functions of random variables. Theorem 1 specifies the probability density of functions of random variables. Theorem 2 gives the proposal ratio in an MCMC algorithm when the Markov chain is formulated using variables **x**, but the proposal is made by changing variables **y**, which are functions of **x**.

Theorem 1 specifies the distribution of functions of random variables. A proof is not provided, as it can be found in any statistics textbook (e.g. Grimmett and Stirzaker 1992, pp. 107–112). Instead a few examples will be discussed to illustrate its use.

Theorem 1. (a) Suppose x is a random variable with density $f(x)$, and $y = y(x)$ and $x = x(y)$ constitute a one-to-one mapping between x and y. Then the random variable y has density

$$f(y) = f(x(y)) \times \left|\frac{dx}{dy}\right|. \tag{A.1}$$

(b) The multivariate version is similar. Suppose random vectors $\mathbf{x} = \{x_1, x_2 \ldots, x_m\}$ and $\mathbf{y} = \{y_1, y_2 \ldots, y_m\}$ constitute a one-to-one mapping through $y_i = y_i(\mathbf{x})$ and $x_i = x_i(\mathbf{y}), i = 1, 2, \ldots, m$, and that $\mathbf{x}$ has density $f(\mathbf{x})$. Then $\mathbf{y}$ has density

$$f(\mathbf{y}) = f(\mathbf{x}(\mathbf{y})) \times |J(\mathbf{y})|, \tag{A.2}$$

where $|J(\mathbf{y})|$ is the absolute value of the Jacobian determinant of the transform

$$J(\mathbf{y}) = \begin{vmatrix} \frac{\partial x_1}{\partial y_1} & \frac{\partial x_1}{\partial y_2} & \cdots & \frac{\partial x_1}{\partial y_m} \\ \frac{\partial x_2}{\partial y_1} & \frac{\partial x_2}{\partial y_2} & \cdots & \frac{\partial x_2}{\partial y_m} \\ \vdots & \vdots & \ddots & \vdots \\ \frac{\partial x_m}{\partial y_1} & \frac{\partial x_m}{\partial y_2} & \cdots & \frac{\partial x_m}{\partial y_m} \end{vmatrix}. \tag{A.3}$$

Example 1. Prior distribution of sequence distance under the JC69 model. Suppose that the probability of different sites p between two sequences is assigned a uniform prior $f(p) = 4/3$, $0 \leq p < 3/4$. What is the corresponding prior distribution of the sequence distance θ? From equation (5.12) (see also equation 1.5), θ and p are related as

$$p = \tfrac{3}{4}\left(1 - e^{-4\theta/3}\right). \tag{A.4}$$

Thus $dp/d\theta = e^{-4\theta/3}$, so that the density of θ is $f(\theta) = \tfrac{4}{3}e^{-4\theta/3}$, $0 \leq \theta < \infty$. Thus θ has an exponential distribution with mean $3/4$. Alternatively, suppose we assign a uniform prior on θ: $f(\theta) = 1/A$, $0 \leq \theta \leq A$, where the upper bound A is a big number (say, 10 changes per site). Then $d\theta/dp = 1/(dp/d\theta) = 1/(1 - 4p/3)$ and

$$f(p) = \tfrac{1}{A}/\left(1 - \tfrac{4}{3}p\right), \qquad 0 \leq p \leq \tfrac{3}{4}\left(1 - e^{-4A/3}\right). \tag{A.5}$$

Because p and θ do not have a linear relationship, it is impossible for them both to have uniform distributions at the same time. Equivalently, we say that the prior is not invariant to reparametrization. □

Example 2. Normal distribution. Suppose z has a standard normal distribution with density

$$\phi(z) = \frac{1}{\sqrt{2\pi}} e^{-z^2/2}. \tag{A.6}$$

Let $x = \mu + \sigma z$, so that $z = (x - \mu)/\sigma$ and $dz/dx = 1/\sigma$. Then x has the density

$$f(x) = \phi\!\left(\tfrac{x-\mu}{\sigma}\right)/\sigma = \frac{1}{\sqrt{2\pi\sigma^2}} \exp\!\left\{-\tfrac{1}{2\sigma^2}(x-\mu)^2\right\}. \tag{A.7}$$

Thus x has a normal distribution with mean μ and variance σ^2. □

Example 3. Multivariate normal distribution. Suppose $z_1, z_2, \ldots, z_p$ are independent standard normal variables, so that the column vector $\mathbf{z} = (z_1, z_2, \ldots, z_p)^\mathrm{T}$ has a standard p-variate normal density

$$\phi_p(\mathbf{z}) = \frac{1}{(2\pi)^{p/2}} \exp\!\left\{-\tfrac{1}{2}(z_1^2 + z_2^2 + \cdots + z_p^2)\right\} = \frac{1}{(2\pi)^{p/2}} \exp\!\left\{-\tfrac{1}{2}\mathbf{z}^\mathrm{T}\mathbf{z}\right\}. \tag{A.8}$$

Let $\mathbf{x} = \boldsymbol{\mu} + A\mathbf{z}$, where A is a $p \times p$ non-singular matrix, so that $\mathbf{z} = A^{-1}(\mathbf{x} - \boldsymbol{\mu})$ and $\partial \mathbf{z}/\partial \mathbf{x} = A^{-1}$. Let $\Sigma = AA^\mathrm{T}$. Then $\mathbf{x}$ has the density

$$f_p(\mathbf{x}) = \phi_p(\mathbf{z}) \cdot \left|\frac{\partial \mathbf{z}}{\partial \mathbf{x}}\right| = \frac{1}{(2\pi)^{p/2}|\Sigma|^{1/2}} \exp\!\left\{-\tfrac{1}{2}(\mathbf{x}-\boldsymbol{\mu})^\mathrm{T}\Sigma^{-1}(\mathbf{x}-\boldsymbol{\mu})\right\}. \tag{A.9}$$

Note that $(A^{-1})^T A^{-1} = (AA^T)^{-1} = \Sigma^{-1}$ and $|AA^T| = |A| \cdot |A^T| = |A|^2$. Equation (A.9) is the density for a multivariate normal distribution with the mean vector μ and variance-covariance matrix Σ. □

Next we describe Theorem 2. In an MCMC algorithm, it is sometimes more convenient to propose changes to the state of the Markov chain using certain transformed variables instead of the original variables. Theorem 2 gives the proposal ratio for such a proposal.

Theorem 2. Suppose the original variables are $\mathbf{x} = \{x_1, x_2 \ldots, x_m\}$, but the proposal ratio is easier to calculate using transformed variables $\mathbf{y} = \{y_1, y_2 \ldots, y_m\}$, where $\mathbf{x}$ and $\mathbf{y}$ constitute a one-to-one mapping through $y_i = y_i(\mathbf{x})$ and $x_i = x_i(\mathbf{y})$, $i = 1, 2, \ldots, m$. Let $\mathbf{x}$ (or $\mathbf{y}$) be the current state of the Markov chain, and $\mathbf{x}^*$ (or $\mathbf{y}^*$) be the proposed state. Then

$$\frac{q(\mathbf{x}|\mathbf{x}^*)}{q(\mathbf{x}^*|\mathbf{x})} = \frac{q(\mathbf{y}|\mathbf{y}^*)}{q(\mathbf{y}^*|\mathbf{y})} \times \frac{|J(\mathbf{y}^*)|}{|J(\mathbf{y})|}. \tag{A.10}$$

The proposal ratio in the original variables $\mathbf{x}$ is the product of the proposal ratio in the transformed variables $\mathbf{y}$ and the ratio of the *Jacobian* determinants. To see this, note

$$q(\mathbf{y}^*|\mathbf{y}) = q(\mathbf{y}^*|\mathbf{x}) = q(\mathbf{x}^*|\mathbf{x}) \times |J(\mathbf{y}^*)|. \tag{A.11}$$

The first equation is because conditioning on $\mathbf{y}$ is equivalent to conditioning on $\mathbf{x}$ due to the one-to-one mapping. The second equation applies Theorem 1 to derive the density of $\mathbf{y}^*$ as a function of $\mathbf{x}^*$. Similarly, $q(\mathbf{y}|\mathbf{y}^*) = q(\mathbf{x}|\mathbf{x}^*) \times |J(\mathbf{y})|$.

Examples of applications of this theorem can be found in Section 5.4.

Appendix B: The delta technique

The delta technique is a general approach to deriving the means, variances, and covariances of functions of random variables. Suppose we are interested in the mean and variance of a function $g(x)$, where the random variable x has mean μ and variance σ^2. Note that if g is not a linear function of x, the mean value of a function is not equal to the function of the mean: $E(g(x)) \neq g(E(x))$. The Taylor expansion of g around the mean μ is

$$g = g(x) = g(\mu) + \frac{dg(\mu)}{dx}(x - \mu) + \frac{d^2 g(\mu)}{2! dx^2}(x - \mu)^2 + \cdots, \tag{B.1}$$

where the function $g(\cdot)$ and the derivatives are all evaluated at $x = \mu$; for example, $dg(\mu)/dx \equiv dg(x)/dx|_{x=\mu}$.

If we take the expectation on both sides and ignore terms of order 3 and higher, we get

$$E(g) \approx g(\mu) + \tfrac{1}{2}\frac{d^2 g(\mu)}{dx^2}\sigma^2, \tag{B.2}$$

because $E(x - \mu) = 0$ and $E(x - \mu)^2 = \sigma^2$. The derivatives are evaluated at $x = \mu$ and are thus constants when we take expectations over x. Similarly, the approximate variance of g is

$$\text{Var}(g) \approx E(g - E(g))^2 \approx \sigma^2 \cdot \left[\frac{dg(\mu)}{dx}\right]^2. \tag{B.3}$$

In a statistical data analysis, x may be a parameter estimate, and g its function. Then μ and σ^2 may be replaced by their estimates from the data set.

The multivariate version is similarly derived using a multivariate Taylor expansion. Suppose $\mathbf{x}$ is a random vector of n variables, and $\mathbf{y} = \mathbf{y}(\mathbf{x})$ is a function of $\mathbf{x}$, with m elements. Then the approximate variance–covariance matrix of $\mathbf{y}$ is given as

$$\text{var}(\mathbf{y}) \approx J \cdot \text{var}(\mathbf{x}) \cdot J^T, \tag{B.4}$$

where $\text{var}(\cdot)$ is the variance–covariance matrix, J is the $m \times n$ Jacobi matrix of the transform

$$J = \begin{pmatrix} \frac{\partial y_1}{\partial x_1} & \frac{\partial y_1}{\partial x_2} & \cdots & \frac{\partial y_1}{\partial x_n} \\ \frac{\partial y_2}{\partial x_1} & \frac{\partial y_2}{\partial x_2} & \cdots & \frac{\partial y_2}{\partial x_n} \\ \vdots & \vdots & \ddots & \vdots \\ \frac{\partial y_m}{\partial x_1} & \frac{\partial y_m}{\partial x_2} & \cdots & \frac{\partial y_m}{\partial x_n} \end{pmatrix}, \tag{B.5}$$

and J^T is its transpose. In particular, the variance of a single-valued function $g(\mathbf{x})$ of $\mathbf{x}$ is approximately

$$\text{var}(g) \approx \sum_{i=1}^{n}\sum_{j=1}^{n} \text{cov}(x_i, x_j)\left(\frac{\partial g}{\partial x_i}\right)\left(\frac{\partial g}{\partial x_j}\right), \tag{B.6}$$

Appendix B: The delta technique • 315

where $\text{cov}(x_i, x_j)$ is the covariance of x_i and x_j if $i \neq j$ or the variance of x_i if $i = j$. The mean of $g(\mathbf{x})$ is approximately

$$E(g) \approx g(\mu_1, \mu_2, \cdots, \mu_n) + \frac{1}{2} \sum_{i=1}^{n} \sum_{j=1}^{n} \text{cov}(x_i, x_j) \frac{\partial^2 g}{\partial x_i \partial x_j}, \quad (B.7)$$

where $\mu_i = E(x_i)$.

Example 1. The variance of the JC69 distance. Suppose x out of n sites are different between two sequences. Equation (1.6) gives the sequence distance under the JC69 model as $\hat{d} = -3/4 \log(4\hat{p}/3)$, where $\hat{p} = x/n$ is the proportion of different sites. This is a binomial proportion, with variance $\text{var}(\hat{p}) = \hat{p}(1-\hat{p})/n$. Consider $\hat{d}$ as a function of $\hat{p}$, and note that $d\hat{d}/d\hat{p} = 1/(1 - 4\hat{p}/3)$. Thus the approximate variance of $\hat{d}$ is $[\hat{p}(1-\hat{p})]/[n(1 - 4\hat{p}/3)^2]$, as given in equation (1.7). □

Example 2. The expectation and variance of the ratio of two random variables. Suppose x and y are two random variables. Let $\mu_x = E(x)$, $\mu_y = E(y)$, $\sigma_x^2 = \text{var}(x)$, $\sigma_y^2 = \text{var}(y)$, and $\sigma_{xy} = \text{cov}(x, y)$. One can then use equations (B.7) and (B.6) to work out the approximate mean and variance of the ratio x/y as

$$E\left(\frac{x}{y}\right) \approx \frac{\mu_x}{\mu_y} - \frac{\sigma_{xy}}{\mu_y^2} + \frac{\mu_x \sigma_y^2}{\mu_y^3}, \quad (B.8)$$

$$\text{var}\left(\frac{x}{y}\right) \approx \frac{\sigma_x^2}{\mu_y^2} - \frac{2\mu_x \sigma_{xy}}{\mu_y^3} + \frac{\mu_x^2 \sigma_y^2}{\mu_y^4}. \quad (B.9)$$

These formulae may be used to derive the mean and variance of the transition/transversion ratio S/V, where S and V are the proportions of sites with transitional and tranversional differences between two sequences (see Subsection 1.2.2). □

Example 3. The variance of the sequence distance under the K80 model. The maximum likelihood estimates of the sequence distance d and the transition/transversion rate ratio κ are given in equation (1.11), as functions of S and V. Note that S and V are multinomial proportions, with variance–covariance matrix

$$\text{var}\begin{pmatrix} S \\ V \end{pmatrix} = \begin{pmatrix} S(1-S)/n & -SV/n \\ -SV/n & V(1-V)/n \end{pmatrix}, \quad (B.10)$$

where n is the number of sites in the sequence. Now consider the estimates $\hat{d}$ and $\hat{\kappa}$ as functions of S and V, so that

$$\text{var}\begin{pmatrix} \hat{d} \\ \hat{\kappa} \end{pmatrix} = J \cdot \text{var}\begin{pmatrix} S \\ V \end{pmatrix} \cdot J^{\text{T}}, \quad (B.11)$$

where J is the Jacobi matrix for the transform:

$$J = \begin{pmatrix} \frac{\partial \hat{d}}{\partial S} & \frac{\partial \hat{d}}{\partial V} \\ \frac{\partial \hat{\kappa}}{\partial S} & \frac{\partial \hat{\kappa}}{\partial V} \end{pmatrix}$$

$$= \begin{pmatrix} \dfrac{1}{1-2S-V} & \dfrac{1}{2(1-2V)} + \dfrac{1}{2(1-2S-V)} \\ -\dfrac{4}{(1-2S-V)\log(1-2V)} & -\dfrac{2}{(1-2S-V)\log(1-2V)} + \dfrac{4\log(1-2S-V)}{(1-2V)(\log(1-2V))^2} \end{pmatrix}.$$

(B.12)

In particular, the variance of $\hat{d}$ can be derived using equation (B.6), as

$$\operatorname{var}(\hat{d}) = \left(\frac{\partial \hat{d}}{\partial S}\right)^2 \operatorname{var}(S) + 2 \cdot \frac{\partial \hat{d}}{\partial S} \cdot \frac{\partial \hat{d}}{\partial V} \cdot \operatorname{cov}(S, V) + \left(\frac{\partial \hat{d}}{\partial V}\right)^2 \operatorname{var}(V) \quad \text{(B.13)}$$

$$= \left[a^2 S + b^2 V - (aS + bV)^2\right]/n,$$

as given in equation (1.13).

Appendix C: Phylogenetics software

Here we give an overview of several widely used programs or software packages in molecular phylogenetics. An almost exhaustive list has been compiled by Joseph Felsenstein, at http://evolution.gs.washington.edu/phylip/software.html.

CLUSTAL (Thompson *et al.* 1994; Higgins and Sharp 1988) is a program for automatic alignment of multiple sequences. It does a crude pairwise alignments using the algorithm of Needleman and Wunsch (1970) to calculate pairwise distances, which are used to reconstruct an NJ tree. It then uses the NJ tree as a guide to progressively align multiple sequences. CLUSTAL exists in two major variants: CLUSTALW with a command-line interface and CLUSTALX with a graphics interface. The program runs on most common platforms. It can be downloaded at, say, ftp://ftp.ebi.ac.uk in directory pub/software.

PHYLIP (Phylogeny Inference Package) is distributed by Joseph Felsenstein. It is a package of about 30 programs for parsimony, distance, and likelihood methods of phylogeny reconstruction. The programs are written in C and can run on virtually any platform. They can deal with various types of data, including DNA and protein sequences. PHYLIP is available from its web site at http://evolution.gs.washington.edu/phylip.html.

PAUP* 4 (Phylogenetic Analysis Using Parsimony *and other methods) is written by David Swofford and distributed by Sinauer Associates. It is a widely used package for phylogenetic analysis of molecular and morphological data using distance, parsimony, and likelihood methods. PAUP* implements efficient heuristic tree search algorithms. Mac versions have a graphics user interface, while command-line portable versions are available for Windows and UNIX. The program web site is http://paup.csit.fsu.edu/, which has links to the Sinauer site, where purchase information is available.

MacClade is written by Wayne Maddison and David Maddison, and published by Sinauer Associates. It is a Macintosh program (including OSX) for ancestral state reconstruction. It can analyse molecular as well as discrete morphological characters to trace changes along branches on the tree. The program web page is at http://www.macclade.org/, which has a link to the Sinauer site.

MEGA3 (Molecular Evolutionary Genetic Analysis) is a Windows program written by S. Kumar, K. Tamura, and M. Nei (2005a). It can be used to conduct distance calculations and phylogeny reconstruction using distance and parsimony methods. It invokes CLUSTAL to perform multiple sequence alignment. The program is distributed at its web site at http://www.megasoftware.net/.

MrBayes is written by John Huelsenbeck and Fredrik Ronquist. It is a program for Bayesian inference of phylogenies from DNA, protein, and codon sequences. The program uses Markov-chain Monte Carlo (MCMC) to search in the tree space and generates the posterior distributions of tree topologies, branch lengths, and other substitution parameters. It can analyse multiple heterogeneous data sets while accommodating their differences. Windows executable and C source codes are available from the program web page at http://mrbayes.csit.fsu.edu/.

PAML (Phylogenetic Analysis by Maximum Likelihood) is my package for likelihood analysis of nucleotide, amino acid, and codon sequences. The programs are not good for making trees, but implement many sophisticated substitution models, and can be used to reconstruct ancestral sequences, detect positive selection, and estimate species divergence times under relaxed molecular clock models. C source code and Windows and Mac executables are available at http://abacus.gene.ucl.ac.uk/software/paml.html.

PHYML, written by Stéphane Guindon and Olivier Gascuel, implements the fast ML search algorithm of Guindon and Gascuel (2003). Executables for various platforms are available at http://atgc.lirmm.fr/phyml/.

TreeView, written by Rod Page, is a program for viewing, editing, and printing trees. The output graphics files can be imported into other software for further editing. Executables for various platforms are available at http://taxonomy.zoology.gla.ac.uk/rod/treeview.html.

References

Adachi, J. and Hasegawa, M. 1996a. Model of amino acid substitution in proteins encoded by mitochondrial DNA. *J. Mol. Evol.* **42**:459–468.

Adachi, J. and Hasegawa, M. 1996b. MOLPHY Version 2.3: Programs for molecular phylogenetics based on maximum likelihood. *Computer Science Monographs*, **28**:1–150. Institute of Statistical Mathematics, Tokyo.

Adachi, J., Waddell, P. J., Martin, W., and Hasegawa, M. 2000. Plastid genome phylogeny and a model of amino acid substitution for proteins encoded by chloroplast DNA. *J. Mol. Evol.* **50**:348–358.

Akaike, H. 1974. A new look at the statistical model identification. *IEEE Trans. Autom. Contr. ACM* **19**:716–723.

Akam, M. 1995. Hox genes and the evolution of diverse body plans. *Philos. Trans. R Soc. Lond. B Biol. Sci.* **349**:313–319.

Akashi, H. 1995. Inferring weak selection from patterns of polymorphism and divergence at "silent" sites in Drosophila DNA. *Genetics* **139**:1067–1076.

Akashi, H. 1999a. Inferring the fitness effects of DNA mutations from polymorphism and divergence data: statistical power to detect directional selection under stationarity and free recombination. *Genetics* **151**:221–238.

Akashi, H. 1999b. Within- and between-species DNA sequence variation and the 'footprint' of natural selection. *Gene* **238**:39–51.

Akashi, H., Goel, P., and John, A. 2006. Ancestral state inference and the study of codon bias evolution: implications for molecular evolutionary analysis of the *Drosophila melanogaster* species subgroup. *Genetics*.

Alba, R., Kelmenson, P. M., Cordonnier-Pratt, M. -M., and Pratt, L. H. 2000. The phytochrome gene family in tomato and the rapid differential evolution of this family in angiosperms. *Mol. Biol. Evol.* **17**:362–373.

Albert, V. A. 2005. *Parsimony, Phylogeny, and Genomics*. Oxford University Press, Oxford.

Alfaro, M. E., Zoller, S., and Lutzoni, F. 2003. Bayes or bootstrap? A simulation study comparing the performance of Bayesian Markov chain Monte Carlo sampling and bootstrapping in assessing phylogenetic confidence. *Mol. Biol. Evol.* **20**:255–266.

Altekar, G., Dwarkadas, S., Huelsenbeck, J. P., and Ronquist, F. 2004. Parallel Metropolis coupled Markov chain Monte Carlo for Bayesian phylogenetic inference. *Bioinformatics* **20**:407–415.

Altschul, S. F., Gish, W., Miller, W. *et al.* 1990. Basic local alignment search tool. *J. Mol. Biol.* **215**:403–410.

Anisimova, A. and Yang, Z. 2004. Molecular evolution of hepatitis delta virus antigen gene: recombination or positive selection? *J. Mol. Evol.* **59**:815–826.

Anisimova, A. and Yang, Z. 2006. Searching for positive selection affecting a few sites and lineages. Submitted.

Anisimova, M., Bielawski, J. P., and Yang, Z. 2001. The accuracy and power of likelihood ratio tests to detect positive selection at amino acid sites. *Mol. Biol. Evol.* **18**:1585–1592.

Anisimova, M., Bielawski, J. P., and Yang, Z. 2002. Accuracy and power of Bayes prediction of amino acid sites under positive selection. *Mol. Biol. Evol.* **19**:950–958.

Anisimova, M., Nielsen, R., and Yang, Z. 2003. Effect of recombination on the accuracy of the likelihood method for detecting positive selection at amino acid sites. *Genetics* **164**:1229–1236.

Aris-Brosou, S. and Yang, Z. 2002. The effects of models of rate evolution on estimation of divergence dates with a special reference to the metazoan 18S rRNA phylogeny. *Syst. Biol.* **51**:703–714.

Atkinson, A. C. 1970. A method of discriminating between models. *J. R. Statist. Soc. B* **32**:323–353.

Bailly, X., Leroy, R., Carney, S. *et al.* 2003. The loss of the hemoglobin H2S-binding function in annelids from sulfide-free habitats reveals molecular adaptation driven by Darwinian positive selection. *Proc. Natl. Acad. Sci. U.S.A.* **100**:5885–5890.

Bargelloni, L., Marcato, S., and Patarnello, T. 1998. Antarctic fish hemoglobins: evidence for adaptive evolution at subzero temperatures. *Proc. Natl. Acad. Sci. U.S.A* **95**:8670–8675.

Barker, D. 2004. LVB: parsimony and simulated annealing in the search for phylogenetic trees. *Bioinformatics* **20**:274–275.

Barkman, T. J. 2003. Evidence for positive selection on the floral scent gene isoeugenol-O-methyltransferase. *Mol. Biol. Evol.* **20**:168–172.

Barrier, M., Robichaux, R. H., and Purugganan, M. D. 2001. Accelerated regulatory gene evolution in an adaptive radiation. *Proc. Natl. Acad. Sci. U.S.A.* **98**:10208–10213.

Barry, D. and Hartigan, J. A. 1987a. Asynchronous distance between homologous DNA sequences. *Biometrics* **43**:261–276.

Barry, D. and Hartigan, J. A. 1987b. Statistical analysis of hominoid molecular evolution. *Statist. Sci.* **2**:191–210.

Baudry, E. and Depaulis, F. 2003. Effect of misoriented sites on neutrality tests with outgroup. *Genetics* **165**:1619–1622.

Baum, J., Ward, R., and Conway, D. 2002. Natural selection on the erythrocyte surface. *Mol. Biol. Evol.* **19**:223–229.

Beerli, P. and Felsenstein, J. 2001. Maximum likelihood estimation of a migration matrix and effective population sizes in n subpopulations by using a coalescent approach. *Proc. Natl. Acad. Sci. U.S.A.* **98**:4563–4568.

Belda, E., Moya, A., and Silva, F. J. 2005. Genome rearrangement distances and gene order phylogeny in γ-proteobacteria. *Mol. Biol. Evol.* **22**:1456–1467.

Benner, S. A. 2001. Natural progression. *Nature* **409**:459.

Benner, S. A. 2002. The past as the key to the present: resurrection of ancient proteins from eosinophils. *Proc. Natl. Acad. Sci. U.S.A.* **99**:4760–4761.

Benton, M. J., Wills, M., and Hitchin, R. 2000. Quality of the fossil record through time. *Nature* **403**:534–538.

Berry, V. and Gascuel, O. 1996. On the interpretation of bootstrap trees: appropriate threshold of clade selection and induced gain. *Mol. Biol. Evol.* **13**:999–1011.

Betran, E. and Long, M. 2003. Dntf-2r, a young Drosophila retroposed gene with specific male expression under positive Darwinian selection. *Genetics* **164**:977–988.

Bielawski, J. P. and Yang, Z. 2001. Positive and negative selection in the DAZ gene family. *Mol. Biol. Evol.* **18**:523–529.

Bielawski, J. P. and Yang, Z. 2004. A maximum likelihood method for detecting functional divergence at individual codon sites, with application to gene family evolution. *J. Mol. Evol.* **59**:121–132.

Bielawski, J. P., Dunn, K., and Yang, Z. 2000. Rates of nucleotide substitution and mammalian nuclear gene evolution: approximate and maximum-likelihood methods lead to different conclusions. *Genetics* **156**:1299–1308.

Bielawski, J. P., Dunn, K. A., Sabehi, G., and Beja, O. 2004. Darwinian adaptation of proteorhodopsin to different light intensities in the marine environment. *Proc. Natl. Acad. Sci. U.S.A.* **101**:14824–14829.

Bierne, N. and Eyre-Walker, A. 2003. The problem of counting sites in the estimation of the synonymous and nonsynonymous substitution rates: implications for the correlation between the synonymous substitution rate and codon usage bias. *Genetics* **165**:1587–1597.

Bininda-Emonds, O. R. P. 2004. *Phylogenetic Supertrees: Combining Information to Reveal the Tree of Life*. Kluwer Academic, Dordrecht.

Birdsey, G. M., Lewin, J., Cunningham, A. A. *et al.* 2004. Differential enzyme targeting as an evolutionary adaptation to herbivory in carnivora. *Mol. Biol. Evol.* **21**:632–646.

Bishop, J. G. 2005. Directed mutagenesis confirms the functional importance of positively selected sites in polygalacturonase inhibitor protein (PGIP). *Mol. Biol. Evol.* **22**:1531–1534.

Bishop, J. G., Dean, A. M., and Mitchell-Olds, T. 2000. Rapid evolution in plant chitinases: molecular targets of selection in plant-pathogen coevolution. *Proc. Natl. Acad. Sci. U.S.A.* **97**:5322–5327.

Bishop, J. G., Ripoll, D. R., Bashir, S. *et al.* 2005. Selection on glycine β-1,3-endoglucanase genes differentially inhibited by a phytophthora glucanase inhibitor protein. *Genetics* **169**:1009–1019.

Bishop, M. J. and Friday, A. E. 1985. Evolutionary trees from nucleic acid and protein sequences. *Proc. R. Soc. Lond. B Biol. Sci.* **226**:271–302.

Bishop, M. J. and Friday, A. E. 1987. Tetropad relationships: the molecular evidence. *Molecules and Morphology in Evolution: Conflict or Compromise?* in (ed. C. Patterson) Cambridge University Press, Cambridge, pp. 123–139.

Bishop, M. J. and Thompson, E. A. 1986. Maximum likelihood alignment of DNA sequences. *J. Mol. Biol.* **190**:159–165.

Bjorklund, M. 1999. Are third positions really that bad? A test using vertebrate cytochrome *b*. *Cladistics* **15**:191–197.

Bjorkman, P. J., Saper, S. A., Samraoui, B. *et al.* 1987a. Structure of the class I histocompatibility antigen, HLA-A2. *Nature* **329**:506–512.

Bjorkman, P. J., Saper, S. A., Samraoui, B. *et al.* 1987b. The foreign antigen binding site and T cell recognition regions of class I histocompatibility antigens. *Nature* **329**:512–518.

Bonhoeffer, S., Holmes, E. C., and Nowak, M. A. 1995. Causes of HIV diversity. *Nature* **376**:125.

Box, G. E. P. 1979. Robustness in the strategy of scientific model building. In *Robustness in Statistics* (ed. R. L. Launer, and G. N. Wilkinson), p. 202. Academic Press, New York.

Box, G. E. P. and Muller, M. E. 1958. A note on the generation of random normal deviates. *Ann. Math. Statist.* **29**:610-611.

Braverman, J. M., Hudson, R. R., Kaplan, N. L. *et al.* 1995. The hitchhiking effect on the site frequency spectrum of DNA polymorphisms. *Genetics* **140**:783–796.

Bremer, K. 1988. The limits of amino acid sequence data in angiosperm phylogenetic reconstruction. *Evolution* **42**:795–803.

Brent, R. P. 1973. *Algorithms for Minimization Without Derivatives*. Prentice-Hall Inc., Englewood Cliffs, NJ.

Brinkmann, H., van der Giezen, M., Zhou, Y. *et al.* 2005. An empirical assessment of long-branch attraction artefacts in deep eukaryotic phylogenomics. *Syst. Biol.* **54**:743–757.

Britten, R. J. 1986. Rates of DNA sequence evolution differ between taxonomic groups. *Science* **231**:1393–1398.

Bromham, L. 2002. Molecular clocks in reptiles: life history influences rate of molecular evolution. *Mol. Biol. Evol.* **19**:302–309.

Bromham, L. and Penny, D. 2003. The modern molecular clock. *Nat. Rev. Genet.* **4**:216–224.

Bromham, L., Rambaut, A., and Harvey, P. H. 1996. Determinants of rate variation in mammalian DNA sequence evolution. *J. Mol. Evol.* **43**:610–621.

Bromham, L., Penny, D., Rambaut, A., and Hendy, M. D. 2000. The power of relative rates tests depends on the data. *J. Mol. Evol.* **50**:296–301.

Brown, W. M., Prager, E. M., Wang, A., and Wilson, A. C. 1982. Mitochondrial DNA sequences of primates: tempo and mode of evolution. *J. Mol. Evol.* **18**:225–239.

Brunet, M., Guy, F., Pilbeam, D. *et al.* 2002. A new hominid from the upper Miocene of Chad, central Africa. *Nature* **418**:145–151.

Bruno, W. J. 1996. Modeling residue usage in aligned protein sequences via maximum likelihood. *Mol. Biol. Evol.* **13**:1368–1374.

Bruno, W. J. and Halpern, A. L. 1999. Topological bias and inconsistency of maximum likelihood using wrong models. *Mol. Biol. Evol.* **16**:564–566.

Bruno, W. J., Socci, N. D., and Halpern, A. L. 2000. Weighted neighbor joining: a likelihood-based approach to distance-based phylogeny reconstruction. *Mol. Biol. Evol.* **17**:189–197.

Bryant, D. 2003. A classication of consensus methods for phylogenetics. In *BioConsensus, DIMACS Series in Discrete Mathematics and Theoretical Computer Science*. (ed. M. Janowitz, F. -J. Lapointe, F. R. McMorris, B. Mirkin, and F. S. Roberts), pp. 163–184. American Mathematical Society, Providence, RI.

Bryant, D. and Waddell, P. J. 1998. Rapid evaluation of least-squares and minimum-evolution criteria on phylogenetic trees. *Mol. Biol. Evol.* **15**:1346–1359.

Buckley, T. R. 2002. Model misspecification and probabilistic tests of topology: evidence from empirical data sets. *Syst. Biol.* **51**:509–523.

Bulmer, M. G. 1990. Estimating the variability of substitution rates. *Genetics* **123**:615–619.

Bulmer, M. S. and Crozier, R. H. 2004. Duplication and diversifying selection among termite antifungal peptides. *Mol. Biol. Evol.* **21**:2256–2264.

Bush, R. M., Fitch, W. M., Bender, C. A., and Cox, N. J. 1999. Positive selection on the H3 hemagglutinin gene of human influenza virus A. *Mol. Biol. Evol.* **16**:1457–1465.

Bustamante, C. D., Wakeley, J., Sawyer, S., and Hartl, D. L. 2001. Directional selection and the site-frequency spectrum. *Genetics* **159**:1779–1788.

Bustamante, C. D., Nielsen, R., Sawyer, S. A. *et al.* 2002. The cost of inbreeding in Arabidopsis. *Nature* **416**:531–534.

Bustamante, C. D., Nielsen, R., and Hartl, D. L. 2003. Maximum likelihood and Bayesian methods for estimating the distribution of selective effects among classes of mutations using DNA polymorphism data. *Theor. Popul. Biol.* **63**:91–103.

Camin, J. H. and Sokal, R. R. 1965. A method for deducing branching sequences in phylogeny. *Evolution* **19**:311–326.

Cao, Y., Adachi, J., Janke, A. *et al.* 1994. Phylogenetic relationships among eutherian orders estimated from inferred sequences of mitochondrial proteins: instability of a tree based on a single gene. *J. Mol. Evol.* **39**:519–527.

Cao, Y., Janke, A., Waddell, P. J. *et al.* 1998. Conflict among individual mitochondrial proteins in resolving the phylogeny of eutherian orders. *J. Mol. Evol.* **47**:307–322.

Cao, Y., Kim, K. S., Ha, J. H., and Hasegawa, M. 1999. Model dependence of the phylogenetic inference: relationship among Carnivores, Perissodactyls and Cetartiodactyls as inferred from mitochondrial genome sequences. *Genes Genet. Syst.* **74**:211–217.

Carlin, B. P. and Louis, T. A. 2000. *Bayes and Empirical Bayes Methods for Data Analysis*. Chapman and Hall, London.

Carroll, S. B. 1995. Homeotic genes and the evolution of the arthropods and chordates. *Nature* **376**:479–485.

Cavalli-Sforza, L. L. and Edwards, A. W. F. 1967. Phylogenetic analysis: models and estimation procedures. *Evolution* **21**:550–570.

Cavender, J. A. 1978. Taxonomy with confidence. *Math. Biosci.* **40**:271–280.

Chang, B. S. and Donoghue, M. J. 2000. Recreating ancestral proteins. *Trends Ecol. Evol.* **15**:109–114.

Chang, J. T. 1996a. Full reconstruction of Markov models on evolutionary trees: identifiability and consistency. *Math. Biosci.* **137**:51–73.

Chang, J. T. 1996b. Inconsistency of evolutionary tree topology reconstruction methods when substitution rates vary across characters. *Math. Biosci.* **134**:189–215.

Charleston, M. A. 1995. Toward a characterization of landscapes of combinatorial optimization problems, with special attention to the phylogeny problem. *J. Comput. Biol.* **2**:439–450.

Chenna, R., Sugawara, H., Koike, T. *et al.* 2003. Multiple sequence alignment with the Clustal series of programs. *Nucleic Acids Res.* **31**:3497–3500.

Chernoff, H. 1954. On the distribution of the likelihood ratio. *Ann. Math. Stat.* **25**:573–578.

Choisy, M., Woelk, C. H., Guegan, J. F., and Robertson, D. L. 2004. Comparative study of adaptive molecular evolution in different human immunodeficiency virus groups and subtypes. *J. Virol.* **78**:1962–1970. [Erratum in *J. Virol.* 2004 **78**:4381–2].

Chor, B. and Snir, S. 2004. Molecular clock fork phylogenies: closed form analytic maximum likelihood solutions. *Syst. Biol.* **53**:963–967.

Chor, B., Holland, B. R., Penny, D., and Hendy, M. D. 2000. Multiple maxima of likelihood in phylogenetic trees: an analytic approach. *Mol. Biol. Evol.* **17**:1529–1541.

Clark, A. G., Glanowski, S., Nielsen, R. *et al.* 2003. Inferring nonneutral evolution from human-chimp-mouse orthologous gene trios. *Science* **302**:1960–1963.

Clark, B. 1970. Selective constraints on amino-acid substitutions during the evolution of proteins. *Nature* **228**:159–160.

Clark, N. L., Aagaard, J. E., and Swanson, W. J. 2006. Evolution of reproductive proteins from animals and plants. *Reproduction* **131**:11–22.

Collins, T. M., Wimberger, P. H., and Naylor, G. J. P. 1994. Compositional bias, character-state bias, and character-state reconstruction using parsimony. *Syst. Biol.* **43**:482–496.

Comeron, J. M. 1995. A method for estimating the numbers of synonymous and nonsynonymous substitutions per site. *J. Mol. Evol.* **41**:1152–1159.

Consuegra, S., Megens, H. -J., Schaschl, H. *et al.* 2005. Rapid evolution of the MHC Class I locus results in different allelic compositions in recently diverged populations of Atlantic salmon. *Mol. Biol. Evol.* **22**:1095–1106.

Cooper, A. and Fortey, R. 1998. Evolutionary explosions and the phylogenetic fuse. *Trends Ecol. Evol.* **13**:151–156.

Cooper, A. and Penny, D. 1997. Mass survival of birds across the Cretaceous-Tertiary boundary: molecular evidence. *Science* **275**:1109–1113.

Cox, D. R. 1961. Tests of separate families of hypotheses. *Proc. 4th Berkeley Symp. Math. Stat. Prob.* **1**:105–123.

Cox, D. R. 1962. Further results on tests of separate families of hypotheses. *J. R. Statist. Soc. B* **24**:406–424.

Cox, D. R. and Hinkley, D. V. 1974. *Theoretical Statistics*. Chapman and Hall, London.

Cummings, M. P., Otto, S. P., and Wakeley, J. 1995. Sampling properties of DNA sequence data in phylogenetic analysis. *Mol. Biol. Evol.* **12**:814–822.

Cutler, D. J. 2000. Understanding the overdispersed molecular clock. *Genetics* **154**:1403–1417.

Dagan, T., Talmor, Y., and Graur, D. 2002. Ratios of radical to conservative amino acid replacement are affected by mutational and compositional factors and may not be indicative of positive Darwinian selection. *Mol. Biol. Evol.* **19**:1022–1025.

Davison, A. C. 2003. *Statistical Models*. Cambridge University Press, Cambridge.

Dawkins, R. and Krebs, J. R. 1979. Arms races between and within species. *Proc. R. Soc. Lond. B. Biol. Sci.* **205**:489–511.

Dayhoff, M. O., Eck, R. V., and Park, C. M. 1972. Evolution of a complex system: the immunoglobulins. Pp. 31–40. *Atlas of protein sequence and structure*, pp. 31–40. National Biomedical Research Foundation, Silver Spring, MD.

Dayhoff, M. O., Schwartz, R. M., and Orcutt, B. C. 1978. A model of evolutionary change in proteins. *Atlas of protein sequence and structure*, Vol 5, Suppl. 3, pp. 345–352. National Biomedical Research Foundation, Washington DC.

DeBry, R. W. 1992. The consistency of several phylogeny-inference methods under varying evolutionary rates. *Mol. Biol. Evol.* **9**:537–551.

DeBry, R. 2001. Improving interpretation of the decay index for DNA sequences. *Syst. Biol.* **50**:742–752.

Deely, J. J. and Lindley, D. V. 1981. Bayes empirical Bayes. *J. Amer. Statist. Assoc.* **76**:833–841.

DeGroot, M. H. and Schervish, M. J. 2002. *Probability and Statistics*. Addison-Wesley, Boston, MA.

Delson, E., Tattersall, I., Van Couvering, J. A., and Brooks, A. S. 2000. In *Encyclopedia of Human Evolution and Prehistory* (ed. E. Delson, I. Tattersall, J. A. Van Couvering, and A. S. Brooks), pp. 166–171. Garland, New York.

Desper, R. and Gascuel, O. 2005. The minimum-evolution distance-based approach to phylogenetic inference. In *Mathematics of Evolution and Phylogeny* (ed. O. Gascuel), pp. 1–32. Oxford University Press, Oxford.

Diggle, P. J. 1990. *Time Series: a Biostatistical Introduction*. Oxford University Press, Oxford.

Doolittle, F. W. 1998. You are what you eat: a gene transfer ratchet could account for bacterial genes in eukaryotic nuclear genomes. *Trends in Genetics* **14**:307–311.

Doolittle, R. F. and Blomback, B. 1964. Amino-acid sequence investigations of fibrinopeptides from various mammals: evolutionary implications. *Nature* **202**:147–152.

Douzery, E. J., Delsuc, F., Stanhope, M. J., and Huchon, D. 2003. Local molecular clocks in three nuclear genes: divergence times for rodents and other mammals and incompatibility among fossil calibrations. *J. Mol. Evol.* **57**:S201–S213.

Drummond, A. J., Nicholls, G. K., Rodrigo, A. G., and Solomon, W. 2002. Estimating mutation parameters, population history and genealogy simultaneously from temporally spaced sequence data. *Genetics* **161**:1307–1320.

Duda, T. F. and Palumbi, S. R. 2000. Evolutionary diversification of multigene families: allelic selection of toxins in predatory cone snails. *Mol. Biol. Evol.* **17**:1286–1293.

Duda, T. F., Jr and Palumbi, S. R. 1999. Molecular genetics of ecological diversification: duplication and rapid evolution of toxin genes of the venomous gastropod *Conus*. *Proc. Natl. Acad. Sci. U.S.A.* **96**:6820–6823.

Duret, L. 2002. Evolution of synonymous codon usage in metazoans. *Curr. Opin. Genet. Dev.* **12**:640–649.

Duret, L., Semon, M., Piganeau, G. *et al.* 2002. Vanishing GC-rich isochores in mammalian genomes. *Genetics* **162**:1837–1847.

Dutheil, J., Pupko, T., Jean-Marie, A., and Galtier, N. 2005. A model-based approach for detecting coevolving positions in a molecule. *Mol. Biol. Evol.* **22**:1919–1928.

Eck, R. V. and Dayhoff, M. O. 1966. Inference from protein sequence comparisons. In *Atlas of protein sequence and structure* (ed. M. O. Dayhoff). National Biomedical Research Foundation, Silver Spring, MD.

Edwards, A. W. F. 1970. Estimation of the branch points of a branching diffusion process (with discussion). *J. R. Statist. Soc. B.* **32**:155–174.

Edwards, A. W. F. 1992. *Likelihood, expanded edition*. Johns Hopkins University Press, London.

Edwards, A. W. F. and Cavalli-Sforza, L. L. 1963. The reconstruction of evolution (abstract). *Ann. Hum. Genet.* **27**:105.

Efron, B. 1979. Bootstrap methods: another look at the jackknife. *Ann. Stat.* **7**:1–26.

Efron, B. 1986. Why isn't everyone a Bayesian? (with discussion). *Am. J. Statist. Assoc.* **40**:1–11.

Efron, B. and Hinkley, D. V. 1978. Assessing the accuracy of the maximum likelihood estimator: observed and expected information. *Biometrika* **65**:457–487.

Efron, B. and Tibshirani, R. J. 1993. *An Introduction to the Bootstrap*. Chapman and Hall, London.

Efron, B., Halloran, E., and Holmes, S. 1996. Bootstrap confidence levels for phylogenetic trees [corrected and republished article originally printed in *Proc. Natl. Acad. Sci. U.S.A.* 1996 **93**:7085–7090]. *Proc. Natl. Acad. Sci. U.S.A.* **93**:13429–13434.

Emes, R. D., Riley, M. C., Laukaitis, C. M. *et al.* 2004. Comparative evolutionary genomics of androgen-binding protein genes. *Genome Res.* **14**:1516–1529.

Enard, W., Przeworski, M., Fisher, S. E. *et al.* 2002. Molecular evolution of FOXP2, a gene involved in speech and language. *Nature* **418**:869–872.

Erixon, P., Svennblad, B., Britton, T., and Oxelman, B. 2003. Reliability of Bayesian posterior probabilities and bootstrap frequencies in phylogenetics. *Syst. Biol.* **52**:665–673.

Evans, P. D., Anderson, J. R., Vallender, E. J. *et al.* 2004. Reconstructing the evolutionary history of microcephalin, a gene controlling human brain size. *Hum. Mol. Genet.* **13**:1139–1145.

Everitt, B. S., Landau, S., and Leese, M. 2001. *Cluster Analysis*. Arnold, London.

Excoffier, L. and Yang, Z. 1999. Substitution rate variation among sites in the mitochondrial hypervariable region I of humans and chimpanzees. *Mol. Biol. Evol.* **16**:1357–1368.

Eyre-Walker, A. 1998. Problems with parsimony in sequences of biased base composition. *J. Mol. Evol.* **47**:686–690.

Fan, C., Purugganan, M. D., Thomas, D. T. *et al.* 2004. Heterogeneous evolution of the Myc-like anthocyanin regulatory gene and its phylogenetic utility in Cornus L. (Cornaceae). *Mol. Phylogenet. Evol.* **33**:580–594.

Fares, M. A. and Wolfe, K. H. 2003. Positive selection and subfunctionalization of duplicated CCT chaperonin subunits. *Mol. Biol. Evol.* **20**:1588–1597.

Fares, M. A., Barrio, E., Sabater-Munoz, B., and Moya, A. 2002. The evolution of the heat-shock protein GroEL from Buchnera, the primary endosymbiont of aphids, is governed by positive selection. *Mol. Biol. Evol.* **19**:1162–1170.

Farris, J. S. 1969. A successive approximation approach to character weighting. *Syst. Zool.* **18**:374–385.

Farris, J. S. 1973. On the use of the parsimony criterion for inferring evolutionary trees. *Syst. Zool.* **22**:250–256.

Farris, J. S. 1977. Phylogenetic analysis under Dollo's law. *Syst. Zool.* **26**:77–88.

Farris, J. S. 1983. The logical basis of phylogenetic analysis. *Advances in Cladistics*. (ed. N. Platnick, and V. Funk), pp. 7–26. Columbia University Press, New York.

Farris, J. S. 1989. The retention index and the rescaled consistency index. *Cladistics* **5**:417–419.

Fay, J. C. and Wu, C. I. 2000. Hitchhiking under positive Darwinian selection. *Genetics* **155**:1405–1413.

Fay, J. C. and Wu, C. -I. 2001. The neutral theory in the genomic era. *Curr. Opinion Genet. Dev.* **11**:642–646.

Fay, J. C. and Wu, C. I. 2003. Sequence divergence, functional constraint, and selection in protein evolution. *Annu. Rev. Genomics Hum. Genet.* **4**:213–235.

Fay, J. C., Wyckoff, G. J., and Wu, C. -I. 2001. Positive and negative selection on the human genome. *Genetics* **158**:1227–1234.

Felsenstein, J. 1973a. Maximum-likelihood estimation of evolutionary trees from continuous characters. *Am. J. Hum. Genet.* **25**:471–492.

Felsenstein, J. 1973b. Maximum likelihood and minimum-steps methods for estimating evolutionary trees from data on discrete characters. *Syst. Zool.* **22**:240–249.

Felsenstein, J. 1978a. The number of evolutionary trees. *Syst. Zool.* **27**:27–33.

Felsenstein, J. 1978b. Cases in which parsimony and compatibility methods will be positively misleading. *Syst. Zool.* **27**:401–410.

Felsenstein, J. 1981. Evolutionary trees from DNA sequences: a maximum likelihood approach. *J. Mol. Evol.* **17**:368–376.

Felsenstein, J. 1983. Statistical inference of phylogenies. *J. R. Statist. Soc. A* **146**:246–272.

Felsenstein, J. 1985a. Confidence limits on phylogenies: an approach using the bootstrap. *Evolution* **39**:783–791.

Felsenstein, J. 1985b. Phylogenies and the comparative method. *Amer. Nat.* **125**:1–15.

Felsenstein, J. 1985c. Confidence limits on phylogenies with a molecular clock. *Evolution* **34**:152–161.

Felsenstein, J. 1988. Phylogenies from molecular sequences: inference and reliability. *Annu. Rev. Genet.* **22**:521–565.

Felsenstein, J. 2001a. Taking variation of evolutionary rates between sites into account in inferring phylogenies. *J. Mol. Evol.* **53**:447–455.

Felsenstein, J. 2001b. The troubled growth of statistical phylogenetics. *Syst. Biol.* **50**:465–467.

Felsenstein, J. 2004. *Inferring Phylogenies*. Sinauer Associates, Sunderland, MA.

Felsenstein, J. and Churchill, G. A. 1996. A hidden Markov model approach to variation among sites in rate of evolution. *Mol. Biol. Evol.* **13**:93–104.

Felsenstein, J. and Kishino, H. 1993. Is there something wrong with the bootstrap on phylogenies? A reply to Hillis and Bull. *Syst. Biol.* **42**:193–200.

Felsenstein, J. and Sober, E. 1986. Parsimony and likelihood: an exchange. *Syst. Zool.* **35**:617–626.

Filip, L. C. and Mundy, N. I. 2004. Rapid evolution by positive Darwinian selection in the extracellular domain of the abundant lymphocyte protein CD45 in primates. *Mol. Biol. Evol.* **21**:1504–1511.
Fisher, R. 1930a. *The Genetic Theory of Natural Selection*. Clarendon Press, Oxford.
Fisher, R. 1930b. Inverse probability. *Proc. Camb. Phil. Soc.* **26**:528–535.
Fisher, R. 1970. *Statistical Methods for Research Workers*. Oliver and Boyd, Edinburgh.
Fitch, W. M. 1971a. Rate of change of concomitantly variable codons. *J. Mol. Evol.* **1**:84–96.
Fitch, W. M. 1971b. Toward defining the course of evolution: minimum change for a specific tree topology. *Syst. Zool.* **20**:406–416.
Fitch, W. M. 1976. Molecular evolutionary clocks. In *Molecular Evolution*. (ed. F. J. Ayala), pp. 160–178. Sinauer Associates, Sunderland, MA.
Fitch, W. M. and Margoliash, E. 1967. Construction of phylogenetic trees. *Science* **155**: 279–284.
Fitch, W. M., Bush, R. M., Bender, C. A., and Cox, N. J. 1997. Long term trends in the evolution of H(3) HA1 human influenza type A. *Proc. Natl. Acad. Sci. U.S.A.* **94**:7712–7718.
Fleissner, R., Metzler, D., and von Haeseler, A. 2005. Simultaneous statistical multiple alignment and phylogeny reconstruction. *Syst. Biol.* **54**:548–561.
Fletcher, R. 1987. *Practical Methods of Optimization*. Wiley, New York.
Foote, M., Hunter, J. P., Janis, C. M., and Sepkoski, J. J. 1999. Evolutionary and preservational constraints on origins of biologic groups: divergence times of eutherian mammals. *Science* **283**:1310–1314.
Ford, M. J. 2001. Molecular evolution of transferrin: evidence for positive selection in salmonids. *Mol. Biol. Evol.* **18**:639–647.
Ford, M. J., Thornton, P. J., and Park, L. K. 1999. Natural selection promotes divergence of transferrin among salmonid species. *Mol. Ecol.* **8**:1055–1061.
Forsberg, R. and Christiansen, F. B. 2003. A codon-based model of host-specific selection in parasites, with an application to the influenza A virus. *Mol. Biol. Evol.* **20**:1252–1259.
Freeland, S. J. and Hurst, L. D. 1998. The genetic code is one in a million. *J. Mol. Evol.* **47**:238–248.
Fu, Y. 1994. Estimating effective population size or mutation rate using the frequencies of mutations of various classes in a sample of DNA sequences. *Genetics* **138**:1375–1386.
Fu, Y. X. and Li, W. H. 1993. Statistical tests of neutrality of mutations. *Genetics* **133**: 693–709.
Fu, Y. -X. 1997. Statistical tests of neutrality of mutations against population growth, hitchhiking and backgroud selection. *Genetics* **147**:915–925.
Fukami, K. and Tateno, Y. 1989. On the maximum likelihood method for estimating molecular trees: uniqueness of the likelihood point. *J. Mol. Evol.* **28**:460–464.
Fukami-Kobayashi, K. and Tateno, Y. 1991. Robustness of maximum likelihood tree estimation against different patterns of base substitutions. *J. Mol. Evol.* **32**:79–91.
Gadagkar, S. R. and Kumar, S. 2005. Maximum likelihood outperforms maximum parsimony even when evolutionary rates are heterotachous. *Mol. Biol. Evol.* **22**:2139–2141.
Galtier, N. 2001. Maximum-likelihood phylogenetic analysis under a covarion-like model. *Mol. Biol. Evol.* **18**:866–873.
Galtier, N. and Gouy, M. 1998. Inferring pattern and process: maximum-likelihood implementation of a nonhomogeneous model of DNA sequence evolution for phylogenetic analysis. *Mol. Biol. Evol.* **15**:871–879.
Galtier, N., Tourasse, N., and Gouy, M. 1999. A nonhyperthermophilic common ancestor to extant life forms. *Science* **283**:220–221.

Gascuel, O. 1994. A note on Sattath and Tversky's, Saitou and Nei's, and Studier and Keppler's algorithms for inferring phylogenies from evolutionary distances. *Mol. Biol. Evol.* **11**: 961–963.

Gascuel, O. 1997. BIONJ: an improved version of the NJ algorithm based on a simple model of sequence data. *Mol. Biol. Evol.* **14**:685–695.

Gascuel, O. 2000. On the optimization principle in phylogenetic analysis and the minimum-evolution criterion. *Mol. Biol. Evol.* **17**:401–405.

Gaucher, E. A. and Miyamoto, M. M. 2005. A call for likelihood phylogenetics even when the process of sequence evolution is heterogeneous. *Mol. Phylogenet. Evol.* **37**:928–931.

Gaut, B. S. and Lewis, P. O. 1995. Success of maximum likelihood phylogeny inference in the four-taxon case. *Mol. Biol. Evol.* **12**:152–162.

Gelfand, A. E. and Smith, A. F. M. 1990. Sampling-based approaches to calculating marginal densities. *J. Amer. Stat. Assoc.* **85**:398–409.

Gelman, A. and Rubin, D. B. 1992. Inference from iterative simulation using multiple sequences (with discussion). *Statist. Sci.* **7**:457–511.

Gelman, A., Roberts, G. O., and Gilks, W. R. 1996. Efficient Metropolis jumping rules. in *Bayesian Statistics 5* (ed. J. M. Bernardo, J. O. Berger, A. P. Dawid, and A. F. M. Smith), pp. 599–607. Oxford University Press, Oxford.

Gelman, S. and Gelman, G. D. 1984. Stochastic relaxation, Gibbs distributions and the Bayes restoration of images. *IEEE Trans. Pattern Anal. Mach. Intel.* **6**:721–741.

Gerrard, D. T. and Filatov, D. A. 2005. Positive and negative selection on mammalian Y chromosomes. *Mol. Biol. Evol.* **22**:1423–1432.

Geyer, C. J. 1991. Markov chain Monte Carlo maximum likelihood. In *Computing Science and Statistics: Proc. 23rd Symp. Interface* (ed. E. M. Keramidas), pp. 156–163. Interface Foundation, Fairfax Station, VA.

Gilks, W. R., Richardson, S., and Spielgelhalter, D. J. 1996. *Markov Chain Monte Carlo in Practice*. Chapman and Hall, London.

Gill, P. E., Murray, W., and Wright, M. H. 1981. *Practical Optimization*. Academic Press, London.

Gillespie, J. H. 1984. The molecular clock may be an episodic clock. *Proc. Natl. Acad. Sci. U.S.A.* **81**:8009–8013.

Gillespie, J. H. 1986a. Rates of molecular evolution. *Ann. Rev. Ecol. Systemat.* **17**:637–665.

Gillespie, J. H. 1986b. Natural selection and the molecular clock. *Mol. Biol. Evol.* **3**:138–155.

Gillespie, J. H. 1991. *The Causes of Molecular Evolution*. Oxford University Press, Oxford.

Gillespie, J. H. 1998. *Population Genetics: a Concise Guide*. Johns Hopkins University Press, Baltimore, MD.

Gogarten, J. P., Kibak, H., Dittrich, P. et al. 1989. Evolution of the vacuolar H^+-ATPase: implications for the origin of eukaryotes. *Proc. Natl. Acad. Sci. U.S.A.* **86**:6661–6665.

Gojobori, T. 1983. Codon substitution in evolution and the "saturation" of synonymous changes. *Genetics* **105**:1011–1027.

Gojobori, T., Li, W. H., and Graur, D. 1982. Patterns of nucleotide substitution in pseudogenes and functional genes. *J. Mol. Evol.* **18**:360–369.

Golding, G. B. 1983. Estimates of DNA and protein sequence divergence: an examination of some assumptions. *Mol. Biol. Evol.* **1**:125–142.

Golding, G. B. and Dean, A. M. 1998. The structural basis of molecular adaptation. *Mol. Biol. Evol.* **15**:355–369.

Goldman, N. 1990. Maximum likelihood inference of phylogenetic trees, with special reference to a Poisson process model of DNA substitution and to parsimony analysis. *Syst. Zool.* **39**:345–361.

Goldman, N. 1993. Statistical tests of models of DNA substitution. *J. Mol. Evol.* **36**: 182–198.

Goldman, N. 1994. Variance to mean ratio, $R(t)$, for Poisson processes on phylogenetic trees. *Mol. Phylogenet. Evol.* **3**:230–239.

Goldman, N. 1998. Phylogenetic information and experimental design in molecular systematics. *Proc. R. Soc. Lond. B Biol. Sci.* **265**:1779–1786.

Goldman, N. and Yang, Z. 1994. A codon-based model of nucleotide substitution for protein-coding DNA sequences. *Mol. Biol. Evol.* **11**:725–736.

Goldman, N., Thorne, J. L., and Jones, D. T. 1998. Assessing the impact of secondary structure and solvent accessibility on protein evolution. *Genetics* **149**:445–458.

Goldman, N., Anderson, J. P., and Rodrigo, A. G. 2000. Likelihood-based tests of topologies in phylogenetics. *Syst. Biol.* **49**:652–670.

Goldstein, D. B. and Pollock, D. D. 1994. Least squares estimation of molecular distance—noise abatement in phylogenetic reconstruction. *Theor. Popul. Biol.* **45**:219–226.

Goloboff, P. A. 1999. Analyzing large data sets in reasonable times: solutions for composite optima. *Cladistics* **15**:415–428.

Goloboff, P. A. and Pol, D. 2005. Parsimony and Bayesian phylogenetics. In *Parsimony, Phylogeny, and Genomics* (ed. V. A. Albert), pp. 148–159. Oxford University Press, Oxford.

Golub, G. H. and Van Loan, C. F. 1996. *Matrix Computations.* Johns Hopkins University Press, Baltimore, MD.

Gonnet, G. H., Cohen, M. A., and Benner, S. A. 1992. Exhaustive matching of the entire protein sequence database. *Science* **256**:1443–1445.

Goodwin, R. L., Baumann, H., and Berger, F. G. 1996. Patterns of divergence during evolution of α_1-Proteinase inhibitors in mammals. *Mol. Biol. Evol.* **13**:346–358.

Götesson, A., Marshall, J. S., Jones, D. A., and Hardham, A. R. 2002. Characterization and evolutionary analysis of a large polygalacturonase gene family in the oomycete pathogen *Phytophthora cinnamomi. Mol. Plant Microbe Interact.* **15**:907–921.

Grantham, R. 1974. Amino acid difference formula to help explain protein evolution. *Science* **185**:862–864.

Graur, D. and Li, W. -H. 2000. *Fundamentals of Molecular Evolution.* Sinauer Associates, Sunderland, MA.

Graur, D. and Martin, W. 2004. Reading the entrails of chickens: molecular timescales of evolution and the illusion of precision. *Trends Genet.* **20**:80–86.

Griffiths, R. C. and Tavaré, S. 1997. Computational methods for the coalescent. In *Progress in Population Genetics and Human Evolution: IMA Volumes in Mathematics and its Applications* (ed. P. Donnelly and S. Tavaré), pp. 165–182. Springer-Verlag, Berlin.

Grimmett, G. R. and Stirzaker, D. R. 1992. *Probability and Random Processes.* Clarendon Press, Oxford.

Gu, X. 2001. Maximum-likelihood approach for gene family evolution under functional divergence. *Mol. Biol. Evol.* **18**:453–464.

Gu, X. and Li, W. -H. 1996. A general additive distance with time-reversibility and rate variation among nucleotide sites. *Proc. Natl. Acad. Sci. U.S.A.* **93**:4671–4676.

Gu, X., Fu, Y. X., and Li, W. H. 1995. Maximum likelihood estimation of the heterogeneity of substitution rate among nucleotide sites. *Mol. Biol. Evol.* **12**:546–557.

Guindon, S. and Gascuel, O. 2003. A simple, fast, and accurate algorithm to estimate large phylogenies by maximum likelihood. *Syst. Biol.* **52**:696–704.

Guindon, S., Rodrigo, A. G., Dyer, K. A., and Huelsenbeck, J. P. 2004. Modeling the site-specific variation of selection patterns along lineages. *Proc. Natl. Acad. Sci. U.S.A.* **101**:12957–12962.

Haldane, J. B. S. 1932. *The Causes of Evolution*. Longmans Green & Co., London.

Harris, H. 1966. Enzyme polymorphism in man. *Proc. R. Soc. Lond. B Biol. Sci.* **164**:298–310.

Hartigan, J. A. 1973. Minimum evolution fits to a given tree. *Biometrics* **29**:53–65.

Hartl, D. L. and Clark, A. G. 1997. *Principles of Population Genetics*. Sinauer Associates, Sunderland, MA.

Hartl, D. L., Moriyama, E. N., and Sawyer, S. A. 1994. Selection intensity for codon bias. *Genetics* **138**:227–234.

Harvey, P. H. and Pagel, M. 1991. *The Comparative Method in Evlutionary Biology*. Oxford University Press, Oxford.

Harvey, P. H. and Purvis, A. 1991. Comparative methods for explaining adaptations. *Nature* **351**:619–624.

Hasegawa, M. and Fujiwara, M. 1993. Relative efficiencies of the maximum likelihood, maximum parsimony, and neihbor joining methods for estimating protein phylogeny. *Mol. Phyl. Evol.* **2**:1–5.

Hasegawa, M. and Kishino, H. 1989. Confidence limits on the maximum-likelihood estimate of the Hominoid tree from mitochondrial DNA sequences. *Evolution* **43**:672–677.

Hasegawa, M. and Kishino, H. 1994. Accuracies of the simple methods for estimating the bootstrap probability of a maximum likelihood tree. *Mol. Biol. Evol.* **11**:142–145.

Hasegawa, M., Yano, T., and Kishino, H. 1984. A new molecular clock of mitochondrial DNA and the evolution of Hominoids. *Proc. Japan Acad. B.* **60**:95–98.

Hasegawa, M., Kishino, H., and Yano, T. 1985. Dating the human–ape splitting by a molecular clock of mitochondrial DNA. *J. Mol. Evol.* **22**:160–174.

Hasegawa, M., Kishino, H., and Saitou, N. 1991. On the maximum likelihood method in molecular phylogenetics. *J. Mol. Evol.* **32**:443–445.

Hasegawa, M., Cao, Y., and Yang, Z. 1998. Preponderance of slightly deleterious polymorphism in mitochondrial DNA: replacement/synonymous rate ratio is much higher within species than between species. *Mol. Biol. Evol.* **15**:1499–1505.

Hasegawa, M., Thorne, J. L., and Kishino, H. 2003. Time scale of eutherian evolution estimated without assuming a constant rate of molecular evolution. *Genes Genet. Syst.* **78**:267–283.

Hastings, W. K. 1970. Monte Carlo sampling methods using Markov chains and their application. *Biometrika* **57**:97–109.

Haydon, D. T., Bastos, A. D., Knowles, N. J., and Samuel, A. R. 2001. Evidence for positive selection in foot-and-mouth-disease virus capsid genes from field isolates. *Genetics* **157**:7–15.

Hedges, S. B. and Kumar, S. 2004. Precision of molecular time estimates. *Trends Genet.* **20**:242–247.

Hedges, S. B., Parker, P. H., Sibley, C. G., and Kumar, S. 1996. Continental breakup and the ordinal diversification of birds and mammals. *Nature* **381**:226–229.

Hein, J., Wiuf, C., Knudsen, B. *et al.* 2000. Statistical alignment: computational properties, homology testing and goodness-of-fit. *J. Mol. Biol.* **302**:265–279.

Hein, J., Jensen, J. L., and Pedersen, C. N. 2003. Recursions for statistical multiple alignment. *Proc. Natl. Acad. Sci. U.S.A.* **100**:14960–14965.

Hein, J., Schieriup, M. H., and Wiuf, C. 2005. *Gene Genealogies, Variation and Evolution: a Primer in Coalescent Theory*. Oxford University Press, Oxford.

Hellberg, M. E. and Vacquier, V. D. 2000. Positive selection and propeptide repeats promote rapid interspecific divergence of a gastropod sperm protein. *Mol. Biol. Evol.* **17**: 458–466.

Hendy, M. D. 2005. Hadamard conjugation: an analytical tool for phylogenetics. In *Mathematics of Evolution and Phylogeny* (ed. O. Gascuel), pp. 143–177. Oxford University Press, Oxford.

Hendy, M. D. and Penny, D. 1982. Branch and bound algorithms ro determine minimum-evolution trees. *Math. Biosci.* **60**:133–142.

Hendy, M. D. and Penny, D. 1989. A framework for the quantitative study of evolutionary trees. *Syst. Zool.* **38**:297–309.

Henikoff, S. and Henikoff, J. 1992. Amino acid substitution matrices from protein blocks. *Proc. Natl. Acad. Sci. U.S.A.* **89**:10915–10919.

Hey, J. and Nielsen, R. 2004. Multilocus methods for estimating population sizes, migration rates and divergence time, with applications to the divergence of *Drosophila pseudoobscura* and *D. persimilis*. *Genetics* **167**:747–760.

Higgins, D. G. and Sharp, P. M. 1988. CLUSTAL: a package for performing multiple sequence alignment on a microcomputer. *Gene* **73**:237–244.

Hillis, D. M. and Bull, J. J. 1993. An empirical test of bootstrapping as a method for assessing confidence in phylogenetic analysis. *Syst. Biol.* **42**:182–192.

Hillis, D. M., Bull, J. J., White, M. E. *et al.* 1992. Experimental phylogenetics: generation of a known phylogeny. *Science* **255**:589–592.

Holbrook, J. D., Birdsey, G. M., Yang, Z. *et al.* 2000. Molecular adaptation of alanine:glyoxylate aminotransferase targeting in primates. *Mol. Biol. Evol.* **17**:387–400.

Holder, M. and Lewis, P. O. 2003. Phylogeny estimation: traditional and Bayesian approaches. *Nat. Rev. Genet.* **4**:275–284.

Holmes, I. 2005. Using evolutionary expectation maximization to estimate indel rates. *Bioinformatics* **21**:2294–2300.

Holmes, S. 2003. Bootstrapping phylogenetic trees: theory and methods. *Stat. Sci.* **18**:241–255.

Horai, S., Hayasaka, K., Kondo, R. *et al.* 1995. Recent African origin of modern humans revealed by complete sequences of hominoid mitochondrial DNAs. *Proc. Natl. Acad. Sci. U.S.A.* **92**:532–536.

Hudson, R. R. 1990. Gene genealogies and the coalescent process. In *Oxford Surveys in Evolutionary Biology* (ed. D. J. Futuyma, and J. D. Antonovics), pp. 1–44. Oxford University Press, New York.

Hudson, R. R. 2001. Two-locus sampling distributions and their application. *Genetics* **159**:1805–1817.

Hudson, R. R., Kreitman, M., and Aguade, M. 1987. A test of neutral molecular evolution based on nucleotide data. *Genetics* **116**:153–159.

Huelsenbeck, J. P. 1995a. The robustness of two phylogenetic methods: four-taxon simulations reveal a slight superiority of maximum likelihood over neighbor joining. *Mol. Biol. Evol.* **12**:843–849.

Huelsenbeck, J. P. 1995b. The performance of phylogenetic methods in simulation. *Syst. Biol.* **44**:17–48.

Huelsenbeck, J. P. 1998. Systematic bias in phylogenetic analysis: is the Strepsiptera problem solved? *Syst. Biol.* **47**:519–537.

Huelsenbeck, J. P. 2002. Testing a covariotide model of DNA substitution. *Mol. Biol. Evol.* **19**:698–707.

Huelsenbeck, J. P. and Bollback, J. P. 2001. Empirical and hierarchical Bayesian estimation of ancestral states. *Syst. Biol.* **50**:351–366.

Huelsenbeck, J. P. and Dyer, K. A. 2004. Bayesian estimation of positively selected sites. *J. Mol. Evol.* **58**:661–672.

Huelsenbeck, J. P. and Lander, K. M. 2003. Frequent inconsistency of parsimony under a simple model of cladogenesis. *Syst Biol* **52**:641–648.

Huelsenbeck, J.P. and Rannala, B. 2004. Frequentist properties of Bayesian posterior probabilities of phylogenetic trees under simple and complex substitution models. *Syst. Biol.* **53**:904–913.

Huelsenbeck, J. P. and Ronquist, F. 2001. MRBAYES: Bayesian inference of phylogenetic trees. *Bioinformatics* **17**:754–755.

Huelsenbeck, J. P., Larget, B., and Swofford, D. 2000a. A compound Poisson process for relaxing the molecular clock. *Genetics* **154**:1879–1892.

Huelsenbeck, J. P., Rannala, B., and Larget, B. 2000b. A Bayesian framework for the analysis of cospeciation. *Evolution* **54**:352–364.

Huelsenbeck, J. P., Rannala, B., and Masly, J. P. 2000c. Accommodating phylogenetic uncertainty in evolutionary studies. *Science* **288**:2349–2350.

Huelsenbeck, J. P., Ronquist, F., Nielsen, R., and Bollback, J. P. 2001. Bayesian inference of phylogeny and its impact on evolutionary biology. *Science* **294**:2310–2314.

Huelsenbeck, J. P., Larget, B., and Alfaro, M. E. 2004. Bayesian phylogenetic model selection using reversible jump Markov chain Monte Carlo. *Mol. Biol. Evol.* **21**:1123–1133.

Hughes, A. L. 1999. *Adaptive Evolution of Genes and Genomes.* Oxford University Press, Oxford.

Hughes, A. L. and Nei, M. 1988. Pattern of nucleotide substitution at major histocompatibility complex class I loci reveals overdominant selection. *Nature* **335**:167–170.

Hughes, A. L., Ota, T., and Nei, M. 1990. Positive Darwinian selection promotes charge profile diversity in the antigen-binding cleft of class I major-histocompatibility-complex molecules. *Mol. Biol. Evol.* **7**:515–524.

Huttley, G. A., Easteal, S., Southey, M. C. *et al.* 2000. Adaptive evolution of the tumour suppressor BRCA1 in humans and chimpanzees. *Nature Genet.* **25**:410–413.

Ina, Y. 1995. New methods for estimating the numbers of synonymous and nonsynonymous substitutions. *J. Mol. Evol.* **40**:190–226.

Ivarsson, Y., Mackey, A. J., Edalat, M. *et al.* 2002. Identification of residues in glutathione transferase capable of driving functional diversification in evolution: a novel approach to protein design. *J. Biol. Chem.* **278**:8733–8738.

Iwabe, N., Kuma, K., Hasegawa, M. *et al.* 1989. Evolutionary relationship of archaebacteria, eubacteria, and eukaryotes inferred from phylogenetic trees of duplicated genes. *Proc. Natl. Acad. Sci. U.S.A.* **86**:9355–9359.

Jeffreys, H. 1939. *Theory of Probability.* Clarendon Press, Oxford.

Jeffreys, H. 1961. *Theory of Probability.* Oxford University Press, Oxford.

Jermann, T. M., Opitz, J. G., Stackhouse, J., and Benner, S. A. 1995. Reconstructing the evolutionary history of the artiodactyl ribonuclease superfamily. *Nature* **374**: 57–59.

Jiggins, F. M., Hurst, G. D. D., and Yang, Z. 2002. Host–symbiont conflicts: positive selection on the outer membrane protein of parasite but not mutualistic Rickettsiaceae. *Mol. Biol. Evol.* **19**:1341–1349.

Jin, L. and Nei, M. 1990. Limitations of the evolutionary parsimony method of phylogenetic analysis [erratum in *Mol. Biol. Evol.* 1990 **7**:201]. *Mol. Biol. Evol.* **7**:82–102.

Johannesson, H., Vidal, P., Guarro, J. *et al.* 2004. Positive directional selection in the proline-rich antigen (PRA) gene among the human pathogenic fungi *Coccidioides immitis*, *C. posadasii* and their closest relatives. *Mol. Biol. Evol.* **21**:1134–1145.

Johnson, M. E., Viggiano, L., Bailey, J. A. *et al.* 2001. Positive selection of a gene family during the emergence of humans and African apes. *Nature* **413**:514–519.

Jones, D. T., Taylor, W. R., and Thornton, J. M. 1992. The rapid generation of mutation data matrices from protein sequences. *CABIOS* **8**:275–282.

Jordan, I. K., Kondrashov, F. A., Adzhubei, I. A. *et al.* 2005. A universal trend of amino acid gain and loss in protein evolution. *Nature* **433**:633–638.

Jukes, T. H. 1987. Transitions, transversions, and the molecular evolutionary clock. *J. Mol. Evol.* **26**:87–98.

Jukes, T. H. and Cantor, C. R. 1969. Evolution of protein molecules. In *Mammalian protein metabolism* (ed. H. N. Munro), pp. 21–123. Academic Press, New York.

Jukes, T. H. and King, J. L. 1979. Evolutionary nucleotide replacements in DNA. *Nature* **281**:605–606.

Kafatos, F. C., Efstratiadis, A., Forget, B. G., and Weissman, S. M. 1977. Molecular evolution of human and rabbit ß-globin mRNAs. *Proc. Natl. Acad. Sci. U.S.A.* **74**:5618–5622.

Kalbfleisch, J. G. 1985. *Probability and Statistical Inference, Vol. 2: Statistical Inference.* Springer-Verlag, New York.

Kalbfleisch, J. G. and Sprott, D. A. 1970. Application of likelihood methods to models involving large numbers of parameters (with discussions). *J. R. Statist. Soc. B* **32**:175–208.

Kao, E. P. C. 1997. *An Introduction to Stochastic Processes*. ITP, Belmont, CA.

Karlin, S. and Taylor, H. M. 1975. *A First Course in Stochastic Processes*. Academic Press, San Diego, CA.

Karn, R. C. and Nachman, M. W. 1999. Reduced nucleotide variability at an androgen-binding protein locus (*Abpa*) in house mice: evidence for positive natural selection. *Mol. Biol. Evol.* **16**:1192–1197.

Katoh, K., Kuma, K., and Miyata, T. 2001. Genetic algorithm-based maximum-likelihood analysis for molecular phylogeny. *J. Mol. Evol.* **53**:477–484.

Keilson, J. 1979. *Markov Chain Models: Rarity and Exponentiality*. Springer-Verlag, New York.

Kelly, C. and Rice, J. 1996. Modeling nucleotide evolution: a heterogeneous rate analysis. *Math. Biosci.* **133**:85–109.

Kelly, F. 1979. *Reversibility and Stochastic Networks*. Springer-Verlag, Berlin.

Kendall, D. G. 1948. On the generalized birth-and-death process. *Ann. Math. Stat.* **19**:1–15.

Kidd, K. K. and Sgaramella-Zonta, L. A. 1971. Phylogenetic analysis: concepts and methods. *Am. J. Hum. Genet.* **23**:235–252.

Kim, J. 1996. General inconsistency conditions for maximum parsimony: effects of branch lengths and increasing numbers of taxa. *Syst. Biol.* **45**:363–374.

Kimura, M. 1968. Evolutionary rate at the molecular level. *Nature* **217**:624–626.

Kimura, M. 1977. Prepondence of synonymous changes as evidence for the neutral theory of molecular evolution. *Nature* **267**:275–276.

Kimura, M. 1980. A simple method for estimating evolutionary rate of base substitution through comparative studies of nucleotide sequences. *J. Mol. Evol.* **16**:111–120.

Kimura, M. 1981. Estimation of evolutionary distances between homologous nucleotide sequences. *Proc. Natl. Acad. Sci. USA* **78**:454–458.

Kimura, M. 1983. *The Neutral Theory of Molecular Evolution*. Cambridge University Press, Cambridge.

Kimura, M. 1987. Molecular evolutionary clock and the neutral theory. *J. Mol. Evol.* **26**:24–33.

Kimura, M. and Ohta, T. 1971. Protein polymorphism as a phase of molecular evolution. *Nature* **229**:467–469.

Kimura, M. and Ohta, T. 1972. On the stochastic model for estimation of mutational distance between homologous proteins. *J. Mol. Evol.* **2**:87–90.

King, C. E. and Jukes, T. H. 1969. Non-Darwinian evolution. *Science* **164**:788–798.

Kirkpatrick, S., Gelatt, C. D., and Vecchi, M. P. 1983. Optimization by simulated annealing. *Science* **220**:671–680.

Kishino, H. and Hasegawa, M. 1989. Evaluation of the maximum likelihood estimate of the evolutionary tree topologies from DNA sequence data, and the branching order in hominoidea. *J. Mol. Evol.* **29**:170–179.

Kishino, H. and Hasegawa, M. 1990. Converting distance to time: application to human evolution. *Methods Enzymol.* **183**:550–570.

Kishino, H., Miyata, T., and Hasegawa, M. 1990. Maximum likelihood inference of protein phylogeny and the origin of chloroplasts. *J. Mol. Evol.* **31**:151–160.

Kishino, H., Thorne, J. L., and Bruno, W. J. 2001. Performance of a divergence time estimation method under a probabilistic model of rate evolution. *Mol. Biol. Evol.* **18**:352–361.

Kitano, T., Sumiyama, K., Shiroishi, T., and Saitou, N. 1998. Conserved evolution of the *Rh50* gene compared to its homologous Rh blood group gene. *Biochem. Biophys. Res. Commun.* **249**:78–85.

Kluge, A. G. and Farris, J. S. 1969. Quantitateive phyletics and the evolution of anurans. *Syst. Zool.* **18**:1–32.

Knoll, A. H. and Carroll, S. B. 1999. Early animal evolution: emerging views from comparative biology and geology. *Science* **284**:2129–2137.

Knudsen, B. and Miyamoto, M. M. 2001. A likelihood ratio test for evolutionary rate shifts and functional divergence among proteins. *Proc. Natl. Acad. Sci. U.S.A.* **98**:14512–14517.

Knuth, D. E. 1997. *The Art of Computer Programming: Fundamental Algorithms*. Addison-Wesley, Reading, MA.

Kocher, T. D. 2004. Adaptive evolution and explosive speciation: the cichlid fish model. *Nature Rev. Genet.* **5**:288–298.

Kolaczkowski, B. and Thornton, J. W. 2004. Performance of maximum parsimony and likelihood phylogenetics when evolution is heterogeneous. *Nature* **431**:980–984.

Kosakovsky Pond, S. L. and Frost, S. D. W. 2005a. A genetic algorithm approach to detecting lineage-specific variation in selection pressure. *Mol. Biol. Evol.* **22**:478–485.

Kosakovsky Pond, S. L. and Frost, S. D. W. 2005b. Not so different after all: a comparison of methods for detecting amino acid sites under selection. *Mol. Biol. Evol.* **22**:1208–1222.

Kosakovsky Pond, S. L. and Muse, S. V. 2004. Column sorting: rapid calculation of the phylogenetic likelihood function. *Syst. Biol.* **53**:685–692.

Kosakovsky Pond, S. L. and Muse, S. V. 2005. Site-to-site variation of synonymous substitution rates. *Mol. Biol. Evol.* **22**:2375–2385.

Koshi, J. M. and Goldstein, R. A. 1996a. Probabilistic reconstruction of ancestral protein sequences. *J. Mol. Evol.* **42**:313–320.

Koshi, J. M. and Goldstein, R. A. 1996b. Correlating structure-dependent mutation matrices with physical-chemical properties. In *Pacific Symposium on Biocomputing '96* (ed. L. Hunter and J. E. Klein), pp. 488–499. World Scientific, Singapore.

Koshi, J. M. Mindell, D. P., and Goldstein, R. A. 1999. Using physical-chemistry-based substitution models in phylogenetic analyses of HIV-1 subtypes. *Mol. Biol. Evol.* **16**:173–179.

Kosiol, C. and Goldman, N. 2005. Different versions of the Dayhoff rate matrix. *Mol. Biol. Evol.* **22**:193–199.

Kreitman, M. 2000. Methods to detect selection in populations with applications to the human. *Annu. Rev. Genomics Hum. Genet.* **1**:539–559.

Kreitman, M. and Akashi, H. 1995. Molecular evidence for natural selection. *Annu. Rev. Ecol. Syst.* **26**:403–422.

Kronmal, R. A. and Peterson, A. V. 1979. On the alias method for generating random variables from a discrete distribution. *Amer. Statist.* **33**:214–218.

Kuhner, M. K. and Felsenstein, J. 1994. A simulation comparison of phylogeny algorithms under equal and unequal evolutionary rates [erratum in *Mol. Biol. Evol.* 1995 **12**:525]. *Mol. Biol. Evol.* **11**:459–468.

Kuhner, M. K., Yamato, J., and Felsenstein, J. 1995. Estimating effective population size and mutation rate from sequence data using Metropolis–Hastings sampling. *Genetics* **140**:1421–1430.

Kumar, S. 2005. Molecular clocks: four decades of evolution. *Nat. Rev. Genet.* **6**:654–662.

Kumar, S. and Hedges, S. B. 1998. A molecular timescale for vertebrate evolution. *Nature* **392**:917–920.

Kumar, S. and Subramanian, S. 2002. Mutation rate in mammalian genomes. *Proc. Natl. Acad. Sci. U.S.A.* **99**:803–808.

Kumar, S., Tamura, K., and Nei, M. 2005a. MEGA3: Integrated software for molecular evolutionary genetics analysis and sequence alignment. *Brief Bioinform.* **5**:150–163.

Kumar, S., Filipski, A., Swarna, V. *et al.* 2005b. Placing confidence limits on the molecular age of the human–chimpanzee divergence. *Proc. Natl. Acad. Sci. U.S.A.* **102**:18842–18847.

Laird, C. D., McConaughy, B. L., and McCarthy, B. J. 1969. Rate of fixation of nucleotide substitutions in evolution. *Nature* **224**:149–154.

Lake, J. A. 1994. Reconstructing evolutionary trees from DNA and protein sequences: paralinear distances. *Proc. Natl. Acad. Sci. U.S.A.* **91**:1455–1459.

Lang, S. 1987. *Linear Algebra*. Springer-Verlag, New York.

Langley, C. H. and Fitch, W. M. 1974. An examination of the constancy of the rate of molecular evolution. *J. Mol. Evol.* **3**:161–177.

Larget, B. and Simon, D. L. 1999. Markov chain Monte Carlo algorithms for the Bayesian analysis of phylogenetic trees. *Mol. Biol. Evol.* **16**:750–759.

Larget, B., Simon, D. L., Kadane, J. B., and Sweet, D. 2005. A Bayesian analysis of metazoan mitochondrial genome arrangements. *Mol. Biol. Evol.* **22**:486–495.

Lee, M. S. Y. 2000. Tree robustness and clade significance. *Syst. Biol.* **49**:829–836.

Lee, Y. and Nelder, J. A. 1996. Hierarchical generalized linear models. *J. R. Statist. Soc. B.* **58**:619–678.

Lee, Y. -H., Ota, T., and Vacquier, V. D. 1995. Positive selection is a general phenomenon in the evolution of abalone sperm lysin. *Mol. Biol. Evol.* **12**:231–238.

Lehmann, P. 2002. Structure and evolution of plant disease resistance genes. *J. Appl. Genet.* **43**:403–414.

Lemmon, A. R. and Milinkovitch, M. C. 2002. The metapopulation genetic algorithm: an efficient solution for the problem of large phylogeny estimation. *Proc. Natl. Acad. Sci. U.S.A.* **99**:10516–10521.

Lemmon, A. R. and Moriarty, E. C. 2004. The importance of proper model assumption in Bayesian phylogenetics. *Syst. Biol.* **53**:265–277.

Leonard, T. and Hsu, J. S. J. 1999. *Bayesian Methods*. Cambridge University Press, Cambridge.

Lewis, P. O. 1998. A genetic algorithm for maximum-likelihood phylogeny inference using nucleotide sequence data. *Mol. Biol. Evol.* **15**:277–283.

Lewis, P. O. 2001. A likelihood approach to estimating phylogeny from discrete morphological character data. *Syst. Biol.* **50**:913–925.

Lewis, P. O., Holder, M. T., and Holsinger, K. E. 2005. Polytomies and Bayesian phylogenetic inference. *Syst. Biol.* **54**:241–253.

Lewontin, R. 1989. Inferring the number of evolutionary events from DNA coding sequence differences. *Mol. Biol. Evol.* **6**:15–32.

Lewontin, R. C. and Hubby, J. L. 1966. A molecular approach to the study of genic heterozygosity in natural populations. II. Amount of variation and degree of heterozygosity in natural populations of *Drosophila pseudoobscura*. *Genetics* **54**:595–609.

Li, S., Pearl, D., and Doss, H. 2000. Phylogenetic tree reconstruction using Markov chain Monte Carlo. *J. Amer. Statist. Assoc.* **95**:493–508.

Li, W. H. and Tanimura, M. 1987. The molecular clock runs more slowly in man than in apes and monkeys. *Nature* **326**:93–96.

Li, W. -H. 1986. Evolutionary change of restriction cleavage sites and phylogenetic inference. *Genetics* **113**:187–213.

Li, W. H. 1989. A statistical test of phylogenies estimated from sequence data. *Mol. Biol. Evol.* **6**:424–435.

Li, W. -H. 1993. Unbiased estimation of the rates of synonymous and nonsynonymous substitution. *J. Mol. Evol.* **36**:96–99.

Li, W. -H. 1997. *Molecular Evolution*. Sinauer Associates, Sunderland, MA.

Li, W. -H. and Gouy, M. 1991. Statistical methods for testing molecular phylogenies. In *Phylogenetic Analysis of DNA Sequences* (ed. M. Miyamoto, and J. Cracraft), pp. 249–277. Oxford University Press, Oxford.

Li, W. H., Tanimura, M., and Sharp, P. M. 1987. An evaluation of the molecular clock hypothesis using mammalian DNA sequences. *J. Mol. Evol.* **25**:330–342.

Li, W. -H., Wu, C. -I., and Luo, C. -C. 1985. A new method for estimating synonymous and nonsynonymous rates of nucleotide substitutions considering the relative likelihood of nucleotide and codon changes. *Mol. Biol. Evol.* **2**:150–174.

Libertini, G. and Di Donato, A. 1994. Reconstruction of ancestral sequences by the inferential method, a tool for protein engineering studies. *J. Mol. Evol.* **39**:219–229.

Lindley, D. V. 1957. A statistical paradox. *Biometrika* **44**:187–192.

Lindley, D. V. 1962. Discussion on "Confidence sets for the mean of a multivariate normal distribution" by C. Stein. *J. R. Statist. Soc. B* **24**:265–296.

Lindley, D. V. and Phillips, L. D. 1976. Inference for a Bernoulli process (a Bayesian view). *Amer. Statist.* **30**:112–119.

Lindsey, J. K. 1974a. Comparison of probability distributions. *J. R. Statist. Soc. B* **36**:38–47.

Lindsey, J. K. 1974b. Construction and comparison of statistical models. *J. R. Statist. Soc. B.* **36**:418–425.

Linhart, H. 1988. A test whether two AIC's differ significantly. *S. Afr. Stat. J.* **22**: 153–161.

Lockhart, P., Novis, P., Milligan, B. G. *et al.* 2006. Heterotachy and tree building: a case study with plastids and Eubacteria. *Mol. Biol. Evol.* **23**:40–45.

Lockhart, P. J., Steel, M. A., Hendy, M. D., and Penny, D. 1994. Recovering evolutionary trees under a more realistic model of sequence evolution. *Mol. Biol. Evol.* **11**:605–612.

Lunter, G. A., Miklos, I., Song, Y. S., and Hein, J. 2003. An efficient algorithm for statistical multiple alignment on arbitrary phylogenetic trees. *J Comput Biol* **10**:869–889.

Lunter, G., Miklos, I., Drummond, A. *et al.* 2005. Bayesian coestimation of phylogeny and sequence alignment. *BMC Bioinformatics* **6**:83.

Lynch, M. and Conery, J. S. 2000. The evolutionary fate and consequences of duplicate genes. *Science* **290**:1151–1155.

Maddison, D. 1991. The discovery and importance of multiple islands of most-parsimonious trees. *Syst. Zool.* **33**:83–103.

Maddison, D. R. and Maddison, W. P. 2000. *MacClade 4: Analysis of Phylogeny and Character Evolution.* Sinauer Associates, Sunderland, MA.

Maddison, W. P. and Maddison, D. R. 1982. *MacClade: Analysis of Phylogeny and Character Evolution.* Sinauer Associates, Sunderland, MA.

Makova, K. D., Ramsay, M., Jenkins, T., and Li, W. H. 2001. Human DNA sequence variation in a 6.6-kb region containing the melanocortin 1 receptor promoter. *Genetics* **158**: 1253–1268.

Malcolm, B. A., Wilson, K. P., Matthews, B. W. *et al.* 1990. Ancestral lysozymes reconstructed, neutrality tested, and thermostability linked to hydrocarbon packing. *Nature* **345**:86–89.

Margoliash, E. 1963. Primary structure and evolution of cytochrome c. *Proc. Natl. Acad. Sci. U.S.A.* **50**:672–679.

Maritz, J. S. and Lwin, T. 1989. *Empirical Bayes Methods.* Chapman and Hall, London.

Martin, A. P. and Palumbi, S. R. 1993. Body size, metabolic rate, generation time, and the molecular clock. *Proc Natl Acad Sci U.S.A.* **90**:4087–4091.

Massingham, T. and Goldman, N. 2005. Detecting amino acid sites under positive selection and purifying selection. *Genetics* **169**:1753–1762.

Maston, G. A. and Ruvolo, M. 2002. Chorionic gonadotropin has a recent origin within primates and an evolutionary history of selection. *Mol. Biol. Evol.* **19**:320–335.

Mateiu, L. M. and Rannala, B. 2006. Inferring complex DNA substitution processes on phylogenies using uniformization and data augmentation. *Syst. Biol.* **55**: 259–269.

Mau, B. and Newton, M. A. 1997. Phylogenetic inference for binary data on dendrograms using Markov chain Monte Carlo. *J. Computat. Graph. Stat.* **6**:122–131.

Mau, B., Newton, M. A., and Larget, B. 1999. Bayesian phylogenetic inference via Markov chain Monte Carlo Methods. *Biometrics* **55**:1–12.

Maynard Smith, J. and Haigh, J. 1974. The hitch-hiking effect of a favorable gene. *Genet. Res.* **23**:23–35.

Mayrose, I., Friedman, N., and Pupko, T. 2005. A gamma mixture model better accounts for among site rate heterogeneity. *Bioinformatics* **21**:151–158.

McDonald, J. H. and Kreitman, M. 1991. Adaptive protein evolution at the *Adh* locus in Drosophila. *Nature* **351**:652–654.

McGuire, G., Denham, M. C., and Balding, D. J. 2001. Models of sequence evolution for DNA sequences containing gaps. *Mol. Biol. Evol.* **18**:481–490.

McVean, G. A. and Charlesworth, D. J. 2000. The effects of Hill–Robertson interference between weakly selected mutations on patterns of molecular evolution and variation. *Genetics* **155**:929–944.

McVean, M., Awadalla, P., and Fearnhead, P. 2002. A coalescent-based method for detecting and estimating recombination from gene sequences. *Genetics* **160**:1231–1241.

Messier, W. and Stewart, C. -B. 1997. Episodic adaptive evolution of primate lysozymes. *Nature* **385**:151–154.

Metropolis, N., Rosenbluth, A. W., Rosenbluth, M. N. *et al.* 1953. Equations of state calculations by fast computing machines. *J. Chem. Physi.* **21**:1087–1092.

Metz, E. C. and Palumbi, S. R. 1996. Positive selection and sequence arrangements generate extensive polymorphism in the gamete recognition protein bindin. *Mol. Biol. Evol.* **13**:397–406.

Metzler, D. 2003. Statistical alignment based on fragment insertion and deletion models. *Bioinformatics* **19**:490–499.

Meyer, A., Kocher, T. D., Basasibwaki, P., and Wilson, A. C. 1990. Monophyletic origin of Lake Victoria cichlid fishes suggested by mitochondrial DNA sequences. *Nature* **347**:550–553.

Mindell, D. P. 1996. Positive selection and rates of evolution in immunodeficiency viruses from humans and chimpanzees. *Proc. Natl. Acad. Sci. U.S.A.* **93**:3284–3288.

Miyata, T. and Yasunaga, T. 1980. Molecular evolution of mRNA: a method for estimating evolutionary rates of synonymous and amino acid substitutions from homologous nucleotide sequences and its applications. *J. Mol. Evol.* **16**:23–36.

Miyata, T., Miyazawa, S., and Yasunaga, T. 1979. Two types of amino acid substitutions in protein evolution. *J. Mol. Evol.* **12**:219–236.

Moler, C. and Van Loan, C. F. 1978. Nineteen dubious ways to compute the exponential of a matrix. *SIAM Review* **20**:801–836.

Mondragon-Palomino, M., Meyers, B. C., Michelmore, R. W., and Gaut, B. S. 2002. Patterns of positive selection in the complete NBS-LRR gene family of *Arabidopsis thaliana*. *Genome Res.* **12**:1305–1315.

Mooers, A. Ø. and Schluter, D. 1999. Reconstructing ancestor states with maximum likelihood: support for one- and two-rate models. *Syst. Biol.* **48**:623–633.

Moore, R. C. and Purugganan, M. D. 2003. The early stages of duplicate gene evolution. *Proc. Natl. Acad. Sci. U.S.A.* **100**:15682–15687.

Moore, R. C. and Purugganan, M. D. 2005. The evolutionary dynamics of plant duplicate genes. *Curr. Opin. Plant Biol.* **8**:122–128.

Morgan, G. J. 1998. Emile Zuckerkandl, Linus Pauling, and the molecular evolutionary clock. *J. Hist. Biol.* **31**:155–178.

Moriyama, E. N. and Powell, J. R. 1997. Synonymous substitution rates in *Drosophila*: mitochondrial versus nuclear genes. *J. Mol. Evol.* **45**:378–391.

Mossel, E. and Vigoda, E. 2005. Phylogenetic MCMC algorithms are misleading on mixtures of trees. *Science* **309**:2207–2209.

Mundy, N. I. and Cook, S. 2003. Positive selection during the diversification of class I vomeronasal receptor-like (V1RL) genes, putative pheromone receptor genes, in human and primate evolution. *Mol. Biol. Evol.* **20**:1805–1810.

Murphy, W. J., Larkin, D. M., der Wind, A. E. -v. *et al.* 2005. Dynamics of mammalian chromosome evolution inferred from multispecies comparative maps. *Science* **309**:613–617.

Muse, S. V. 1996. Estimating synonymous and nonsynonymous substitution rates. *Mol. Biol. Evol.* **13**:105–114.

Muse, S. V. and Gaut, B. S. 1994. A likelihood approach for comparing synonymous and nonsynonymous nucleotide substitution rates, with application to the chloroplast genome. *Mol. Biol. Evol.* **11**:715–724.

Muse, S. V. and Gaut, B. S. 1997. Comparing patterns of nucleotide substitution rates among chloroplast loci using the relative ratio test. *Genetics* **146**:393–399.

Muse, S. V. and Weir, B. S. 1992. Testing for equality of evolutionary rates. *Genetics* **132**: 269–276.

Nachman, M. W., Boyer, S., and Aquadro, C. F. 1996. Non-neutral evolution at the mitochondrial NADH dehydrogenase subunit 3 gene in mice. *Proc. Natl. Acad. Sci. U.S.A.* **91**:6364–6368.

Needleman, S. G. and Wunsch, C. D. 1970. A general method applicable to the search for similarities in the amino acid sequence of two proteins. *J. Mol. Biol.* **48**:443–453.

Nei, M. 1987. *Molecular Evolutionary Genetics*. Columbia University Press, New York.

Nei, M. 1996. Phylogenetic analysis in molecular evolutionary genetics. *Annu. Rev. Genet.* **30**:371–403.

Nei, M. and Gojobori, T. 1986. Simple methods for estimating the numbers of synonymous and nonsynonymous nucleotide substitutions. *Mol. Biol. Evol.* **3**:418–426.

Nei, M., Stephens, J. C., and Saitou, N. 1985. Methods for computing the standard errors of branching points in an evolutionary tree and their application to molecular data from humans and apes. *Mol. Biol. Evol.* **2**:66–85.

Nielsen, R. 1997. Site-by-site estimation of the rate of substitution and the correlation of rates in mitochondrial DNA. *Syst. Biol.* **46**:346–353.

Nielsen, R. 2001a. Mutations as missing data: inferences on the ages and distributions of nonsynonymous and synonymous mutations. *Genetics* **159**:401–411.

Nielsen, R. 2001b. Statistical tests of selective neutrality in the age of genomics. *Heredity* **86**:641–647.

Nielsen, R. and Wakeley, J. 2001. Distinguishing migration from isolation: a Markov chain Monte Carlo approach. *Genetics* **158**:885–896.

Nielsen, R. and Yang, Z. 1998. Likelihood models for detecting positively selected amino acid sites and applications to the HIV-1 envelope gene. *Genetics* **148**:929–936.

Nielsen, R., Bustamante, C., Clark, A. G. et al. 2005. A scan for positively selected genes in the genomes of humans and chimpanzees. *PLoS Biol.* **3**:e170.

Nixon, K. C. 1999. The parsimony ratchet, a new method for rapid parsimony analysis. *Cladistics* **15**:407–414.

Norris, J. R. 1997. *Markov Chains*. Cambridge University Press, Cambridge.

Nylander, J. A. A., Ronquist, F., Huelsenbeck, J. P., and Nieves-Aldrey, J. L. 2004. Bayesian phylogenetic analysis of combined data. *Syst. Biol.* **53**:47–67.

O'Hagan, A. and Forster, J. 2004. *Kendall's Advanced Theory of Statistics: Bayesian Inference*. Arnold, London.

Ohno, S. 1970. *Evolution by Gene Duplication*. Springer-Verlag, New York.

Ohta, T. 1973. Slightly deleterious mutant substitutions in evolution. *Nature* **246**:96–98.

Ohta, T. 1992. Theoretical study of near neutrality. II. Effect of subdivided population structure with local extinction and recolonization. *Genetics* **130**:917–923.

Ohta, T. 1995. Synonymous and nonsynonymous substitutions in mammalian genes and the nearly neutral theory. *J. Mol. Evol.* **40**:56–63.

Ohta, T. 2002. Near-neutrality in evolution of genes and gene regulation. *Proc. Natl. Acad. Sci. U.S.A.* **99**:16134–16137.

Ohta, T. and Gillespie, J. H. 1996. Development of neutral and nearly neutral theories. *Theor. Popul. Biol.* **49**:128–142.

Ohta, T. and Kimura, M. 1971. On the constancy of the evolutionary rate of cistrons. *J. Mol. Evol.* **1**:18–25.

Ohta, T. and Tachida, H. 1990. Theoretical study of near neutrality. I. Heterozygosity and rate of mutant substitution. *Genetics* **126**:219–229.

Olsen, G. J., Matsuda, H., Hagstrom, R., and Overbeek, R. 1994. fastDNAML: a tool for construction of phylogenetic trees of DNA sequences using maximum likelihood. *Comput. Appl. Biosci.* **10**:41–48.

Opazo, J. C., Palma, R. E., Melo, F., and Lessa, E. P. 2005. Adaptive evolution of the insulin gene in caviomorph rodents. *Mol. Biol. Evol.* **22**:1290–1298.

Osawa, S. and Jukes, T. H. 1989. Codon reassignment (codon capture) in evolution. *J. Mol. Evol.* **28**:271–278.

Ota, S. and Li, W. H. 2000. NJML: a hybrid algorithm for the neighbor-joining and maximum-likelihood methods. *Mol. Biol. Evol.* **17**:1401–1409.

Pagel, M. 1994. Detecting correlated evolution on phylogenies: a general method for the comparative analysis of discrete characters. *Proc. R. Soc. Lond. B Biol. Sci.* **255**:37–45.

Pagel, M. 1999. The maximum likelihood approach to reconstructing ancestral character states of discrete characters on phylogenies. *Syst. Biol.* **48**:612–622.

Pagel, M. and Meade, A. 2004. A phylogenetic mixture model for detecting pattern-heterogeneity in gene sequence or character-state data. *Syst. Biol.* **53**:571–581.

Pagel, M., Meade, A., and Barker, D. 2004. Bayesian estimation of ancestral character states on phylogenies. *Syst. Biol.* **53**:673–684.

Palmer, C. A., Watts, R. A., Gregg, R. G. *et al.* 2005. Lineage-specific differences in evolutionary mode in a salamander courtship pheromone. *Mol. Biol. Evol.* **22**:2243–2256.

Palumbi, S. R. 1994. Genetic divergence, reproductive isolation and marine speciation. *Annu. Rev. Ecol. Syst.* **25**:547–572.

Pamilo, P. and Bianchi, N. O. 1993. Evolution of the *Zfx* and *Zfy* genes—rates and interdependence between the genes. *Mol. Biol. Evol.* **10**:271–281.

Pamilo, P. and O'Neill, R. W. 1997. Evolution of *Sry* genes. *Mol. Biol. Evol.* **14**:49–55.

Pauling, L. and Zuckerkandl, E. 1963. Chemical paleogenetics: molecular "restoration studies" of extinct forms of life. *Acta Chem. Scand.* **17**:S9–S16.

Pechirra, P., Nunes, B., Coelho, A. *et al.* 2005. Molecular characterization of the HA gene of influenza type B viruses. *J. Med. Virol.* **77**:541–549.

Penny, D. and Hendy, M. D. 1985. The use of tree comparison metrics. *Syst. Zool.* **34**:75–82.

Perler, F., Efstratiadis, A., Lomedica, P. *et al.* 1980. The evolution of genes: the chicken preproinsulin gene. *Cell* **20**:555–566.

Perna, N. T. and Kocher, T. D. 1995. Unequal base frequencies and the estimation of substitution rates. *Mol. Biol. Evol.* **12**:359–361.

Pfister, P. and Rodriguez, I. 2005. Olfactory expression of a single and highly variable V1r pheromone receptor-like gene in fish species. *Proc. Natl. Acad. Sci. U.S.A.* **102**:5489–5494.

Philippe, H., Zhou, Y., Brinkmann, H. *et al.* 2005. Heterotachy and long-branch attraction in phylogenetics. *BMC Evol. Biol.* **5**:50.

Pickett, K. M. and Randle, C. P. 2005. Strange Bayes indeed: uniform topological priors imply non-uniform clade priors. *Mol. Phylogenet. Evol.* **34**:203–211.

Polley, S. D. and Conway, D. J. 2001. Strong diversifying selection on domains of the *Plasmodium falciparum* apical membrane antigen 1 gene. *Genetics* **158**:1505–1512.

Posada, D. and Buckley, T. R. 2004. Model selection and model averaging in phylogenetics: advantages of Akaike Informtaion Criterion and Bayesian approaches over likelihood ratio tests. *Syst. Biol.* **53**:793–808.

Posada, D. and Crandall, K. A. 1998. MODELTEST: testing the model of DNA substitution. *Bioinformatics* **14**:817–818.

Posada, D. and Crandall, K. 2001. Simple (wrong) models for complex trees: a case from retroviridae. *Mol. Biol. Evol.* **18**:271–275.

Prince, V. E. and Pickett, F. B. 2002. Splitting pairs: the diverging fates of duplicated genes. *Nat. Rev. Genet.* **3**:827–837.

Pupko, T., Pe'er, I., Shamir, R., and Graur, D. 2000. A fast algorithm for joint reconstruction of ancestral amino acid sequences. *Mol. Biol. Evol.* **17**:890–896.

Pupko, T., Pe'er, I., Hasegawa, M. *et al.* 2002a. A branch-and-bound algorithm for the inference of ancestral amino-acid sequences when the replacement rate varies among sites: application to the evolution of five gene families. *Bioinformatics* **18**:1116–1123.

Pupko, T., Huchon, D., Cao, Y. *et al.* 2002b. Combining multiple data sets in a likelihood analysis: which models are the best? *Mol. Biol. Evol.* **19**:2294–2307.

Raaum, R. L., Sterner, K. N., Noviello, C. M. *et al.* 2005. Catarrhine primate divergence dates estimated from complete mitochondrial genomes: concordance with fossil and nuclear DNA evidence. *J. Human Evol.* **48**:237–257.

Rambaut, A. 2000. Estimating the rate of molecular evolution: incorporating non-comptemporaneous sequences into maximum likelihood phylogenetics. *Bioinformatics* **16**:395–399.

Rambaut, A. and Bromham, L. 1998. Estimating divergence dates from molecular sequences. *Mol. Biol. Evol.* **15**:442–448.

Rambaut, A. and Grassly, N. C. 1997. Seq-Gen: an application for the Monte Carlo simulation of DNA sequence evolution along phylogenetic trees. *Comput. Appl. Biosci.* **13**:235–238.

Rand, D., Dorfsman, M., and Kann, L. 1994. Neutral and nonneutral evolution of *Drosophila* mitochondrial DNA. *Genetics* **138**:741–756.

Rand, D. M., Weinreich, D. M., and Cezairliyan, B. O. 2000. Neutrality tests of conservative-radical amino acid changes in nuclear- and mitochondrially-encoded proteins. *Gene* **261**:115–125.

Rannala, B. 2002. Identifiability of parameters in MCMC Bayesian inference of phylogeny. *Syst. Biol.* **51**:754–760.

Rannala, B. and Yang, Z. 1996. Probability distribution of molecular evolutionary trees: a new method of phylogenetic inference. *J. Mol. Evol.* **43**:304–311.

Rannala, B. and Yang, Z. 2003. Bayes estimation of species divergence times and ancestral population sizes using DNA sequences from multiple loci. *Genetics* **164**:1645–1656.

Rannala, B. and Yang, Z. 2006. Inferring speciation times under an episodic molecular clock in preparation.

Ranwez, V. and Gascuel, O. 2002. Improvement of distance-based phylogenetic methods by a local maximum likelihood approach using triplets. *Mol. Biol. Evol.* **19**:1952–1963.

Redelings, B. D. and Suchard, M. A. 2005. Joint Bayesian estimation of alignment and phylogeny. *Syst. Biol.* **54**:401–418.

Ren, F., Tanaka, T. and Yang, Z. 2005. An empirical examination of the utility of codon-substitution models in phylogeny reconstruction. *Syst. Biol.* **54**:808–818.

Ripley, B. 1987. *Stochastic Simulation*. Wiley, New York.
Robbins, H. 1955. An empirical Bayes approach to statistics. *Proc. 3rd Berkeley Symp. Math. Stat. Prob.* **1**:157–164.
Robbins, H. 1983. Some thoughts on empirical Bayes estimation. *Ann. Statist.* **1**:713–723.
Robert, C. P. and Casella, G. 2004. *Monte Carlo Statistical Methods*. Springer-Verlag, New York.
Robinson, D. F. and Foulds, L. R. 1981. Comparison of phylogenetic trees. *Math. Biosci.* **53**:131–147.
Rodriguez, F., Oliver, J. F., Marin, A., and Medina, J. R. 1990. The general stochastic model of nucleotide substitutions. *J. Theor. Biol.* **142**:485–501.
Rodriguez-Trelles, F., Tarrio, R., and Ayala, F. J. 2003. Convergent neofunctionalization by positive Darwinian selection after ancient recurrent duplications of the xanthine dehydrogenase gene. *Proc. Natl. Acad. Sci. U.S.A.* **100**:13413–13417.
Rogers, J. S. 1997. On the consistency of maximum likelihood estimation of phylogenetic trees from nucleotide sequences. *Syst. Biol.* **46**:354–357.
Rogers, J. S. and Swofford, D. L. 1998. A fast method for approximating maximum likelihoods of phylogenetic trees from nucleotide sequences. *Syst. Biol.* **47**:77–89.
Rogers, J. S. and Swofford, D. L. 1999. Multiple local maxima for likelihoods of phylogenetic trees: a simulation study. *Mol. Biol. Evol.* **16**:1079–1085.
Rokas, A., Kruger, D., and Carroll, S. B. 2005. Animal evolution and the molecular signature of radiations compressed in time. *Science* **310**:1933–1938.
Ronquist, F. 1998. Fast Fitch-parsimony algorithms for large data sets. *Cladistics* **14**:387–400.
Ronquist, F. and Huelsenbeck, J. P. 2003. MrBayes 3: Bayesian phylogenetic inference under mixed models. *Bioinformatics* **19**:1572–1574.
Rooney, A. P. and Zhang, J. 1999. Rapid evolution of a primate sperm protein: relaxation of functional constraint or positive Darwinian selection? *Mol. Biol. Evol.* **16**:706–710.
Ross, R. 1997. *Simulation*. Academic Press, London.
Ross, S. 1996. *Stochastic Processes*. Springer-Verlag, New York.
Roth, C., Betts, M. J., Steffansson, P. *et al.* 2005. The Adaptive Evolution Database (TAED): a phylogeny based tool for comparative genomics. *Nucl. Acids Res.* **33**:D495–D497.
Rubin, D. B. and Schenker, N. 1986. Efficiently simulating the coverage properties of interval estimates. *Appl. Statist.* **35**:159–167.
Russo, C. A., Takezaki, N., and Nei, M. 1996. Efficiencies of different genes and different tree-building methods in recovering a known vertebrate phylogeny. *Mol. Biol. Evol.* **13**:525–536.
Rzhetsky, A. and Nei, M. 1992. A simple method for estimating and testing minimum-evolution trees. *Mol. Biol. Evol.* **9**:945–967.
Rzhetsky, A. and Nei, M. 1993. Theoretical foundation of the minimum-evolution method of phylogenetic inference. *Mol. Biol. Evol.* **10**:1073–1095.
Rzhetsky, A. and Nei, M. 1994. Unbiased estimates of the number of nucleotide substitutions when substitution rate varies among different sites. *J. Mol. Evol.* **38**:295–299.
Rzhetsky, A. and Sitnikova, T. 1996. When is it safe to use an oversimplified substitution model in tree-making? *Mol. Biol. Evol.* **13**:1255–1265.
Saitou, N. 1988. Property and efficiency of the maximum likelihood method for molecular phylogeny. *J. Mol. Evol.* **27**:261–273.
Saitou, N. and Imanishi, T. 1989. Relative efficiencies of the Fitch-Margoliash, maximum parsimony, maximum likelihood, minimum evolution, and neighbor joining methods of phylogenetic tree construction in obtaining the correct tree. *Mol. Biol. Evol.* **6**:514–525.

Saitou, N. and Nei, M. 1986. The number of nucleotides required to determine the branching order of three species, with special reference to the human-chimpanzee-gorilla divergence. *J. Mol. Evol.* **24**:189–204.

Saitou, N. and Nei, M. 1987. The neighbor-joining method: a new method for reconstructing phylogenetic trees. *Mol. Biol. Evol.* **4**:406–425.

Salter, L. A. 2001. Complexity of the likelihood surface for a large DNA dataset. *Syst. Biol.* **50**:970–978.

Salter, L. A. and Pearl, D. K. 2001. Stochastic search strategy for estimation of maximum likelihood phylogenetic trees. *Syst. Biol.* **50**:7–17.

Sanderson, M. J. 1997. A nonparametric approach to estimating divergence times in the absence of rate constancy. *Mol. Biol. Evol.* **14**:1218–1232.

Sanderson, M. J. 2002. Estimating absolute rates of molecular evolution and divergence times: a penalized likelihood approach. *Mol. Biol. Evol.* **19**:101–109.

Sanderson, M. J. and Kim, J. 2000. Parametric phylogenetics? *Syst. Biol.* **49**:817–829.

Sankoff, D. 1975. Minimal mutation trees of sequences. *SIAM J. Appl. Math.* **28**:35–42.

Sarich, V. M. and Wilson, A. C. 1967. Rates of albumin evolution in primates. *Proc. Natl. Acad. Sci. U.S.A.* **58**:142–148.

Sarich, V. M. and Wilson, A. C. 1973. Generation time and genomic evolution in primates. *Science* **179**:1144–1147.

Savage, L. J. 1962. *The Foundations of Statistical Inference*. Metheun & Co., London.

Sawyer, K. R. 1984. Multiple hypothesis testing. *J. R. Statist. Soc. B* **46**:419–424.

Sawyer, S. A. and Hartl, D. L. 1992. Population genetics of polymorphism and divergence. *Genetics* **132**:1161–1176.

Sawyer, S. L. Emerman, M., and Malik, H. S. 2004. Ancient adaptive evolution of the primate antiviral DNA-editing enzyme APOBEC3G. *PLoS Biol.* **2**:E275.

Sawyer, S. L., Wu, L. I., Emerman, M., and Malik, H. S. 2005. Positive selection of primate TRIM5α identifies a critical species-specific retroviral restriction domain. *Proc. Natl. Acad. Sci. U.S.A.* **102**:2832–2837.

Scheffler, K. and Seoighe, C. 2005. A Bayesian model comparison approach to inferring positive selection. *Mol. Biol. Evol.* **22**:2531–2540.

Schluter, D. 1995. Uncertainty in ancient phylogenies. *Nature* **377**:108–110.

Schluter, D. 2000. *The Ecology of Adaptive Radiation*. Oxford University Press, Oxford.

Schmidt, H. A., Strimmer, K., Vingron, M., and von Haeseler, A. 2002. TREE-PUZZLE: maximum likelihood phylogenetic analysis using quartets and parallel computing. *Bioinformatics* **18**:502–504.

Schoeniger, M. and von Haeseler, A. 1993. A simple method to improve the reliability of tree reconstructions. *Mol. Biol. Evol.* **10**:471–483.

Schott, J. R. 1997. *Matrix Analysis for Statistics*. Wiley, New York.

Schultz, T. R. and Churchill, G. A. 1999. The role of subjectivity in reconstructing ancestral character states: a Bayesian approach to unknown rates, states, and transformation asymmetries. *Syst. Biol.* **48**:651–664.

Schwarz, G. 1978. Estimating the dimension of a model. *Ann. Statist.* **6**:461–464.

Self, S. G. and Liang, K. -Y. 1987. Asymptotic properties of maximum likelihood estimators and likelihood ratio tests under nonstandard conditions. *J. Am. Stat. Assoc.* **82**:605–610.

Shackelton, L. A., Parrish, C. R., Truyen, U., and Holmes, E. C. 2005. High rate of viral evolution associated with the emergence of carnivore parvovirus. *Proc. Natl. Acad. Sci. U.S.A.* **102**:379–384.

Shapiro, B., Rambaut, A., and Drummond, A. J. 2006. Choosing appropriate substitution models for the phylogenetic analysis of protein-coding sequences. *Mol. Biol. Evol.* **23**:7–9.

Sharp, P. M. 1997. In search of molecular Darwinism. *Nature* **385**:111–112.

Shields, D. C., Harmon, D. L., and Whitehead, A. S. 1996. Evolution of hemopoietic ligands and their receptors: influence of positive selection on correlated replacements throughout ligand and receptor proteins. *J. Immunol.* **156**:1062–1070.

Shimodaira, H. 2002. An approximately unbiased test of phylogenetic tree selection. *Syst. Biol.* **51**:492–508.

Shimodaira, H. and Hasegawa, M. 1999. Multiple comparisons of log-likelihoods with applications to phylogenetic inference. *Mol. Biol. Evol.* **16**:1114–1116.

Shimodaira, H. and Hasegawa, M. 2001. CONSEL: for assessing the confidence of phylogenetic tree selection. *Bioinformatics* **17**:1246–1247.

Shindyalov, I. N., Kolchanov, N. A., and Sander, C. 1994. Can three-dimensional contacts in protein structures be predicted by analysis of correlated mutations? *Protein Eng.* **7**:349–358.

Shriner, D., Nickle, D. C., Jensen, M. A., and Mullins, J. I. 2003. Potential impact of recombination on sitewise approaches for detecting positive natural selection. *Genet. Res.* **81**:115–121.

Siddall, M. E. 1998. Success of parsimony in the four-taxon case: long branch repulsion by likelihood in the Farris zone. *Cladistics* **14**:209–220.

Silverman, B. W. 1986. *Density Estimation for Statistics and Data Analysis*. Chapman and Hall, London.

Simes, R. J. 1986. An improved Bonferroni procedure for multiple tests of significance. *Biometrika* **73**:751–754.

Simonsen, K. L., Churchill, G. A., and Aquadro, C. F. 1995. Properties of statistical tests of neutrality for DNA polymorphism data. *Genetics* **141**:413–429.

Sitnikova, T., Rzhetsky, A., and Nei, M. 1995. Interior-branch and bootstrap tests of phylogenetic trees. *Mol. Biol. Evol.* **12**:319–333.

Slowinski, J. B. and Arbogast, B. S. 1999. Is the rate of molecular evolution inversely related to body size? *Syst. Biol.* **48**:396–399.

Smith, A. B. and Peterson, K. J. 2002. Dating the time of origin of major clades: molecular clocks and the fossil record. *Ann. Rev. Earth Planet. Sci.* **30**:65–88.

Sober, E. 1988. *Reconstructing the Past: Parsimony, Evolution, and Inference*. MIT Press, Cambridge, MA.

Sober, E. 2004. The contest between parsimony and likelihood. *Syst. Biol.* **53**:644–653.

Sokal, R. R. and Sneath, P. H. A. 1963. *Numerical Taxonomy*. W.H. Freeman and Co., San Francisco, CA.

Sourdis, J. and Nei, M. 1988. Relative efficiencies of the maximum parsimony and distance-matrix methods in obtaining the correct phylogenetic tree. *Mol. Biol. Evol.* **5**:298–311.

Spady, T. C., Seehausen, O., Loew, E. R. *et al.* 2005. Adaptive molecular evolution in the opsin genes of rapidly speciating cichlid species. *Mol. Biol. Evol.* **22**:1412–1422.

Spencer, M., Susko, E., and Roger, A. J. 2005. Likelihood, parsimony, and heterogeneous evolution. *Mol. Biol. Evol.* **22**:1161–1164.

Springer, M. S., Murphy, W. J., Eizirik, E., and O'Brien, S. J. 2003. Placental mammal diversification and the Cretaceous–Tertiary boundary. *Proc. Natl. Acad. Sci. U.S.A.* **100**:1056–1061.

Stackhouse, J., Presnell, S. R., McGeehan, G. M. *et al.* 1990. The ribonuclease from an ancient bovid ruminant. *FEBS Lett.* **262**:104–106.

Steel, M. A. 1994a. The maximum likelihood point for a phylogenetic tree is not unique. *Syst. Biol.* **43**:560–564.

Steel, M. A. 1994b. Recovering a tree from the leaf colourations it generates under a Markov model. *Appl. Math. Lett.* **7**:19–24.

Steel, M. A. and Penny, D. 2000. Parsimony, likelihood, and the role of models in molecular phylogenetics. *Mol. Biol. Evol.* **17**:839–850.

Stein, C. 1956. Inadmissibility of the usual estimator for the mean of a multivariate normal distribution. *Proc. Third Berkeley Symp. Math. Stat. Prob.* **1**:197–206.

Stein, C. 1962. Confidence sets for the mean of a multivariate normal distribution. *J. R. Statist. Soc. B.* **24**:265–296.

Steiper, M. E., Young, N. M., and Sukarna, T. Y. 2004. Genomic data support the hominoid slowdown and an Early Oligocene estimate for the hominoid-cercopithecoid divergence. *Proc. Natl. Acad. Sci. U.S.A.* **101**:17021–17026.

Stephens, M. and Donnelly, P. 2000. Inference in molecular population genetics (with discussions). *J. R. Statist. Soc. B* **62**:605–655.

Stewart, C. -B., Schilling, J. W., and Wilson, A. C. 1987. Adaptive evolution in the stomach lysozymes of foregut fermenters. *Nature* **330**:401–404.

Stotz, H. U., Bishop, J. G., Bergmann, C. W. *et al.* 2000. Identification of target amino acids that affect interactions of fungal polygalacturonases and their plant inhibitors. *Mol. Physiol. Plant Pathol.* **56**:117–130.

Strimmer, K. and von Haeseler, A. 1996. Quartet puzzling: a quartet maximum-likelihood method for reconstructing tree topologies. *Mol. Biol. Evol.* **13**:964–969.

Stuart, A., Ord, K., and Arnold, S. 1999. *Kendall's Advanced Theory of Statistics*. Arnold, London.

Studier, J. A. and Keppler, K. J. 1988. A note on the neighbor-joining algorithm of Saitou and Nei. *Mol. Biol. Evol.* **5**:729–731.

Su, C., Nguyen, V. K., and Nei, M. 2002. Adaptive evolution of variable region genes encoding an unusual type of immunoglobulin in Camelids. *Mol. Biol. Evol.* **19**:205–215.

Suchard, M. A., Weiss, R. E., and Sinsheimer, J. S. 2001. Bayesian selection of continuous-time Markov chain evolutionary models. *Mol. Biol. Evol.* **18**:1001–1013.

Suchard, M. A., Kitchen, C. M., Sinsheimer, J. S., and Weiss, R. E. 2003. Hierarchical phylogenetic models for analyzing multipartite sequence data. *Syst. Biol.* **52**:649–664.

Sullivan, J. and Swofford, D. L. 2001. Should we use model-based methods for phylogenetic inference when we know that assumptions about among-site rate variation and nucleotide substitution pattern are violated? *Syst. Biol.* **50**:723–729.

Sullivan, J., Holsinger, K. E., and Simon, C. 1995. Among-site rate variation and phylogenetic analysis of 12S rRNA in sigmodontine rodents. *Mol. Biol. Evol.* **12**:988–1001.

Sullivan, J., Swofford, D. L., and Naylor, G. J. P. 1999. The effect of taxon-sampling on estimating rate heterogeneity parameters on maximum-likelihood models. *Mol. Biol. Evol.* **16**:1347–1356.

Sutton, K. A. and Wilkinson, M. F. 1997. Rapid evolution of a homeodomain: evidence for positive selection. *J. Mol. Evol.* **45**:579–588.

Suzuki, Y. 2004. New methods for detecting positive selection at single amino acid sites. *J. Mol. Evol.* **59**:11–19.

Suzuki, Y. and Gojobori, T. 1999. A method for detecting positive selection at single amino acid sites. *Mol. Biol. Evol.* **16**:1315–1328.

Suzuki, Y., Glazko, G. V., and Nei, M. 2002. Overcredibility of molecular phylogenies obtained by Bayesian phylogenetics. *Proc. Natl. Acad. Sci. U.S.A.* **99**:16138–16143.

Swanson, W. J. and Vacquier, V. D. 2002a. The rapid evolution of reproductive proteins. *Nature Rev. Genet.* **3**:137–144.

Swanson, W. J. and Vacquier, V. D. 2002b. Reproductive protein evolution. *Ann. Rev. Ecol. Systemat.* **33**:161–179.

Swanson, W. J., Yang, Z., Wolfner, M. F., and Aquadro, C. F. 2001a. Positive Darwinian selection in the evolution of mammalian female reproductive proteins. *Proc. Natl. Acad. Sci. U.S.A.* **98**:2509–2514.

Swanson, W. J., Clark, A. G., Waldrip-Dail, H. M. *et al.* 2001b. Evolutionary EST analysis identifies rapidly evolving male reproductive proteins in Drosophila. *Proc. Natl. Acad. Sci. U.S.A.* **98**:7375–7379.

Swofford, D. L. 2000. *PAUP*: Phylogenetic Analysis by Parsimony*, Version 4. Sinauer Associates, Sunderland, MA.

Swofford, D. L., Waddell, P. J., Huelsenbeck, J. P. *et al.* 2001. Bias in phylogenetic estimation and its relevance to the choice between parsimony and likelihood methods. *Syst. Biol.* **50**:525–539.

Tajima, F. 1983. Evolutionary relationship of DNA sequences in finite populations. *Genetics* **105**:437–460.

Tajima, F. 1989. Statistical method for testing the neutral mutation hypothesis by DNA polymorphism. *Genetics* **123**:585–595.

Tajima, F. 1993. Simple methods for testing the molecular evolutionary clock hypothesis. *Genetics* **135**:599–607.

Tajima, F. and Takezaki N. 1994. Estimation of evolutionary distance for reconstructing molecular phylogenetic trees. *Mol. Biol. Evol.* **11**:278–286.

Tajima, F. and Nei, M. 1982. Biases of the estimates of DNA divergence obtained by the restriction enzyme technique. *J. Mol. Evol.* **18**:115–120.

Takahata, N. 1986. An attempt to estimate the effective size of the ancestral species common to two extant species from which homologous genes are sequenced. *Genet. Res.* **48**:187–190.

Takahata, N., Satta, Y., and Klein, J. 1995. Divergence time and population size in the lineage leading to modern humans. *Theor. Popul. Biol.* **48**:198–221.

Takezaki, N. and Gojobori, T. 1999. Correct and incorrect vertebrate phylogenies obtained by the entire mitochondrial DNA sequences. *Mol. Biol. Evol.* **16**:590–601.

Takezaki, N. and Nei, M. 1994. Inconsistency of the maximum parsimony method when the rate of nucleotide substitution is constant. *J. Mol. Evol.* **39**:210–218.

Takezaki, N., Rzhetsky, A., and Nei, M. 1995. Phylogenetic test of the molecular clock and linearized trees. *Mol. Biol. Evol.* **12**:823–833.

Tamura, K. 1992. Estimation of the number of nucleotide substitutions when there are strong transition/transversion and G+C content biases. *Mol. Biol. Evol.* **9**:678–687.

Tamura, K. and Nei, M. 1993. Estimation of the number of nucleotide substitutions in the control region of mitochondrial DNA in humans and chimpanzees. *Mol Biol Evol* **10**:512–526.

Tanaka, T. and Nei, M. 1989. Positive darwinian selection observed at the variable-region genes of immunoglobulins. *Mol. Biol. Evol.* **6**:447–459.

Tateno, Y., Takezaki, N., and Nei, M. 1994. Relative efficiencies of the maximum-likelihood, neighbor-joining, and maximum-parsimony methods when substitution rate varies with site. *Mol. Biol. Evol.* **11**:261–277.

Tavaré, S. 1986. Some probabilistic and statistical problems on the analysis of DNA sequences. *Lect. Math. Life Sci.* **17**:57–86.

Tavaré, S., Marshall, C. R., Will, O. et al. 2002. Using the fossil record to estimate the age of the last common ancestor of extant primates. *Nature* **416**:726–729.

Templeton, A. R. 1983. Phylogenetic inference from restriction endonuclease cleavage site maps with particular reference to the evolution of man and the apes. *Evolution* **37**:221–224.

Terai, Y., Mayer, W. E., Klein, J. et al. 2002. The effect of selection on a long wavelength-sensitive (LWS) opsin gene of Lake Victoria cichlid fishes. *Proc. Natl. Acad. Sci. U.S.A.* **99**:15501–15506.

Thomas, J. H., Kelley, J. L., Robertson, H. M. et al. 2005. Adaptive evolution in the SRZ chemoreceptor families of *Caenorhabditis elegans* and *Caenorhabditis briggsae*. *Proc. Natl. Acad. Sci. U.S.A.* **102**:4476–4481.

Thomas, J. W., Touchman, J. W., Blakesley, R. W. et al. 2003. Comparative analyses of multi-species sequences from targeted genomic regions. *Nature* **424**:788–793.

Thompson, E. A. 1975. *Human Evolutionary Trees*. Cambridge University Press, Cambridge.

Thompson, J. D., Higgins, D. G., and Gibson, T. J. 1994. CLUSTAL W: improving the sensitivity of progressive multiple sequence alignment through sequence weighting, position-specific gap penalties and weight matrix choice. *Nucleic Acids Res.* **22**:4673–4680.

Thorne, J. L. and Kishino, H. 1992. Freeing phylogenies from artifacts of alignment. *Mol. Biol. Evol.* **9**:1148–1162.

Thorne, J. L. and Kishino, H. 2002. Divergence time and evolutionary rate estimation with multilocus data. *Syst. Biol.* **51**:689–702.

Thorne, J. L. and Kishino, H. 2005. Estimation of divergence times from molecular sequence data. In *Statistical Methods in Molecular Evolution*. (ed. R. Nielsen), pp. 233–256. Springer-Verlag, New York.

Thorne, J. L., Kishino, H., and Felsenstein, J. 1991. An evolutionary model for maximum likelihood alignment of DNA sequences [erratum in *J. Mol. Evol.* 1992 **34**:91]. *J. Mol. Evol.* **33**:114–124.

Thorne, J. L., Kishino, H., and Felsenstein, J. 1992. Inching toward reality: an improved likelihood model of sequence evolution. *J. Mol. Evol.* **34**:3–16.

Thorne, J. L., Goldman, N., and Jones, D. T. 1996. Combining protein evolution and secondary structure. *Mol. Biol. Evol.* **13**:666–673.

Thorne, J. L., Kishino, H., and Painter, I. S. 1998. Estimating the rate of evolution of the rate of molecular evolution. *Mol. Biol. Evol.* **15**:1647–1657.

Thornton, J. 2004. Resurrecting ancient genes: experimental analysis of extinct molecules. *Nat. Rev. Genet.* **5**:366–375.

Thornton, J. W., Need, E., and Crews, D. 2003. Resurrecting the ancestral steroid receptor: ancient origin of estrogen signaling. *Science* **301**:1714–1717.

Tillier, E. R. M. 1994. Maximum likelihood with multiparameter models of substitution. *J. Mol. Evol.* **39**:409–417.

Ting, C. T., Tsaur, S. C., Wu, M. L., and Wu, C. I. 1998. A rapidly evolving homeobox at the site of a hybrid sterility gene. *Science* **282**:1501–1504.

Torgerson, D. G. and Singh, R. S. 2004. Rapid evolution through gene duplication and sub-functionalization of the testes-specific a4 proteasome subunits in Drosophila. *Genetics* **168**:1421–1432.

Tsaur, S. C. and Wu, C. -I. 1997. Positive selection and the molecular evolution of a gene of male reproduction, *Acp26Aa* of *Drosophila*. *Mol. Biol. Evol.* **14**:544–549.

Tucker, A. 1995. *Applied Combinatorics*. Wiley, New York.

Tuff, P. and Darlu, P. 2000. Exploring a phylogenetic approach for the detection of correlated substitutions in proteins. *Mol. Biol. Evol.* **17**:1753–1759.

Tuffley, C. and Steel, M. 1997. Links between maximum likelihood and maximum parsimony under a simple model of site substitution. *Bull. Math. Biol.* **59**:581–607.

Tuffley, C. and Steel, M. 1998. Modeling the covarion hypothesis of nucleotide substitution. *Math. Biosci.* **147**:63–91.

Twiddy, S. S., Woelk, C. H., and Holmes, E. C. 2002. Phylogenetic evidence for adaptive evolution of dengue viruses in nature. *J. Gen. Virol.* **83**:1679–1689.

Tzeng, Y. H., Pan, R., and Li, W. H. 2004. Comparison of three methods for estimating rates of synonymous and nonsynonymous nucleotide substitutions. *Mol. Biol. Evol.* **21**:2290–2298.

Ugalde, J. A., Chang, B. S. W., and Matz, M. V. 2004. Evolution of coral pigments recreated. *Science* **305**:1433.

Vacquier, V. D., Swanson, W. J., and Lee, Y. -H. 1997. Positive Darwinian selection on two homologous fertilization proteins: what is the selective pressure driving their divergence? *J. Mol. Evol.* **44**:S15–S22.

Vallender, E. J. and Lahn, B. T. 2004. Positive selection on the human genome. *Hum. Mol. Genet.* **13**:R245–R254.

Vinh, Y. and von Haeseler, A. 2004. IQPNNI: Moving fast through tree space and stopping in time. *Mol. Biol. Evol.* **21**:1565–1571.

Vuong, Q. H. 1989. Likelihood ratio tests for model selection and non-nested hypotheses. *Econometrica* **57**:307–333.

Waddell, P. J. and Steel, M. A. 1997. General time-reversible distances with unequal rates across sites: mixing gamma and inverse Gaussian distributions with invariant sites. *Mol. Phylogenet. Evol.* **8**:398–414.

Waddell, P. J., Penny, D., and Moore, T. 1997. Hadamard conjugations and modeling sequence evolution with unequal rates across sites [erratum in *Mol. Phylogenet. Evol.* 1997 **8**:446]. *Mol. Phylogenet. Evol.* **8**:33–50.

Wakeley, J. 1994. Substitution-rate variation among sites and the estimation of transition bias. *Mol. Biol. Evol* **11**:436–442.

Wald, A. 1949. Note on the consistency of the maximum likelihood estimate. *Ann. Math. Statist.* **20**:595–601.

Walker, A. J. 1974. New fast method for generating discrete random numbers with arbitrary frequency distributions. *Electron. Lett.* **10**:127–128.

Wallace, D. L. 1980. The Behrens-Fisher and Fieller-Creasy problems. In *R.A. Fisher: An Appreciation* (ed. S. Fienberg, J. Gani, J. Kiefer, and K. Krickeberg) pp. 119–147. Springer-Verlag, New York.

Walsh, J. B. 1995. How often do duplicated genes evolve new functions? *Genetics* **139**:421–428.

Wang, X. and Zhang, J. 2004. Rapid evolution of mammalian X-linked testis-expressed homeobox genes. *Genetics* **167**:879–888.

Wang, Y. -Q. and Su, B. 2004. Molecular evolution of microcephalin, a gene determining human brain size. *Hum. Mol. Genet.* **13**:1131–1137.

Ward, T. J., Honeycutt, R. L., and Derr, J. N. 1997. Nucleotide sequence evolution at the κ-casein locus: evidence for positive selection within the family Bovidae. *Genetics* **147**:1863–1872.

Ward, T. J., Bielawski, J. P., Kistler, H. C. *et al.* 2002. Ancestral polymorphism and adaptive evolution in the trichothecene mycotoxin gene cluster of phytopathogenic Fusarium. *Proc. Natl. Acad. Sci. U.S.A.* **99**:9278–9283.

Waterston, R. H., Lindblad-Toh, K., Birney, E. *et al.* 2002. Initial sequencing and comparative analysis of the mouse genome. *Nature* **420**:520–562.

Watterson, G. A. 1975. On the number of segregating sites in genetical models without recombination. *Theor. Popul. Biol.* **7**:256–276.

Weerahandi, S. 1993. Generalized confidence intervals. *J. Amer. Statist. Assoc.* **88**:899–905.

Weerahandi, S. 2004. *Generalized Inference in Repeated Measures: Exact Methods in MANOVA and Mixed Models.* Wiley, New York.

Whelan, S. and Goldman, N. 2001. A general empirical model of protein evolution derived from multiple protein families using a maximum likelihood approach. *Mol. Biol. Evol.* **18**:691–699.

Whelan, S., Liò, P., and Goldman, N. 2001. Molecular phylogenetics: state of the art methods for looking into the past. *Trends Genet.* **17**:262–272.

White, H. 1982. Maximum likelihood estimation of misspecified models. *Econometrica* **50**:1–25.

Wiley, E. O. 1981. *Phylogenetics. The Theory and Practice of Phylogenetic Systematics.* John Wiley & Sons, New York.

Wilkinson, M., Lapointe, F. -J., and Gower, D. J. 2003. Branch lengths and support. *Syst. Biol.* **52**:127–130.

Willett, C. S. 2000. Evidence for directional selection acting on pheromone-binding proteins in the genus Choristoneura. *Mol. Biol. Evol.* **17**:553–562.

Williamson, S. and Orive, M. E. 2002. The genealogy of a sequence subject to purifying selection at multiple sites. *Mol. Biol. Evol.* **19**:1376–1384.

Wilson, A. C., Carlson, S. S., and White, T. J. 1977. Biochemical evolution. *Ann. Rev. Biochem.* **46**:573–639.

Wilson, D. J. and McVean, G. 2006. Estimating diversifying selection and functional constraint in the presence of recombination. *Genetics* **172**:1411–1425.

Wilson, I. J., Weal, M. E., and Balding, D. J. 2003. Inference from DNA data: population histories, evolutionary processes and forensic match probabilities. *J. R. Statist. Soc. A* **166**:155–201.

Wong, W. S. W., Yang, Z., Goldman, N., and Nielsen, R. 2004. Accuracy and power of statistical methods for detecting adaptive evolution in protein coding sequences and for identifying positively selected sites. *Genetics* **168**:1041–1051.

Wray, G. A., Levinton, J. S., and Shapiro, L. H. 1996. Molecular evidence for deep Precambrian divergences. *Science* **274**:568–573.

Wright, F. 1990. The 'effective number of codons' used in a gene. *Gene* **87**:23–29.

Wright, S. 1931. Evolution in Mendelian populations. *Genetics* **16**:97–159.

Wu, C. -I. and Li, W. -H. 1985. Evidence for higher rates of nucleotide substitution in rodents than in man. *Proc. Natl. Acad. Sci. U.S.A.* **82**:1741–1745.

Wu, Q. 2005. Comparative genomics and diversifying selection of the clustered vertebrate protocadherin genes. *Genetics* **169**:2179–2188.

Wyckoff, G. J., Wang, W., and Wu, C. -I. 2000. Rapid evolution of male reproductive genes in the descent of man. *Nature* **403**:304–309.

Xia, X. and Xie, Z. 2001. DAMBE: Data analysis in molecular biology and evolution. *J. Hered.* **92**:371–373.

Yamaguchi-Kabata, Y. and Gojobori, T. 2000. Reevaluation of amino acid variability of the human immunodeficiency virus type 1 gp120 envelope glycoprotein and prediction of new discontinuous epitopes. *J. Virol.* **74**:4335–4350.

Yang, W., Bielawski, J. P., and Yang, Z. 2003. Widespread adaptive evolution in the human immunodeficiency virus type 1 genome. *J. Mol. Evol.* **57**:57:212–221.

Yang, Z. 1993. Maximum-likelihood estimation of phylogeny from DNA sequences when substitution rates differ over sites. *Mol. Biol. Evol.* **10**:1396–1401.

Yang, Z. 1994a. Maximum likelihood phylogenetic estimation from DNA sequences with variable rates over sites: approximate methods. *J. Mol. Evol.* **39**:306–314.

Yang, Z. 1994b. Estimating the pattern of nucleotide substitution. *J. Mol. Evol.* **39**:105–111.

Yang, Z. 1994c. Statistical properties of the maximum likelihood method of phylogenetic estimation and comparison with distance matrix methods. *Syst. Biol.* **43**:329–342.

Yang, Z. 1995a. A space-time process model for the evolution of DNA sequences. *Genetics* **139**:993–1005.

Yang, Z. 1995b. Evaluation of several methods for estimating phylogenetic trees when substitution rates differ over nucleotide sites. *J. Mol. Evol.* **40**:689–697.

Yang, Z. 1996a. Phylogenetic analysis using parsimony and likelihood methods. *J. Mol. Evol.* **42**:294–307.

Yang, Z. 1996b. Maximum-likelihood models for combined analyses of multiple sequence data. *J. Mol. Evol.* **42**:587–596.

Yang, Z. 1996c. Among-site rate variation and its impact on phylogenetic analyses. *Trends Ecol. Evol.* **11**:367–372.

Yang, Z. 1997a. PAML: a program package for phylogenetic analysis by maximum likelihood. *Comput. Appl. Biosci.* **13**:555–556 (http://abacus.gene.ucl.ac.uk/software/paml.html).

Yang, Z. 1997b. How often do wrong models produce better phylogenies? *Mol. Biol. Evol.* **14**:105–108.

Yang, Z. 1998a. Likelihood ratio tests for detecting positive selection and application to primate lysozyme evolution. *Mol. Biol. Evol.* **15**:568–573.

Yang, Z. 1998b. On the best evolutionary rate for phylogenetic analysis. *Syst. Biol.* **47**:125–133.

Yang, Z. 2000a. Complexity of the simplest phylogenetic estimation problem. *Proc. R. Soc. B Biol. Sci.* **267**:109–116.

Yang, Z. 2000b. Maximum likelihood estimation on large phylogenies and analysis of adaptive evolution in human influenza virus A. *J. Mol. Evol.* **51**:423–432.

Yang, Z. 2002. Likelihood and Bayes estimation of ancestral population sizes in Hominoids using data from multiple loci. *Genetics* **162**:1811–1823.

Yang, Z. 2004. A heuristic rate smoothing procedure for maximum likelihood estimation of species divergence times. *Acta Zoologica Sinica* **50**:645–656.

Yang, Z. and Kumar, S. 1996. Approximate methods for estimating the pattern of nucleotide substitution and the variation of substitution rates among sites. *Mol. Biol. Evol.* **13**:650–659.

Yang, Z. and Nielsen, R. 1998. Synonymous and nonsynonymous rate variation in nuclear genes of mammals. *J. Mol. Evol.* **46**:409–418.

Yang, Z. and Nielsen, R. 2000. Estimating synonymous and nonsynonymous substitution rates under realistic evolutionary models. *Mol. Biol. Evol.* **17**:32–43.

Yang, Z. and Nielsen, R. 2002. Codon-substitution models for detecting molecular adaptation at individual sites along specific lineages. *Mol. Biol. Evol.* **19**:908–917.

Yang, Z. and Rannala, B. 1997. Bayesian phylogenetic inference using DNA sequences: a Markov chain Monte Carlo Method. *Mol. Biol. Evol.* **14**:717–724.

Yang, Z. and Rannala, B. 2005. Branch-length prior influences Bayesian posterior probability of phylogeny. *Syst. Biol.* **54**:455–470.

Yang, Z. and Rannala, B. 2006. Bayesian estimation of species divergence times under a molecular clock using multiple fossil calibrations with soft bounds. *Mol. Biol. Evol.* **23**: 212–226.

Yang, Z. and Roberts, D. 1995. On the use of nucleic acid sequences to infer early branchings in the tree of life. *Mol. Biol. Evol.* **12**:451–458.

Yang, Z. and Swanson, W. J. 2002. Codon-substitution models to detect adaptive evolution that account for heterogeneous selective pressures among site classes. *Mol. Biol. Evol.* **19**:49–57.

Yang, Z. and Wang, T. 1995. Mixed model analysis of DNA sequence evolution. *Biometrics* **51**:552–561.

Yang, Z. and Yoder, A. D. 2003. Comparison of likelihood and Bayesian methods for estimating divergence times using multiple gene loci and calibration points, with application to a radiation of cute-looking mouse lemur species. *Syst. Biol.* **52**:705–716.

Yang, Z., Kumar, S., and Nei, M. 1995a. A new method of inference of ancestral nucleotide and amino acid sequences. *Genetics* **141**:1641–1650.

Yang, Z., Lauder, I. J., and Lin, H. J. 1995b. Molecular evolution of the hepatitis B virus genome. *J. Mol. Evol.* **41**:587–596.

Yang, Z., Goldman, N., and Friday, A. E. 1995c. Maximum likelihood trees from DNA sequences: a peculiar statistical estimation problem. *Syst. Biol.* **44**:384–399.

Yang, Z., Nielsen, R., and Hasegawa, M. 1998. Models of amino acid substitution and applications to mitochondrial protein evolution. *Mol. Biol. Evol.* **15**:1600–1611.

Yang, Z., Nielsen, R., Goldman, N., and Pedersen, A. -M. K. 2000. Codon-substitution models for heterogeneous selection pressure at amino acid sites. *Genetics* **155**:431–449.

Yang, Z., Wong, W. S. W., and Nielsen, R. 2005. Bayes empirical Bayes inference of amino acid sites under positive selection. *Mol. Biol. Evol.* **22**:1107–1118.

Yoder, A. D. and Yang, Z. 2000. Estimation of primate speciation dates using local molecular clocks. *Mol. Biol. Evol.* **17**:1081–1090.

Yokoyama, S. 2002. Molecular evolution of color vision in vertebrates. *Gene* **300**:69–78.

York, T. L., Durrett, R., and Nielsen, R. 2002. Bayesian estimation of the number of inversions in the history of two chromosomes. *J. Comp. Biol.* **9**:805–818.

Yu, N., Zhao, Z., Fu, Y. X. *et al.* 2001. Global patterns of human DNA sequence variation in a 10-kb region on chromosome 1. *Mol. Biol. Evol.* **18**:214–222.

Zakon, H. H., Lu, Y., Zwickl, D. J., and Hillis, D. M. 2006. Sodium channel genes and the evolution of diversity in communication signals of electric fishes: convergent molecular evolution. *Proc. Natl. Acad. Sci. U.S.A.* **103**:3675–3680.

Zanotto, P. M., Kallas, E. G., Souza, R. F., and Holmes, E. C. 1999. Genealogical evidence for positive selection in the *nef* gene of HIV-1. *Genetics* **153**:1077–1089.

Zardoya, R. and Meyer, A. 1996. Phylogenetic performance of mitochondrial protein-coding genes in resolving relationships among vertebrates. *Mol. Biol. Evol.* **13**:933–942.

Zhang, J. 2000. Rates of conservative and radical nonsynonymous nucleotide substitutions in mammalian nuclear genes. *J. Mol. Evol.* **50**:56–68.

Zhang, J. 2003. Evolution of the human ASPM gene, a major determinant of brain size. *Genetics* **165**:2063–2070.

Zhang, J. 2004. Frequent false detection of positive selection by the likelihood method with branch-site models. *Mol. Biol. Evol.* **21**:1332–1339.

Zhang, J. and Nei, M. 1997. Accuracies of ancestral amino acid sequences inferred by the parsimony, likelihood, and distance methods. *J. Mol. Evol.* **44**:S139–146.

Zhang, J. and Nei, M. 2000. Positive selection in the evolution of mammalian interleukin-2 genes. *Mol. Biol. Evol.* **17**:1413–1416.

Zhang, J. and Webb, D. M. 2004. Rapid evolution of primate antiviral enzyme APOBEC3G. *Hum. Mol. Genet.* **13**:1785–1791.

Zhang, J., Kumar, S., and Nei, M. 1997. Small-sample tests of episodic adaptive evolution: a case study of primate lysozymes. *Mol. Biol. Evol.* **14**:1335–1338.

Zhang, J., Nielsen, R., and Yang, Z. 2005. Evaluation of an improved branch-site likelihood method for detecting positive selection at the molecular level. *Mol. Biol. Evol.* **22**:2472–2479.

Zhang, J., Rosenberg, H. F., and Nei, M. 1998. Positive Darwinian selection after gene duplication in primate ribonuclease genes. *Proc. Natl. Acad. Sci. U.S.A.* **95**:3708–3713.

Zhang, J., Zhang, Y. P., and Rosenberg, H. F. 2002. Adaptive evolution of a duplicated pancreatic ribonuclease gene in a leaf-eating monkey. *Nat. Genet.* **30**:411–415.

Zhang, L., Gaut, B. S., and Vision, T. J. 2001. Gene duplication and evolution. *Science* **293**:1551.

Zhao, Z., Jin, L., Fu, Y. X. *et al.* 2000. Worldwide DNA sequence variation in a 10-kilobase noncoding region on human chromosome 22. *Proc. Natl. Acad. Sci. U.S.A.* **97**:11354–11358.

Zharkikh, A. 1994. Estimation of evolutionary distances between nucleotide sequences. *J. Mol. Evol.* **39**:315–329.

Zharkikh, A. and Li, W. -H. 1993. Inconsistency of the maximum parsimony method: the case of five taxa with a molecular clock. *Syst. Biol.* **42**:113–125.

Zharkikh, A. and Li, W. -H. 1995. Estimation of confidence in phylogeny: the complete-and-partial bootstrap technique. *Mol. Phylogenet. Evol.* **4**:44–63.

Zhu, L. and Bustamante, C. D. 2005. A composite likelihood approach for detecting directional selection from DNA sequence data. *Genetics* **170**:1411–1421.

Zhu, S., Bosmans, F., and Tytgat, J. 2004. Adaptive evolution of scorpion sodium channel toxins. *J. Mol. Evol.* **58**:145–153.

Zuckerkandl, E. 1964. Further principles of chemical paleogenetics as applied to the evolution of hemoglobin. In *Protides of the Biological Fluids* (ed. H. Peeters). pp. 102–109. Elsevier, Amsterdam.

Zuckerkandl, E. and Pauling, L. 1962. Molecular disease, evolution, and genetic heterogeneity. In *Horizons in Biochemistry* (ed. M. Kasha, and B. Pullman), pp. 189–225. Academic Press, New York.

Zuckerkandl, E. and Pauling, L. 1965. Evolutionary divergence and convergence in proteins. In *Evolving Genes and Proteins* (ed. V. Bryson, and H. J. Vogel), pp. 97–166. Academic Press, New York.

Index

acceptance proportion, 169, 184
acceptance ratio, 162, 313
adaptive evolution, 259–292
additive-tree method, 90
agglomerative method, 83
AIC *See* Akaike information criterion
Akaike information criterion, 140
alias method, 299–301
alignment gap, 107
amino acid exchangeability, 41
amino acid model, 40–45, 119
ancestral polymorphism, 80
ancestral reconstruction, 94–98, 119–128, 264
   bias, 126–128, 264, 272
   empirical Bayes, 121–124
   hierarchical Bayesian, 123
   joint, 122
   likelihood, 121–124
   marginal, 121
   morphological characters, 124–126
   parsimony, 93–98, 120, 123
antigen-recognition site, 271
apes, 108, 180
ARS *See* antigen-recognition site
AU test 212
auto-discrete-gamma model, 115

Bayes's theorem, 146
Bayesian information criterion, 141
Bayesian method, 145–184
   ancestral reconstruction, 121–124
   coalescent model, 181–183
   molecular clock dating, 245–257
   tree reconstruction, 174–180
BFGS, 133
BIC *See* Bayesian information criterion
bifurcating tree, 75
binary tree *See* bifurcating tree
binomial distribution, 39, 153, 297

bipartition, 77
birth–death process, 249, 305
birthday problem, 305
blocking, 164
bootstrap, 176, 178, 293, 296
Box–Muller transformation, 296
branch swapping, 84
Bremer support *See* decay index
burn-in, 174

Chapman-Kolmogorov theorem, 8
character length, 94, 216
chi-square distribution, 25, 138
   mixture, 140, 281
cladistics 198
cluster method, 83
coalescent, 181, 305
codon model, 48, 119, 284
   branch-site, 279
   covarion, 282
   site, 273
   switching, 282
   branch, 270
   clade, 281
combined analysis, 116, 175, 257
composition method, 298
compound distribution *See* mixture model
computer simulation, 143, 293–307
confidence interval, 24, 151, 241
consensus tree
   majority-rule, 79, 176
   strict, 79
consistency, 24, 98, 307
consistency index, 213
covarion model, 115, 282
coverage probability, 173
credibility interval, 148
   equal-tail, 148, 174
   HPD, 149
cross-validation, 232

Dayhoff, 46
decay index, 213
deletion, 108
delta technique, 9, 11, 15, 313–316
detailed balance, 33, 41
Dirichlet distribution, 248
discrete distribution, 275, 297
discrete uniform distribution, 296
discrete gamma model, 112
divergence, 265
divisive method, 83
dynamic-programming algorithm, 95–98, 102, 122

efficiency, 24, 187, 192–196
eigenvalue, 13, 68
eigenvector, 13, 68
empirical Bayes method, 114, 121–128, 159, 276
 Bayes, 276
 naive, 276
 nonparametric, 273
ENC *See* effective number of codons
error
 random, 186
 systematic, 186
expected information, 24, 187
exponential distribution, 296, 303

fiducial probability, 236
Fisher information, *see* expected information
Fisher, Ronald A., 145
fixed-sites model, 272, 304
fossil, 235
 uncertainty, 235–245, 248–251
four-fold degenerate site, 54, 63, 64
frequentist statistics, 145, 151
Fu and Li's *D*, 264

Galton, Francis, 145
gamma model, 19–22, 35, 44–46, 110–115
gene tree, 80
general time-reversible model, 33–37, 41, 134, 143, 232
generator matrix, 30
genetic algorithm, 89, 136
genetic hitchhiking, 265, 286
genome rearrangement, 309
geometric Brownian motion, 232, 247–248
Gibbs sampler, 166
golden section search, 129

gradient, 132
GTR *See* general time reversible model

Hadamard conjugation, 105
Hastings ratio *See* proposal ratio
Hessian matrix, 132
heterotachy, 197
hidden Markov model, 115
hierarchical Bayesian method, 123, 276
HKA test *See* Hudson–Kreitman–Aquade test
homoplasy, 213
HPD interval, 149
Hudson-Kreitman-Aquade test, 267
hypothesis testing, 153

i.i.d. model, 114, 117, 143, 302, 304, 305
importance sampling, 293
inconsistency
 parsimony, 99, 144
indel, 108
index
 consistency, 213
 dispersion, 228
 decay, 213
 retention, 213
infinite-sites model, 262, 286
infinite-sites theory, 257
information
 expected, 24, 187
 observed, 24
ingroup, 73
insertion, 108
interior branch test, 210
interval of uncertainty, 129
invariable-sites model, 111
invariance, 24
inversion method, 295, 297

JC69, 4–9, 21–25, 149–150
JTT, 41, 46
jump chain, 303
jumping kernel *See* proposal density

K80, 10-11, 17, 20–22, 25–30
K-H test, *See* Kishino–Hasegawa test
Kishino–Hasegawa test, 211
K-T boundary, 225
Kullback–Leibler divergence, 192

Index

labelled history, 174, 181, 305
least squares, 90, 109, 269
   generalized, 92, 93
   ordinary, 92
   weighted, 92
likelihood, 22–30
   integrated, 28, 30, 240
   marginal, 28
   penalized, 233
   profile, 27
   pseudo-, 27
   relative, 27
likelihood equation, 27, 128
likelihood function, 22, 100
likelihood interval, 25, 241
likelihood ratio test, 25, 137
   molecular clock, 226
   positive selection, 270, 273, 279
limiting distribution, 7
Lindley's paradox, 157
line search, 129
log likelihood function, 23
log normal distribution, 113, 232, 248
long-branch attraction, 98, 144, 306
LRT *See* likelihood ratio test
LS *See* least squares

major histocompatibility complex, 271, 276
mammals, 254
MAP tree, 176
Markov-chain, 30–33
   hidden, 115
Markov chain Monte Carlo, 159–167, 175, 181, 245, 293
   convergence, 172
   mixing, 172
   monitoring, 171
   proposals, 167–171
Markov process *See* Markov chains
Markovian property, 4
maximum likelihood estimate, 22
maximum likelihood method, 22–30, 100–144
   ancestral reconstruction, 120
   computation, 128
   distance estimation, 22, 46
   estimation of d_S and d_N 59
   molecular clock dating, 228–245
   tree reconstruction, 100
McDonald–Kreitman test, 265
maximum likelihood tree, 82
maximum parsimony likelihood, 199

maximum parsimony tree, 82
MCMC *See* Markov chain Monte Carlo
MCMCMC, 166
Metropolis–Hastings algorithm, 161
   sliding window, 168, 169
   single-component, 164
MHC *See* major histocompatibility complex
minimum-evolution method 82, 93
missing data, 107
mitochondrial protein, 108, 116, 214
mixture model, 298, 299
   d_N/d_S for sites, 275
   branch-length prior, 179
   rates for sites, 111, 114
   simulation, 298
ML *See* maximum likelihood method
MLE, *see* maximum likelihood estimate
model robustness, 61, 137, 142, 293
model selection, 137
Model Test, 141
molecular clock, 106, 223–228
   local, 230
   global, 228
   over-dispersed, 238
molecular-clock dating
   calibration, 229, 248
   rate smoothing, 231
   Bayesian, 245–257
   likelihood, 228–245
molecular evolutionary clock *See* molecular clock
molecular-clock rooting, 73
Monte Carlo integration, 160, 293, 306
Monte Carlo simulation, *See* computer simulation
morphological character, 124–126
most parsimonious tree, *See* maximum parsimony tree
multifurcating tree, 75
multinomial distribution, 32, 298
   alias method, 301
multiple comparison, *see* multiple testing
multiple genes, 116, 175, 232, 247, 257, 272
multiple hits, *See* multiple substitutions
multiple optima, 87, 135
multiple substitutions, 3, 52
multiple testing, 212, 271, 273, 281
multiplication-congruent method, 294

nearly neutral theory, 261
negative binomial distribution, 39, 153
neighbour joining, 81, 92
neutral theory, 260
Newton's method, 129, 132

Newton–Raphson method, *see* Newton's method
Neyman, Jerzy, 146
NG86, 50
NJ, *see* Neighbor joining
nocommon mechanism model, 200
nonhomogeneous model, 118
noninformative prior, 154
nonlinear programming, *see* optimisation
nonstationary model, 118
nonsynonymous site, 49, 54, 58
nonsynonymous substitution, 40
nonsynonymous/synonymous rate ratio, 48, 268, 273, 279
normal distribution, 24, 140, 151, 157, 158, 160, 168, 169, 296
nuisance parameter, 27, 149

objective function, 128
Ockman's razor *See* principle of parsimony
optimization, 128
    multivariate, 131
    univariate, 129
outgroup rooting, 73
overall type-1 error, 213

p53 gene, 47
PAM, 41
parsimony method, 93–99, 198–206
    ancestral reconstruction, 94
    positive selection, 269
    tree reconstruction, 93
parsimony principle, 141, 204
parsimony-informative site, 95
partition distance, 77
Pearson, Egon, 146
Pearson, Karl, 145
phytochrome gene, 270, 282
Poisson distribution, 45, 232
Poisson random field, 267
polymorphism, 265
polytomy, 75
posterior clade probability, 176
posterior distribution, 145
post-trial evaluation, 152
Potential scale reduction statistic, 173
pre-trial evaluation, 152
prior, 158–159
    branch lengths, 175
    conjugate, 158
    controversy, 154
    diffuse, 158
    hyper, 158
    non-informative, 158
    sensitivity, 157
    tree topology, 175
    vague, 158
prior distribution *See* prior
proposal, 162
proposal density, 162
proposal ratio, 163, 167, 313
pruning algorithm, 102
pseudo-random number, 294
pulley principle, 106
purine, 10
pyrimidine, 10

quartet-puzzling, 137
quartet-dating, 230

random number, 294
random-number generator, 294
random-sites model, 273, 279, 304
rate smoothing, 231
rbcL gene, 53, 56, 60, 64, 66, 138
recombination, 285
regulatory gene, 291
relative-rate test, 233
relative-ratio test, 272
RELL bootstrap, 214, 207
reproductive proteins, 290
retention index, 213
REV *See* general time-reversible model
reversibility, 14, 68, 106
root, 106
rooted tree, 73

safe-guided Newton algorithm, 133
saturation, 38
segregating sites, 262
selection
    negative, 260
    positive, 260, 286
selection bias, 212
separate analysis, 118
sequential addition *See* stepwise addition
S-H test *See* Shimodaira–Hasegawa test
Shimodaira–Hasegawa test, 212
simulated annealing, 89
site configuration *See* site pattern
site length *See* character length

site pattern, 14, 95
site-frequency spectrum, 264, 267
  folded, 264
sliding window, 168–169
species tree, 80
spectral decomposition, 13
split *See* bipartition
SPR, 86
star decomposition, 83
stationary distribution, 8
steady-state distribution *See* stationary distribution
steepest ascent, 132
steepest descent, 132
step matrix, 95
stepwise addition, 75, 83, 84
stochastic simulation *See* computer stimulation
substitution-rate matrix, 6, 30
supermatrix, 118
supertree, 118
synonymous site, 49, 54, 58
synonymous substitution, 40

Tajima's *D*, 262
TBR, 86
thinning, 174
transformation, 295
transition, 10
transition probability, 6
transition-probability matrix, 6, 31
  computation, 68
transition/transversion rate ratio, 10, 17, 27, 48, 54, 306

transversion, 10
transversion parsimony, 95
tree rearrangement *See* branch swapping
tree search, 136
  exhaustive, 83
  heuristic, 83
  stochastic, 89
  multiple peaks, 88
tree space, 87
  local peak, 87, 167
tree-reconstruction method, 81
  Bayesian method, 174
  distance method, 89
  likelihood method, 100
  minimum-evolution method, 92
  optimality-based, 82
  parsimony method, 93

unbiased estimate, 24
uniform distribution, 295
unrooted tree, 73
UPGMA, 81

WAG, 42
waiting time, 303
weighted parsimony, 95
winning-sites test, 213

YN00, 65
Yule process, 305